NEUROMETHODS

Series Editor
Wolfgang Walz
University of Saskatchewan
Saskatoon, SK, Canada

For further volumes:
http://www.springer.com/series/7657

Neuromethods publishes cutting-edge methods and protocols in all areas of neuroscience as well as translational neurological and mental research. Each volume in the series offers tested laboratory protocols, step-by-step methods for reproducible lab experiments and addresses methodological controversies and pitfalls in order to aid neuroscientists in experimentation. *Neuromethods* focuses on traditional and emerging topics with wide-ranging implications to brain function, such as electrophysiology, neuroimaging, behavioral analysis, genomics, neurodegeneration, translational research and clinical trials. *Neuromethods* provides investigators and trainees with highly useful compendiums of key strategies and approaches for successful research in animal and human brain function including translational "bench to bedside" approaches to mental and neurological diseases.

Lesion-to-Symptom Mapping

Principles and Tools

Edited by

Dorian Pustina

Experimental Neuroimaging, CHDI Management/CHDI Foundation, Princeton, NJ, USA

Daniel Mirman

Department of Psychology, University of Edinburgh, Edinburgh, UK

Humana Press

Editors
Dorian Pustina
Experimental Neuroimaging
CHDI Management/CHDI Foundation
Princeton, NJ, USA

Daniel Mirman
Department of Psychology
University of Edinburgh
Edinburgh, UK

ISSN 0893-2336 ISSN 1940-6045 (electronic)
Neuromethods
ISBN 978-1-0716-2227-8 ISBN 978-1-0716-2225-4 (eBook)
https://doi.org/10.1007/978-1-0716-2225-4

This Springer imprint is published by the registered company Springer Science+Business Media, LLC part of Springer Nature.
The registered company address is: 1 New York Plaza, New York, NY 10004, U.S.A.

Preface to the Series

Experimental life sciences have two basic foundations: concepts and tools. The Neuromethods series focuses on the tools and techniques unique to the investigation of the nervous system and excitable cells. It will not, however, shortchange the concept side of things as care has been taken to integrate these tools within the context of the concepts and questions under investigation. In this way, the series is unique in that it not only collects protocols but also includes theoretical background information and critiques which led to the methods and their development. Thus, it gives the reader a better understanding of the origin of the techniques and their potential future development. The Neuromethods series strikes a balance between recent and exciting developments like those concerning new animal models of disease, imaging, *in vivo* methods, and more established techniques, including, for example, immunocytochemistry and electrophysiological technologies. New trainees in neurosciences still need a sound footing in these older methods in order to apply a critical approach to their results.

Under the guidance of its founders, Alan Boulton and Glen Baker, the Neuromethods series has been a success since its first volume published through Humana Press in 1985. The series continues to flourish through many changes over the years. It is now published under the umbrella of Springer Protocols. While methods involving brain research have changed a lot since the series started, the publishing environment and technology have changed even more radically. Neuromethods has the distinct layout and style of the Springer Protocols program, designed specifically for readability and ease of reference in a laboratory setting.

The careful application of methods is potentially the most important step in the process of scientific inquiry. In the past, new methodologies led the way in developing new disciplines in the biological and medical sciences. For example, physiology emerged out of anatomy in the nineteenth century by harnessing new methods based on the newly discovered phenomenon of electricity. Nowadays, the relationships between disciplines and methods are more complex. Methods are now widely shared between disciplines and research areas. New developments in electronic publishing make it possible for scientists that encounter new methods to quickly find sources of information electronically. The design of individual volumes and chapters in this series takes this new access technology into account. Springer Protocols makes it possible to download single protocols separately. In addition, Springer makes its print-on-demand technology available globally. A print copy can therefore be acquired quickly and for a competitive price anywhere in the world.

Wolfgang Walz

Preface

The investigation of relationships between brain lesions and behavioral symptoms has increased rapidly in the past 20 years (see Fig. 1). Yet, lesion studies are not a new approach, they have been around for centuries. From the animal ablation experiments of Jean Pierre Flourens [1] to the deficits reported in humans by Marc Dax [2] and Paul Broca [3], lesions have played a key role in shaping our understanding of functional organization of the brain. Why, then, are lesion studies growing more in the last couple of decades? We can think of four main reasons. First, lesion studies make contact with the real-world problems of understanding, diagnosing, and treating clinical deficits that result from brain injury and neurodegenerative diseases. Lesion studies provide a way to answer fundamental questions about the functional organization of the brain with stronger potential for impact on health and well-being, which is appealing to researchers as well as funding agencies. Second, lesion studies help make the highly valuable jump from correlative to causative relationships. The investigation of the regional specialization in the last decades has relied heavily on functional imaging methods (i.e., fMRI or PET), but these correlative methods have not provided conclusive answers on which brain region is critical for performing a task. Third, the methods of lesion investigation have improved substantially in recent years. This has led to more precise lesion definition, more expertise in image processing, more advanced statistical models, and larger datasets. Fourth, a new field has gradually emerged, network neuroscience, which is introducing itself in lesion investigations [4]. Network analyses go beyond the topological mapping of individual brain regions and consider the networks that perform the tasks as well as the role of brain regions in the network. In this view, a lesion disturbs nodes and edges of the network, and the deficits arise from network anomalies. Classical cases have been revisited and were found to carry important network disruptions [5].

The Reason for This Book

The field of lesion symptom mapping (LSM) has reached a methodological maturity that demands proper collection of the current knowledge in a single volume. The overall goal of this book is to provide comprehensive coverage of the conceptual and practical principles of LSM methods. The primary target audience are researchers who already have some familiarity with human neuroscience and statistics, but who are only getting started with LSM and want a full explanation of the steps and considerations involved in LSM. Nonetheless, the book tackles some difficult methodological issues in enough detail to be useful to expert researchers as well.

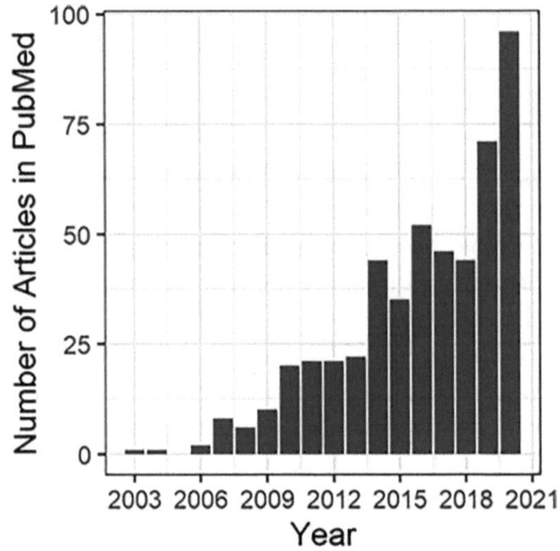

Fig. 1 Number of articles in PubMed with "lesion symptom mapping" or "lesion behavior/behaviour mapping" in the title or abstract for each year from 2003 to 2020

This book focuses on methods and includes illustrative examples; it does not contain cutting-edge results, which are more appropriately found in peer-reviewed journals. Rather, the goal here is to cover the core conceptual issues and connect them with available tools, facilitating real-world application and setting up the next generation of innovation in LSM.

The Book Creation Process

A quick understanding of the book creation process can help the undecided reader form an idea on what to expect of this book. In the early phases of planning, the editors created a table of contents with all the chapters and the topics to cover. The chapter authors were invited to contribute, with each chapter being offered to researchers with the appropriate expertise. A timeline was defined for each chapter, and the editors followed up until all chapters were completed.

Book chapters are not typically peer reviewed. However, the chapters of this book were commented on by the two editors and at least one independent expert. The feedback was provided to the authors in semi-anonymous form, and the authors were invited to address the comments where possible. The commenting and revision process was somewhat laborious and slow, but we think it substantially increased the quality of the chapters.

We limited the amount of repetitive information across chapters in two ways. First, the authors were given the table of contents from the beginning, so they could understand which topics are already covered in other chapters. Second, the editors read all the chapters in sequence prior to submitting to the publisher to make sure the information flow is natural and not repetitive. Corrections were made to improve the overall book readability.

The final result, this book you have in your hands, is a carefully organized volume written by international leaders in the field – senior scientists who have worked on LSM for many years – and by mid-career scientists who are leading the development of new LSM methods.

The Final Chapter Structure

The book has 15 chapters and two appendices:

- The first four chapters explain brain lesions and preparing them for analysis: specifically, what is a lesion (Chap. 1), how to manually draw ("segment") lesions (Chap. 2), how to obtain lesion maps automatically (Chap. 3), and what are some methods of lesion registration in template space (Chap. 4).
- The next three chapters describe the most typical lesion to symptom mapping methods: first by introducing voxel-based lesion symptom mapping (Chap. 5), then by discussing statistical considerations of that approach (Chap. 6), and then expanding to describe methods applied to progressive neurodegenerative lesions (Chap. 7).
- The next three chapters focus on network-based lesion symptom mapping: first by introducing the need for a hodologic approach (Chap. 8) and then by describing the methods to map lesions of structural (Chap. 9) and functional (Chap. 10) networks.
- The next three chapters deal with broader conceptual issues in lesion symptom mapping methods. Chapter 11 discusses limitations of the current statistical approaches and their potential alleviation with multivariate statistical methods that allow testing all the voxels at once rather than testing each voxel separately. Chapter 12 considers methods for predicting behavioral deficits based on the lesion information, and Chap. 13 describes how to select or build behavioral tests for use in lesion symptom mapping analyses.
- The last two chapters explain how to map brain functions using transient lesions via brain stimulation during surgical procedures (Chap. 14) or through non-invasive stimulation over the scalp (Chap. 15).

In addition to the chapters, the book includes practical, hands-on mini tutorials on implementing the different analyses procedures using freely available software. These tutorials are collected in two appendices: one focused on conducting lesion-symptom mapping analyses (Appendix A) and one focused on automated lesion segmentation (Appendix B). The appendices are written by young researchers who are particularly familiar with the software and are ideally positioned to train other users.

Data Availability

One of the recurrent themes in the book is the use of two publicly available datasets across all chapters:

1. A dataset of 131 lesion masks generously provided by the Moss Rehabilitation Research Institute. A simulated behavioral score was added based on the lesion load of an inferior frontal cortex region (see [6]). This dataset was used by chapters that talk about lesion to symptom mapping analyses, and all the examples of Appendix A. The data can be downloaded via Github (https://github.com/dorianps/LESYMAP/tree/master/inst/extdata).
2. A single subject from the ATLAS dataset (Native space ID: c0005s0007t01, Left MCA embolism, lesion size 56ml, [7]) is used to show examples of lesion drawing or automated segmentation. This case was used in chapters that explain lesion properties and automated segmentation methods, as well as in Appendix B. The ATLAS dataset can be downloaded at http://fcon_1000.projects.nitrc.org/indi/retro/atlas.html).

You may notice that figures and examples often refer to one of the above datasets. Using the same datasets allows the reader to follow the same thread from chapter to chapter, and perhaps to follow along or familiarize with the examples.

A Note on Terminology

Sometimes authors use different terms to refer to the same – or closely related – concept or method. Even basic terms like "lesion to symptom mapping" may not be consistent between chapters. Rather than asking the authors to harmonize the terminology – a feat that requires broad community consensus – we clarify here the meaning of some equivalent terms.

- "Lesion-Symptom Mapping" or "Lesion-to-Symptom Mapping" (LSM) refers to the same substantial analyses as "Lesion-Behavior Mapping" and "Lesion-Function Mapping." The method was initially built to study clinical deficits (i.e., symptoms), although many analyses are performed on behavioral scores that do not necessarily constitute a clinically significant symptom, and, therefore, some authors prefer to use "behavior" or "function" instead. "Voxel-based" LSM (VLSM) emphasizes use of voxel-level lesion information; multivariate LSM methods often omit the "V" because multivariate analyses use all voxels at once. Voxel-based lesion-behavior mapping is typically abbreviated as VLBM, and can be considered a synonym of VLSM. N.b., VLBM should not be confused with "Voxel-based morphometry" (VBM) – a method that uses graded tissue concentration information and is not specific to lesion research.
- "Connectome-based" LSM (CLSM) uses connection damage information rather than voxel damage information.
- A "lesion map" is typically a 3D binary image with all the voxels set to 0 except the lesioned voxels, which are set to 1. The lesion may be labeled with other values than 1, for example, the software MRIcron labels the lesions with the value 255. The lesion map is sometimes called "lesion mask," "lesion image," or "segmentation image" (i.e., because the lesion is "segmented" from the full brain image). The lesion map may also be called a "label image" because the lesion area is "labeled" with a different value than the background (i.e., lesion = 1 or 255, background = 0). Please note that the lesion map/mask/segmentation is just a binary mask with no brain structure to see, unlike the MRI or CT structural image on which the lesion map was drawn.
- Lesion "normalization," "registration," or "transfer[ing]" in template space all refer to image processing procedures for aligning an individual's lesion map with a template brain. This is necessary for conducting group-level analyses to make sure that a voxel corresponds to the same (or expected) brain structure across individuals. The term "normalization" is sometimes misleading in neuroscience papers because it can refer both to the process of registration in template space and to the mathematical scaling of voxel values (e.g., voxel values can be "normalized" in the 0 to 1000 range).

Book Limitations

The appendices provide examples with several, widely used software packages, but not all available software options are covered. In part, this is because software options change all the time, new ones become available and older ones become obsolete. The appendices are there only to jump-start the hands-on experience of young researchers.

We attempted to cover most of the LSM topics without making the chapters and the book too long or too discursive. The authors were instructed to keep the chapters short, educative, and avoid repeating information. It is possible that some methodological topics might be missing, while some opinions or techniques may not be widely accepted by the community. This is science after all. If we all knew what to do and agreed unanimously on everything, there would be no research. This said, the editors are interested in knowing your opinion on what can be improved in future editions of the book; feel free to send your feedback to the editors.

Princeton, NJ, USA *Dorian Pustina*
Edinburgh, UK *Daniel Mirman*

Acknowledgments

We sincerely thank all of the authors for their generous contributions – this book would not have been possible without them.

References

1. Pearce JM (2009) Marie-Jean-Pierre Flourens (1794–1867) and cortical localization. Eur Neurol 61(5):311–314. https://doi.org/10.1159/000206858
2. Joynt RJ, Benton AL (1964) The memoir of Marc Dax on aphasia. Neurology 14:851–854. https://doi.org/10.1212/wnl.14.9.851
3. Rorden C, Karnath HO (2004) Using human brain lesions to infer function: a relic from a past era in the fMRI age? Nat Rev Neurosci 5:813–819. https://doi.org/10.1038/nrn1521
4. Bassett DS, Sporns O (2017) Network neuroscience. Nat Neurosci 20(3):353–364. https://doi.org/10.1038/nn.4502
5. Dronkers NF, Plaisant O, Iba-Zizen MT, Cabanis EA (2007) Paul Broca's historic cases: high resolution MR imaging of the brains of Leborgne and Lelong. Brain J Neurol 130(Pt 5):1432–1441. https://doi.org/10.1093/brain/awm042
6. Pustina D, Avants B, Faseyitan OK, Medaglia JD, Coslett HB (2017) Improved accuracy of lesion to symptom mapping with multivariate sparse canonical correlations. Neuropsychologia. https://doi.org/10.1016/j.neuropsychologia.2017.08.027
7. Liew SL, Anglin JM, Banks NW, Sondag M, Ito KL, Kim H et al (2018) A large, open source dataset of stroke anatomical brain images and manual lesion segmentations. Sci Data 5:180011. https://doi.org/10.1038/sdata.2018.11

Contents

Preface to the Series v
Preface vii
Contributors xv

1 Defining the Lesion for Lesion-Symptom Mapping 1
Shannon M. Sheppard, Andrea L. C. Schneider, and Argye E. Hillis

2 Manual Lesion Segmentation 27
Casey Ferrara, Branch Coslett, and Laurel Buxbaum

3 Automated Lesion Segmentation 53
Joseph C. Griffis and Dorian Pustina

4 Mapping the Spatial Distribution of Lesions in Stroke: Effect of Diffeomorphic Registration Strategy in the ATLAS Dataset 79
Brian B. Avants and Nicholas J. Tustison

5 Voxel-Based Lesion Symptom Mapping 95
Juliana V. Baldo, Maria V. Ivanova, Timothy J. Herron, Stephen M. Wilson, and Nina F. Dronkers

6 Statistical Considerations in Voxel-Based Lesion-Behavior Mapping 119
Christoph Sperber and Hans-Otto Karnath

7 Voxel-Based Brain-Behavior Mapping in Neurodegenerative Diseases 135
Sladjana Lukic, Valentina Borghesani, Maria Luisa Gorno-Tempini, and Giovanni Battistella

8 Lesion-Network Mapping: From a Topologic to Hodologic Approach 149
Amy Kuceyeski and Aaron Boes

9 Connectome-Based Lesion-Symptom Mapping Using Structural Brain Imaging 167
Ezequiel Gleichgerrcht, Janina Wilmskoetter, and Leonardo Bonilha

10 Lesion Network Mapping Using Resting-State Functional Connectivity MRI 181
Juho Joutsa, R. Ryan Darby, and Michael D. Fox

11 Multivariate Lesion-Behavior Mapping 199
Yee-Haur Mah, Ashwani Jha, Tianbo Xu, and Parashkev Nachev

12 Lesion-Based Prediction and Predictive Inference 219
Daniel Mirman and Melissa Thye

13 Selecting and Handling Behavioral Measures for Lesion-Symptom Mapping 237
Peter E. Turkeltaub, Andrew T. DeMarco, and Maryam Ghaleh

14 Lesion-Behavior Awake Mapping with Direct Cortical and Subcortical Stimulation 257
Stephanie K. Ries, Kesshi Jordan, Robert T. Knight, and Mitchel Berger

15 Transcranial Magnetic Stimulation Mapping for Perceptual and Cognitive Functions 271
Samantha Strong and Edward H. Silson

Correction to: Voxel-Based Lesion Symptom Mapping C1

Appendix 1: Lesion-Symptom Mapping—Tools and Examples 289

Appendix 2: Automated Segmentation—Tools and Examples 337

Index 347

Contributors

BRIAN B. AVANTS • *Department of Radiology and Medical Imaging, University of Virginia, Charlottesville, VA, USA*
JULIANA V. BALDO • *Center for Language, Imaging, Mind & Brain (CLIMB), VA Northern California Health Care System, Martinez, CA, USA*
GIOVANNI BATTISTELLA • *Department of Neurology, Memory and Aging Center, Weill Institute for Neurosciences, University of California, San Francisco, CA, USA*
MITCHEL BERGER • *Department of Neurological Surgery, University of California San Francisco, San Francisco, CA, USA*
AARON BOES • *University of Iowa, Iowa City, IA, USA; UIHC, Iowa City, IA, USA*
LEONARDO BONILHA • *Department of Neurology, Medical University of South Carolina, Charleston, SC, USA*
VALENTINA BORGHESANI • *Université de Montreal, QC, Canada*
LAUREL BUXBAUM • *Moss Rehabilitation Research Institute, Elkins Park, PA, USA*
BRANCH COSLETT • *University of Pennsylvania, Philadelphia, PA, USA*
R. RYAN DARBY • *Vanderbilt University Medical Center, Nashville, TN, USA*
ANDREW T. DEMARCO • *Center for Brain Plasticity and Recovery, Georgetown University Medical Center, Washington, DC, USA*
NINA F. DRONKERS • *Center for Language, Imaging, Mind & Brain (CLIMB), VA Northern California Health Care System, Martinez, CA, USA; Aphasia Recovery Lab, University of California, Berkeley, CA, USA; Department of Neurology, University of California, Davis, CA, USA*
CASEY FERRARA • *University of Chicago, Chicago, IL, USA*
MICHAEL D. FOX • *Brigham and Women's Hospital, Harvard Medical School, Boston, MA, USA*
MARYAM GHALEH • *Center for Brain Plasticity and Recovery, Georgetown University Medical Center, Washington, DC, USA*
EZEQUIEL GLEICHGERRCHT • *Department of Neurology, Medical University of South Carolina, Charleston, SC, USA*
MARIA LUISA GORNO-TEMPINI • *Department of Neurology, Memory and Aging Center, Weill Institute for Neurosciences, University of California, San Francisco, CA, USA*
JOSEPH C. GRIFFIS • *Washington University, St. Louis, MO, USA*
TIMOTHY J. HERRON • *Center for Language, Imaging, Mind & Brain (CLIMB), VA Northern California Health Care System, Martinez, CA, USA*
ARGYE E. HILLIS • *Johns Hopkins University, Baltimore, MD, USA*
MARIA V. IVANOVA • *Center for Language, Imaging, Mind & Brain (CLIMB), VA Northern California Health Care System, Martinez, CA, USA; Aphasia Recovery Lab, University of California, Berkeley, CA, USA; Center for Language and Brain, National Research University Higher School of Economics, Moscow, Russia*
ASHWANI JHA • *UCL Queen Square Institute of Neurology, University College London, London, UK*
KESSHI JORDAN • *Department of Neurology, University of California San Francisco, San Francisco, CA, USA*
JUHO JOUTSA • *University of Turku and Turku University Hospital, Turku, Finland*

HANS-OTTO KARNATH • *Center of Neurology, Division of Neuropsychology, Hertie-Institute for Clinical Brain Research, University of Tübingen, Tübingen, Germany; Department of Psychology, University of South Carolina, Columbia, SC, USA*
ROBERT T. KNIGHT • *Department of Psychology, University of California Berkeley, Berkeley, CA, USA; Helen Wills Neuroscience Institute, University of California Berkeley, Berkeley, CA, USA*
AMY KUCEYESKI • *Weill Cornell Medicine, New York, NY, USA*
SLADJANA LUKIC • *Department of Neurology, Memory and Aging Center, Weill Institute for Neurosciences, University of California, San Francisco, CA, USA*
YEE-HAUR MAH • *King's College London, London, UK*
DANIEL MIRMAN • *Department of Psychology, University of Edinburgh, Edinburgh, UK*
PARASHKEV NACHEV • *UCL Queen Square Institute of Neurology, University College London, London, UK*
DORIAN PUSTINA • *Experimental Neuroimaging, CHDI Management/CHDI Foundation, Princeton, NJ, USA*
STEPHANIE K. RIES • *School of Speech, Language, and Hearing Sciences, San Diego State University, San Diego, CA, USA; Center for Clinical and Cognitive Neuroscience, San Diego State University, San Diego, CA, USA; Joint Doctoral Program in Language and Communicative Disorders, San Diego State University and the University of California San Diego, San Diego, CA, USA*
ANDREA L. C. SCHNEIDER • *University of Pennsylvania, Philadelphia, PA, USA*
SHANNON M. SHEPPARD • *Department of Communications Sciences & Disorders, Chapman University, Orange, CA, USA; Johns Hopkins University, Baltimore, MD, USA*
EDWARD H. SILSON • *Department of Psychology, School of Philosophy, Psychology & Language Sciences, University of Edinburgh, Edinburgh, UK*
CHRISTOPH SPERBER • *Center of Neurology, Division of Neuropsychology, Hertie-Institute for Clinical Brain Research, University of Tübingen, Tübingen, Germany*
SAMANTHA STRONG • *Optometry and Vision Science, Aston University, Birmingham, UK*
MELISSA THYE • *Department of Psychology, University of Edinburgh, Edinburgh, UK*
PETER E. TURKELTAUB • *Center for Brain Plasticity and Recovery, Georgetown University Medical Center, Washington, DC, USA; Research Division, MedStar National Rehabilitation Hospital, Washington, DC, USA*
NICHOLAS J. TUSTISON • *Department of Radiology and Medical Imaging, University of Virginia, Charlottesville, VA, USA*
JANINA WILMSKOETTER • *Department of Neurology, Medical University of South Carolina, Charleston, SC, USA*
STEPHEN M. WILSON • *Language Neuroscience Laboratory, Department of Hearing and Speech Sciences, Vanderbilt University Medical Center, Nashville, TN, USA*
TIANBO XU • *UCL Queen Square Institute of Neurology, University College London, London, UK*

Chapter 1

Defining the Lesion for Lesion-Symptom Mapping

Shannon M. Sheppard, Andrea L. C. Schneider, and Argye E. Hillis

Abstract

Brain lesions can have many causes including injury, disease, and infections. Lesion-symptom mapping is a tool used by investigators to understand the relationship between brain structure and function. There are many different types of brain lesions with varying characteristics that researchers must consider when deciding which participants to include to best answer their specific research questions of interest. In this chapter, we discuss the different types of lesions that may be used in brain mapping research, the characteristics of those lesions including the stability of lesions over time, and appropriateness and challenges specific to each lesion type for behavior mapping at various time points. Different types of lesions discussed include lesions resulting from ischemic or hemorrhagic stroke, traumatic brain injury, neurodegenerative disease, brain tumors, surgical resection of lesions, brain abscesses, and transient lesions from transcranial magnetic stimulation (TMS). As the brain will reorganize during spontaneous recovery processes and rehabilitation following an insult, the importance of considering not only which types of lesions to study, but also the time point at which they are studied, is also discussed.

Key words Brain lesions, Lesion-symptom mapping, Stroke, Neurodegenerative disease, Traumatic brain injury

1 Introduction

Brain lesions can have many causes, including injury, disease, and infections. Lesion-symptom mapping is a tool used by investigators to understand the relationship between brain structure and function. There are many different types of brain lesions with varying characteristics that researchers must consider when deciding which participants to include to best answer their specific research questions of interest. One important point to consider is that the brain will reorganize during spontaneous recovery processes and rehabilitation following an insult, so investigators must also carefully consider not only which types of lesions to study but also the time point at which they are studying them. In this chapter, we discuss the different types of lesions that may be used in brain mapping research, the characteristics of those lesions including the stability of lesions over time, and appropriateness and

Dorian Pustina and Daniel Mirman (eds.), *Lesion-to-Symptom Mapping: Principles and Tools*, Neuromethods, vol. 180, https://doi.org/10.1007/978-1-0716-2225-4_1, © Springer Science+Business Media, LLC, part of Springer Nature 2022

challenges specific to each lesion type for behavior mapping at various time points.

2 Defining Different Types of Lesions

2.1 Lesions Resulting from Stroke

The main categories of strokes are ischemic strokes and hemorrhagic strokes. Ischemic strokes result from clots or atherosclerotic plaque that occludes or severely narrows an artery, limiting blood flood to brain tissue. Transient ischemic attacks (TIAs) also result from limited blood flow due to occluded or narrow arteries, but the symptoms of TIAs are short-lived, and they do not cause permanent brain damage. Hemorrhages occur when weakened blood vessels rupture. Secondary intracerebral hemorrhage can occur when ischemic strokes undergo hemorrhagic conversion when the blood vessels around the primary ischemic infarct lose the ability to retain blood. Defining lesions at the acute stage varies across etiologies. At the chronic stage, while ischemic strokes and primary intracerebral hemorrhage both produce focal lesions that can be defined using identical methods, subarachnoid hemorrhage can cause more diffuse damage at the chronic stage. Of all strokes, 87% are ischemic strokes, 10% are intracerebral hemorrhagic strokes, and 3% are subarachnoid hemorrhages. Subarachnoid hemorrhages consist of bleeding in the space between the brain parenchyma and the arachnoid mater (part of the meninges) and can cause both focal deficits and more general deficits, such as reduced level of consciousness. Subdural and epidural hematomas are caused by bleeding outside of the arachnoid mater, and do not typically cause focal neurological deficits (but do cause increased intracranial pressure), so they will be considered only under traumatic brain injury (the most common cause). Defining lesions for these different types of strokes at acute and chronic stages will be discussed below.

2.2 Acute Stage of Stroke

2.2.1 Ischemic Stroke Resulting from Arterial Occlusion, Stenosis, or Embolism

Ischemic stroke results from inadequate blood flow to tissue, causing subsequent tissue death. After an ischemic stroke, there is often some degree of dynamic reorganization of language and cognitive networks, such that undamaged areas assume the functions of the damaged areas. However, it is difficult at later times to determine how much reorganization has taken place and how beneficial reorganization has been for restoration of function. Functional imaging, including functional magnetic resonance imaging (fMRI) and positron emission tomography (PET) research, has demonstrated that areas activated during language tasks change over time as a function of natural recovery and behavioral treatment [1, 2]. Therefore, when considering participants to include, it is important to determine whether it would be best to include participants at the acute phase of stroke (before reorganization or recovery) or the chronic phase of recovery (generally at least 6 months post-stroke) when structure-function relationships have more or less stabilized in the newly reorganized brain. This decision largely depends on the question to be studied and access to patients.

An advantage of studying acute stroke is that it allows one to evaluate areas of the brain that are normally critical for a given function, before reorganization. Another advantage is that it allows one to study the effects of very small lesions; patients with small lesions typically recover completely before 6 months after stroke. However, one important challenge in studying acute stroke is that the lesion itself may be evolving. Deficits at the acute stage are often due to both the infarct and the area of low blood flow (hypoperfusion) surrounding the infarct. This area, where there is adequate blood flow for neurons to survive, but not enough for neurons to function, is often known as the "ischemic penumbra." In this area, restoration of blood flow not only prevents the area from progressing to infarct but also restores function of the area. Therefore, to capture the entire area of dysfunctional brain tissue, it is essential to include both structural imaging and blood flow imaging (e.g., with perfusion-weighted MRI (PWI), arterial spin labeling (ASL) perfusion MRI, CT perfusion (CTP), or PET). There can be an area of significant hypoperfusion beyond the infarct that contributes to the deficit for days to weeks after stroke onset [3, 4] (Fig. 1). However, with each of these methods of perfusion imaging, it is important to define the threshold of blood flow change that actually causes

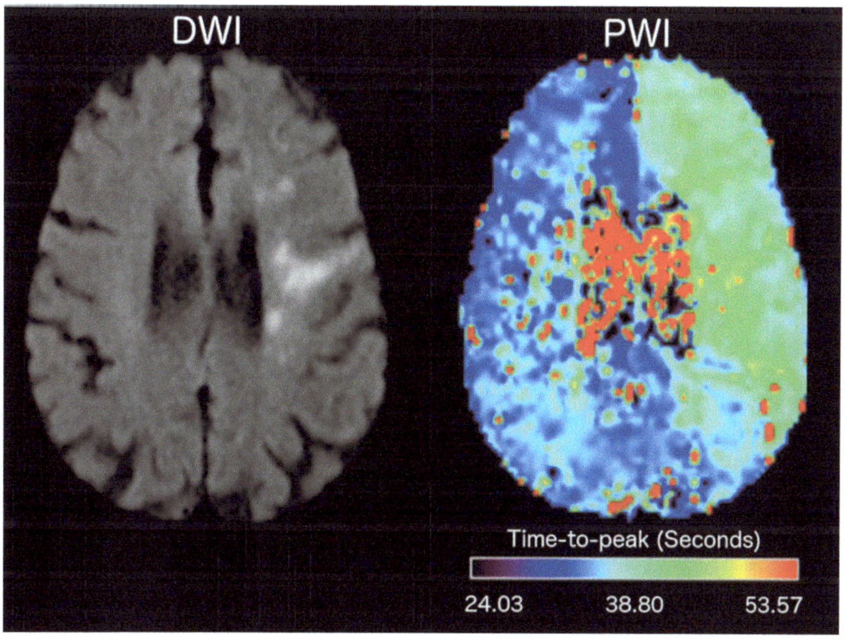

Fig. 1 Diffusion-weighted image (DWI, left) showing core area of densely ischemic tissue unlikely to survive (white), and perfusion-weighted image (PWI, right) showing the larger area of poor perfusion in green

deficits. For example, a delay in blood flow of just 1–2 s PWI or CTP time to peak (TTP) arrival or time to drain (TTD) does not cause dysfunction of the neurons. A delay of at least 4 s is necessary to cause functional deficits [3, 5], and this threshold (3.9–6.6 s delay) corresponds to PET measures of rCBF of <20 mL/100 g/min [6]. Early studies of primates established that when rCBF dropped below 20 mL/100 g/min, neurons ceased to function, but did not start to die until rCBF dropped below 10 mL/100 g/min [7].

The structural lesion may also be difficult to define in acute stroke. The infarct is not visible on computed tomography (CT) for several hours (up to 24 h) and is best seen on diffusion-weighted image (DWI) sequences of MRI. On DWI, acute lesions will appear bright white and can be distinguished from old chronic lesions, which generally appear dark (Fig. 2). The age of the lesion should be confirmed with the apparent diffusion coefficient (ADC) sequence, on which acute lesions (<10 days) appear dark (low ADC) and chronic lesions (>14 days) appear light (high ADC because CSF has filled the area where brain tissue once existed), as shown in Fig. 2. The DWI trace image is derived from the ADC map, as it is easier to visualize bright white (infarct on DWI trace images) on gray brain than dark (infarct on ADC) on gray brain. Fluid-attenuated inversion recovery (FLAIR) imaging is a T2-weighted sequence that can be used to detect ischemic changes after the first few hours of stroke, although it cannot be used to detect hyperacute ischemic lesions in the initial hours following a stroke [8, 9]. Signal change in FLAIR scans is very subtle during the first several hours of acute ischemia, but as they become more

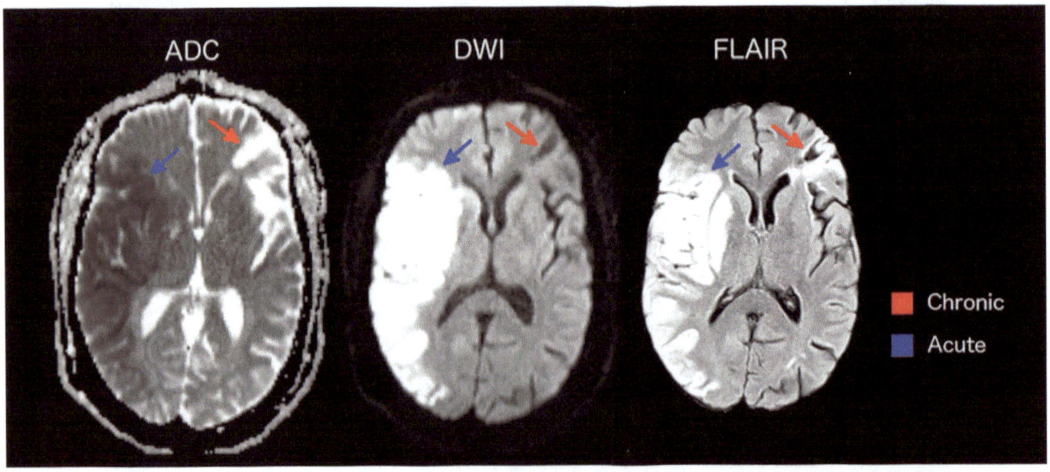

Fig. 2 Chronic vs. acute lesions. ADC (left), DWI (center), and FLAIR (left). Blue arrows point to acute infarct (dark on ADC, bright on DWI and FLAIR). Red arrows point to older infarct (bright on ADC, dark on DWI and FLAIR). Note the chronic lesion is visualized with greater clarity on FLAIR image, and the bright area surrounding the dark chronic lesion indicates hypoperfused tissue surrounding the lesion

pronounced, lesions appear white. DWI is preferred to FLAIR in the first few days after stroke, because the lesions are more easily identified.

Functional imaging can be difficult to interpret in acute stroke because areas around the infarct are likely to be compromised due to low blood flow, which may or may not proceed to infarct. These areas may not show a blood-oxygen-level-dependent (BOLD) effect in fMRI studies even in the presence of neural activation [10, 11] because of an inability to show a hemodynamic response to activation. That is, the BOLD effect depends on the fact that in normal tissue, there is an increase in blood flow, out of proportion to oxygen consumption, in response to activation of an area. In hypoperfused tissue, the vessels are often maximally dilated in an effort to perfuse the tissue and cannot provide increased blood flow. In fact, there can be a reverse BOLD effect because oxygen consumption by activated neurons exceeds blood flow. Recent studies show the promise of oxygen extraction MRI, which could show areas of activation in poorly perfused tissue [12]. Other investigators have shown the potential usefulness of individualized modelling of the hemodynamic response in areas of compromised blood flow [13, 14]. This individualized hemodynamic response modelling may be useful in task-based fMRI studies in all stages of stroke recovery.

2.2.2 Acute Primary Intracerebral Hemorrhage

Primary intracerebral hemorrhage occurs when a blood vessel within the brain bursts, causing blood to leak inside the brain. Intracranial pressure rises due to the blood that is released into the brain cavity and due to the development of cerebral edema, which increases progressively in the hours and days after injury. The primary injury results from hematoma formation and its expansion within the brain, which causes damage to brain tissue close to the site of hemorrhage. The secondary injury arises from cerebral edema (due to inflammation, mechanical pressure, thrombin production, and so on) and thus results in more diffuse damage throughout the brain depending on how well intracerebral pressure is managed and reduced. This diffuse damage can cause many general deficits; therefore, patients with primary intracerebral hemorrhage are typically not ideal candidates for mapping structure to function in the brain at the acute stage. However, at the chronic stage of recovery, researchers can choose to enroll these patients since they have focal lesions that can be isolated without the confound of general deficits caused by cerebral edema.

The appearance of the hematoma on imaging changes rapidly in the first couple of weeks after a stroke. Imaging primary intracerebral hemorrhage at the hyperacute stage, within several hours of stroke, often includes noncontrast CT (NCCT) with or without MRI. NCCT can be used to quantify the volume of the hematoma and to monitor the evolution of the hemorrhage over several days

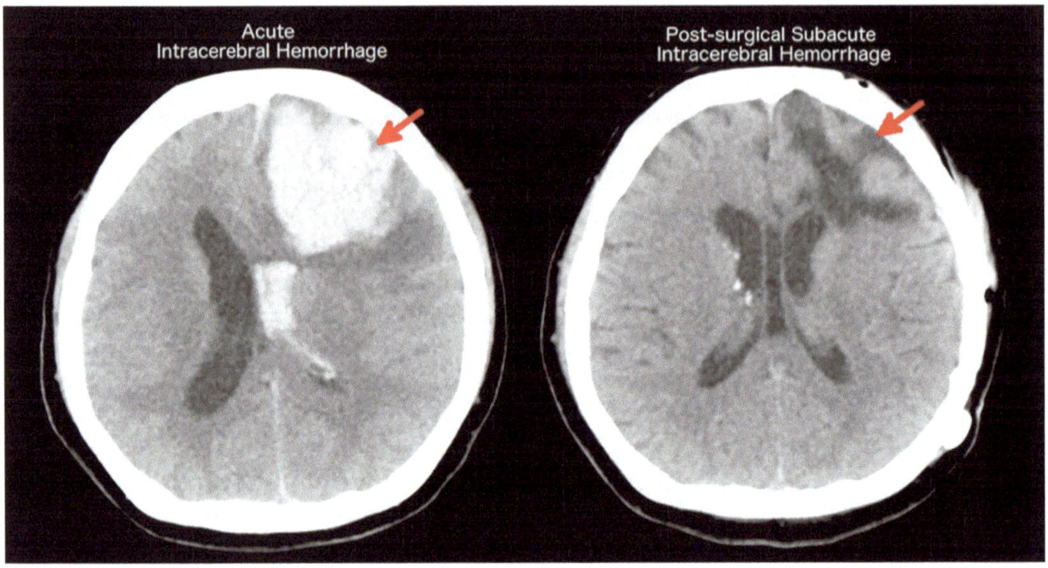

Fig. 3 Intracerebral hemorrhage on noncontrast CT (NCCT). Acute left frontal intracerebral hemorrhage (left) and subacute lesion 2 weeks postsurgery (right). Arrows point to acute intracerebral hemorrhage with extension into the lateral ventricle (bright on NCCT) and resolved hemorrhage 2 weeks postsurgery with infarcted tissue (dark on NCCT)

[15]. Primary intracerebral hemorrhage appears hyperdense (bright white) on NCCT in the first 48 h of stroke and then evolves to a hypodense (dark) region of edema with mass effect in the next several days (Fig. 3). Finally, the lesion shrinks and becomes less intense and more uneven with a ring-like appearance 3–20 days post-onset [16].

MRI can also be used for acute primary intracerebral hemorrhage imaging. The most sensitive sequences to both old and new blood are susceptibility-weighted imaging (SWI) and gradient recalled echo (GRE), but these sequences do not distinguish between old and new blood. The combination of T1 and T2 can determine the age of the hemorrhage. MRI is also helpful for distinguishing primary intracerebral hemorrhage from ischemic strokes followed by hemorrhagic transformation. Moreover, there is typically a larger area of cerebral edema surrounding the hematoma in primary intracerebral hemorrhage, which is well-visualized by MRI.

2.2.3 Acute Venous Infarct

Venous infarction is a relatively rare form of stroke [17] that results from a blood clot in the brain's venous sinuses, which blocks blood from draining out of the brain. The subsequent increase in pressure can cause cerebral hemorrhage. Acute clinical presentation varies widely with this form of stroke and can include both focal deficits and more general deficits, particularly when extensive cerebral edema causes mass effect and hydrocephalus. These patients are not typically candidates for brain mapping because it is difficult to tease apart focal deficits caused by the infarct from deficits caused by cerebral edema. Imaging used for diagnosis at the acute stage often includes contrast-enhanced CT, as well as MRI. CT venography and MR venography are also being used as diagnostic tools, particularly in cases where venous infarct is suspected but cannot be imaged with CT or MRI alone [18]. Once patients are properly diagnosed and treated, they typically experience a good recovery and therefore are not often used in chronic stroke studies.

2.3 Chronic Stage of Recovery After Stroke

2.3.1 Focal Lesions

Studying participants in the chronic stage after an ischemic stroke or primary intracerebral hemorrhage allows one to identify areas where damage often results in a failure to recover a particular function. Advantages of studying patients long after stroke include the fact that they are more medically stable and often able to tolerate longer assessment periods. Nevertheless, many lesion factors must be considered when using chronic lesions for lesion-deficit mapping. CT is adequate and equivalent to MRI for delineating large, chronic lesions. MRI scans are preferable for identifying smaller chronic lesions and white matter changes (leukoaraiosis) because MRIs have better image quality due to better signal-to-noise ratio (SNR) and contrast resolution (Fig. 2). T2 and FLAIR images are ideal for visualizing and quantifying white matter changes, which have recently been shown to contribute to the functional impairment [19] and impair recovery after stroke [20, 21].

Diffusion tensor imaging (DTI) can be used to assess how white matter tracts have been damaged by stroke by detecting disruptions of specific tracts (tractography) or reduction in measures, such as fractional anisotropy. DTI can also be used to investigate Wallerian degeneration, which is the secondary degeneration of white matter fibers distant to the primary lesion occurring from 2 months to many years post-stroke [22, 23]. Furthermore, DTI allows for the investigation of damage to networks in the brain rather than focusing solely on damage to specific regions in isolation. For example, recent research in patients with impaired language processing following a stroke has focused on using DTI to study how lesions in white matter tracts that are part of the language network (e.g., tracts connecting language areas such as Broca's area, Wernicke's area, etc.) impact various aspects of language processing [24].

Functional imaging studies (described elsewhere in this book) typically require a high-resolution T1 image, such as a high-resolution three-dimensional (3D) magnetization-prepared rapid acquisition with gradient echo (MPRAGE) or spoiled gradient recalled acquisition (SPGR), for adequate normalization of the functional images to a common space or template. Furthermore, to interpret activation patterns, it is essential to define the lesion.

The total "lesion" may include areas of hypoperfusion, as described above in acute stroke. Chronic hypoperfusion can also affect the BOLD response to activation [11, 25, 26], which can be minimized by examining individual hemodynamic response functions (HRF) rather than relying on canonical HRF in statistical analyses of fMRI data. An alternative approach is to identify any significant hypoperfusion extending beyond the infarct, using CT perfusion, MR dynamic contrast perfusion (PWI), or arterial spin labeling (ASL) as described earlier. Areas of hypoperfusion as defined by a specific threshold can be considered to be part of the lesion itself. For example, a 4-s delay in time to peak (TTP) arrival of contrast compared to the homologous region in the unimpaired hemisphere on dynamic contrast PWI can be considered to be dysfunctional tissue and can be included as part of the lesion [27]. ASL can also be used to calibrate fMRI BOLD responses [28, 29].

It is also unclear how diaschisis impacts the BOLD signal. Diaschisis refers to a reduction of blood flow and metabolism in cortical areas distant from the lesion. It can be helpful to use converging evidence from different imaging methods to determine the relationship between blood flow, the BOLD signal, and neuronal activity. For example, Fair and colleagues [30] investigated how chronic diaschisis measured by resting PET would affect task-evoked BOLD responses during a word-stem completion task in a block design fMRI study. The study included three patients who all had a single left hemisphere stroke involving the inferior frontal gyrus (IFG) and the operculum and were 6 months post-stroke. BOLD responses were only minimally affected in areas with chronic diaschisis.

One variable that is rarely considered adequately in lesion-deficit mapping studies of chronic stroke is the timing of the imaging to define the lesion and the behavioral assessment. It is frequently assumed that both the lesion and the behavior are stable at the chronic stage. While both the lesion and the behavior are more stable long after stroke than acutely after stroke, both frequently change even more than 1-year post-stroke [31, 32]. If imaging substantially precedes behavior testing, some participants will likely have developed some "silent" (asymptomatic) stroke or extension of the infarct in the interim between scanning and behavioral assessment. More importantly, behavior can substantially change even in the chronic stage, often in response to language therapy or other intervention. Therefore, if lesion-symptom mapping were to be repeated with the same individuals 1 year later (perhaps before and after treatment), different results would be obtained if some of the patients showed substantial recovery by the second time point [29]. Ideally, participants should be homogeneous with respect to time post-onset and interventions, so that one could conclude that lesion-symptom mapping revealed areas where damage is associated with incomplete recovery of behavior X by Y months post-onset when individuals have received treatment Z. However, given that it is nearly impossible to identify individuals who have undergone identical interventions over a year, lesion-symptom mapping in chronic stroke should be interpreted with these caveats in mind. The variability in time post-onset and interventions likely explains much of the variability in results across studies [32]. Only the most robust lesion-deficit associations will be consistently obtained, given the noise introduced by the variability in recovery. However, in datasets with very chronic lesions (often several years post stroke), there may be very little variability due to time post-onset, as there is generally less recovery years after stroke.

3 Traumatic Brain Injury

Traumatic brain injury (TBI) is a structural brain injury that results from an external physical force which causes disruption of the normal architecture and function of the brain [33]. Brain lesions associated with TBI are heterogeneous and occur both as a result of the primary injury sustained at the time of impact and as a result of secondary injuries from subsequent hypoxia, hypercarbia, hypotension, and brain swelling that occur after the initial head injury [31].

3.1 Defining Lesions Due to TBI for the Purposes of Brain Mapping

Traditional lesion-symptom mapping in TBI patients is complicated by the fact that the lesions that are clearly visible on CT and/or MRI are typically associated with less visible, more diffuse injury as described above. However, many investigators have studied the cognitive or language deficits associated with disruptions of white matter tracts caused by TBI, typically in the chronic stage of recovery [34–36]. The idea is that acute microbleeds and DAI cause disruptions of white matter tracts that are later measurable with DTI, including fiber tracking, fractional anisotropy (FA), mean diffusivity (MD), and radial diffusivity (RD) [37].

3.2 Lesions of Primary Brain Injury

Primary brain injury from TBI is characterized by three broad types of brain lesions: hematomas, contusions, and diffuse axonal injury [31]. Hematomas can be subdivided into epidural, subdural, and subarachnoid hemorrhages and are well visualized on CT. Epidural hematomas, where bleeding occurs between the skull and the dura mater, typically consist of arterial blood, but sometimes venous blood, in the space between the dura and inner skull. Epidural hematomas are biconvex (lens-shaped) hyperdensities that are external to the brain parenchyma (extra-axial) and are often associated with skull fractures, which are well visualized on CT [38]. Subdural hematomas are located in the space between the dura and arachnoid and are typically caused when the veins in the meninges rupture due to trauma. On head CT, subdural hematomas are crescent-shaped, extra-axial hyperdensities. Over the course of weeks, epidural and subdural hematomas decrease in attenuation from hyperattenuating to isoattenuating [38]. In TBI, subarachnoid hemorrhage is typically located over the periphery of the cerebral hemispheres and is often accompanied by intraventricular hemorrhage. Subarachnoid hemorrhages appear as increased CT attenuation along sulci, Sylvian fissures, and basilar cisterns. In addition to head CT, SWI or GRE and FLAIR MRI sequences are also reliable for detection of extra-axial hematomas [39].

Contusions are bruises or lacerations of the brain parenchyma which occur when the brain forcibly impacts the irregular surface of the skull at the site of trauma (coup injury) or opposite the site of trauma (contrecoup injury) [38]. Contusions can be focal or multifocal and in mild injury are typically localized to the cortex, especially the inferior frontal lobes and temporal lobes. In more severe injuries, contusions may extend to subcortical white matter. Contusions may be hemorrhagic or non-hemorrhagic in nature [38]. Between 25% and 35% of hemorrhagic contusions expand in size over several hours after trauma. Approximately 15% of non-hemorrhagic contusions progress to microhemorrhages or hemorrhagic contusions over the first 6–9 h after injury [40]. On CT, hemorrhagic contusions are characterized as intraparenchymal increased attenuation with or without surrounding edema. Smaller hemorrhagic contusions are often not seen on head CT but are seen on MRI, SWI, and GRE sequences [41]. Non-hemorrhagic contusions restrict diffusion and are best identified as bright on DWI sequences and dark on apparent diffusion coefficient (ADC) sequences in the acute setting, like acute stroke. Non-hemorrhagic contusions are hyperintense on FLAIR sequences in the subacute period and transform to encephalomalacia over months to years [38].

Diffuse axonal injury (DAI) is caused by angular or rotational acceleration, causing shear strain and stretching of axons [42]. MRI is the imaging modality of choice for assessing DAI. DAI can be characterized by both hemorrhagic and non-hemorrhagic lesions. DAI lesions are small (i.e., 1–15 mm), ovoid lesions that are parallel to axonal tracts [38]. Hemorrhagic DAI lesions can be seen on GRE and SWI sequences as an increase in magnetic susceptibility (seen as black on light-gray brain tissue), and non-hemorrhagic DAI lesions restrict diffusion and are bright on DWI and dark on ADC MRI sequences [38]. Imaging grading of DAI is based on the histopathological grading system proposed by Adams and colleagues [43]. DAI is defined by the following grades: Grade 1, DAI lesions in the cerebral hemispheres, particularly at the gray-white junction in the frontal and temporal lobes; Grade 2, DAI lesions in the corpus callosum; and Grade 3, DAI lesions in the dorsolateral brainstem [42]. DAI lesions can also be assessed on MRI DTI sequences, which is a measure of white matter microstructural integrity. In the immediate time after TBI with DAI, there is a reduction in fractional anisotropy (FA) and an increase in mean diffusivity (MD), suggesting impaired white matter microstructural integrity; these alterations in FA and MD persist for at least 6–12 months postinjury [44].

3.3 Lesions of Secondary Brain Injury

Secondary brain injury occurs over the hours and days following a primary brain injury and are due to subsequent hypoxia, hypercarbia, hypotension, and brain swelling, which result in a pathophysiologic cascade of events, including excitatory neurotransmitter release, free radical generation, calcium-mediated damage, gene activation, mitochondrial dysfunction, and inflammatory responses [45]. This cascade can lead to astrocyte swelling, cerebral edema, and neuronal and glial cell necrosis [38].

On imaging, secondary brain injuries are characterized by ischemia, edema, herniation, and hydrocephalus. TBI-related ischemia is often a result of cardiac/vascular compromise, including hypoxia and hypotension [46]. TBI-related ischemia is seen on CT with loss of gray-white differentiation, gray-white reversal (i.e., the white matter appears denser than the gray matter), and pseudo-subarachnoid hemorrhage, where the cisterns appear hyperdense to the brain [38]. On MRI, ischemic lesions restrict diffusion and are bright on DWI and dark on ADC MRI sequences.

Cerebral edema is commonly seen surrounding primary brain injury lesions and is characterized by hyperintensities on FLAIR MRI sequences. Cerebral edema can lead to mass effect, which can be seen by effacement of basal cisterns and midline shift, and herniation (i.e., the displacement of brain parenchyma into a different brain compartment) [38]. Multiple types of herniation can occur depending on the location of the TBI and can be seen on both CT and MRI, although MRI is better for posterior fossa due to beam hardening artifact in this area on CT [46] Subfalcine herniation occurs when structures, such as the cingulate gyrus, anterior cerebral artery, and lateral ventricle, are displaced under the falx cerebri. Transtentorial herniation results in brain parenchyma descending from the supratentorial to infratentorial compartment or vice versa. Uncal herniation is a specific type of transtentorial herniation that occurs with downward displacement of the medial temporal lobes, which puts pressure on the brainstem. Tonsillar herniation occurs with displacement of the cerebellar tonsils through the foramen magnum [47]. Cerebral edema and herniation can lead to hydrocephalus, which is seen on CT or MRI imaging as enlargement of the ventricular system [38].

4 Neurodegenerative Diseases

Neurodegenerative diseases cause the progressive loss of nerve cells, which can result in language deficits as in primary progressive aphasia (PPA), as well as dementia or motor deficits such as ataxia and Parkinsonism. Many different diseases fall under the umbrella of neurodegenerative disorders. Alzheimer's disease and Parkinson's disease are the most common, but they also include other disorders such as frontotemporal lobar degeneration (FTLD), Lewy body dementia, Huntington's disease, and amyotrophic lateral sclerosis. Each of these diseases is characterized by a different pattern of neuronal loss. For example, Alzheimer's disease is typically associated with atrophy in the medial temporal lobe (MTL) and particularly the hippocampus and entorhinal cortex [37], while FTLD is characterized by the progressive loss of neurons in the frontal and temporal lobes [48]. These diseases are associated with the aggregation of misfolded proteins with progressive loss of neurons. Several different types of proteins have been implicated, including tau, Pick bodies, β-amyloid, α-synuclein, and prions. Patients with neurodegenerative diseases are typically good candidates for lesion-symptom mapping, although since they are experiencing progressive neuronal loss, it is vital that imaging and behavioral testing are performed very close to one another in time. Also, neurodegenerative diseases typically result in relatively diffuse lesions rather than focal lesions, such as typically seen in stroke. Regional volumetric loss can be mapped to behavior using voxel-based morphometry (*see* Chapter 7).

Areas of atrophy can be imaged using high-resolution structural MRI techniques, which allow for the examination of gray and white matter loss, as well as overall brain volume loss. Many voxel-based morphometry studies have identified areas of focal atrophy associated with deterioration in a variety of functions and behaviors [49–55]. Studies have variously identified areas of volume loss or areas of cortical thinning. Although there are theoretical reasons for believing areas of cortical thinning might be more sensitive to change in behavior, particularly in frontotemporal dementia [56–58], several studies have found that the degree of volume loss in regions of interest correlated more strongly with behavior (e.g., naming accuracy) than cortical thinning in the same areas [59, 60].

Even neurotypical controls have a wide range of brain shapes and sizes, so comparing diseased brains to a "normal brain" to identify areas of atrophy can be problematic, even if age- and sex-matched controls are used. One way to avoid this problem is to track atrophy by conducting longitudinal studies, in which imaging is acquired at multiple time points so atrophy can be quantified in comparison to each individual's baseline scan. For example, Faria and colleagues [61] investigated auditory word comprehension deficits in three variants of PPA, a neurological syndrome where two variants are associated with FTLD and the third with Alzheimer's disease. The authors acquired cognitive testing and MPRAGE scans at least 9 months apart and examined the relationship between word comprehension deficits and focal atrophy. Rather than co-registering brains to a neurotypical template, areas of longitudinal atrophy were obtained using an "intra-subject" registration by mapping the second time point to the first time point. Using this technique, the authors determined that focal atrophy in left and right middle temporal cortex, left angular gyrus, and right inferior temporal cortex between the two time points most strongly correlated with the degree of decline in word comprehension over those two time points in PPA.

On MRI, atrophy is characterized by prominent cerebral sulci and enlarged ventricles (Fig. 4). Relatively focal atrophy can be identified by comparing volumes of structures to age-matched controls, or by measuring change in volume of each voxel within subjects over time, with repeat imaging [61]. DTI can be used to detect white matter abnormalities in neurodegenerative disorders, as it is sensitive to microstructural white matter alterations. For example, low fractional anisotropy (one measure from DTI) may indicate various types of damage such as demyelination, gliosis, or axonal degeneration [62]. DTI can be helpful both for lesion-symptom mapping and for differentiating between dementia

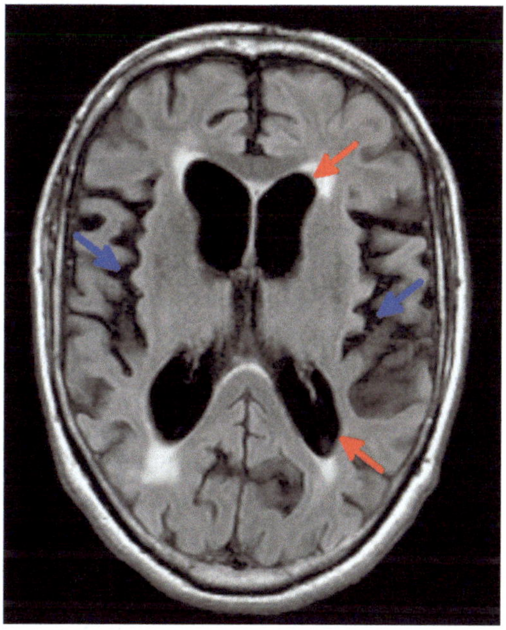

Fig. 4 Neurodegeneration. FLAIR scan demonstrates prominent cerebral sulci (blue arrows) and enlarged ventricles (red arrows) that are markers of neurodegenerative diseases

variants. For example, Mahoney and colleagues [63] discovered that measures of fractional anisotropy were optimal for differentiating a group of patients with the behavioral variant of FTD from those with Alzheimer's disease. ASL MRI and FDG PET have also become increasingly helpful for clinical differential diagnosis by showing areas of slow blood flow (ASL) or reduced metabolism (FDG PET). PET and MRI are becoming increasingly specific as more advanced labeling techniques and tracers have continued to be developed. For example, rather than relying on biopsy or autopsy, amyloid MRI and tau PET imaging can now be used to measure early biomarkers of disease and track disease progression longitudinally [64]. Functional imaging may also be used for investigating cognitive decline associated with neurodegenerative disease. Changes in resting-state (task-free) functional MRI within various language and cognitive networks can be investigated in association with deficits [65–67]. Neuronal loss is associated with reduced functional connectivity between network regions and can be detected even when atrophy is not prominent [14, 68]. However, researchers should be aware that brain atrophy can potentially bias both FDG PET and fMRI measures [69], so results should be interpreted with these considerations in mind.

5 Brain Tumors

Brain tumors can be classified as either primary tumors, where the abnormal growth of cells originates in the brain itself, or metastatic (secondary) tumors that began as a malignant tumor in another part of the body and subsequently spread to the brain. There are many types of primary brain tumors which can grow from the glial cells (glioma), nerve cells (neuroma), the dura (meningioma), or the pituitary gland (craniopharyngioma or pituitary adenoma). Brain tumors can affect behavior through several mechanisms, including damage to the brain cells themselves and dysfunction due to increased intracranial pressure. Intracranial pressure increases as a tumor grows and brain tissue is displaced, sometimes blocking the flow of cerebral spinal flow (CSF), resulting in cerebral edema with mass effect.

Several characteristics of brain tumors must be considered to determine the appropriateness for mapping affected brain regions to behaviors. One important consideration is the speed at which the tumor is growing. Tumors that are slow-growing are likely not ideal candidates for brain mapping because brain networks have an abundance of time to reorganize to incorporate alternative brain regions or tracts, which is not the case with fast-growing tumors. Thus, patients with slow-growing tumors may not have many (or any) deficits because time has allowed for brain reorganization. As discussed in the "Lesions Resulting from Stroke" section above, when a brain region is damaged, critical language and cognitive functions will often at least partially be co-opted by undamaged areas. Since timing is critical, investigators not only should consider tumor growth rate but should also ensure neuroimaging and behavioral testing are obtained at the same time to capture the impact of the tumor on functioning. Additionally, patients with metastatic tumors may be too sick to participate in behavioral testing, so it may be best to only include patients with primary tumors.

CT may be used to visualize brain tumors, but MRI is preferable. It is common to use FLAIR, pre- and post-gadolinium T1 sequences, as well as high-resolution 3D T2 gradient echo sequences like SWI [70]. Structural MRI can be used to identify the lesion location, including differentiating extra-axial tumors outside the brain parenchyma from intra-axial tumors in the parenchyma, and evaluate mass effect. FLAIR is helpful for visualizing the tumor as well as peritumoral edema. Comparing pre- and post-contrast T1 scans allows researchers to distinguish between enhancing portions of the tumor (uptake of gadolinium in the lesion) and non-enhancing portions (no gadolinium uptake) (Fig. 5). SWI evaluates susceptibility differences between tissues and is particularly sensitive to differences in calcium, ferritin (the primary form of iron stored inside cells), and deoxygenated blood [71]. Therefore, it is particularly helpful for detecting hemorrhage and micro-hemorrhage. DWI can also be used in brain tumor imaging to calculate ADC value, which reflects the degree of diffusion of water molecules throughout various tissues. ADC values vary in different tumor types; for example, lower ADC values are often found in higher-grade gliomas when compared to lower-grade gliomas [72]. DTI is also beneficial particularly when studying patients who are undergoing radiation therapy, because even in cases where white matter appears normal, radiation can impact fractional anisotropy in white matter tracts [73].

Functional imaging can be used for evaluating activation associated with tasks in participants with brain tumors, but researchers should be cautious in interpreting results. Brain tumors can lead to neurovascular uncoupling, where the presence of brain tumors affects the association between increased brain activation and blood flow [74]. This can potentially invalidate fMRI results because accurate fMRI data relies on the strong relationship between increased brain activation and blood flow to activated neural regions. Direct cortical stimulation (DCS) provides an alternative for functional mapping in participants with brain tumors because it allows for direct temporary inhibition of the brain tissue that is electrically "stimulated," and neurovascular uncoupling is not a concern [75–81]. However, DCS can only be done during awake craniotomy or via an implanted subdural grid of electrodes and covers limited brain areas (areas being considered for resection,

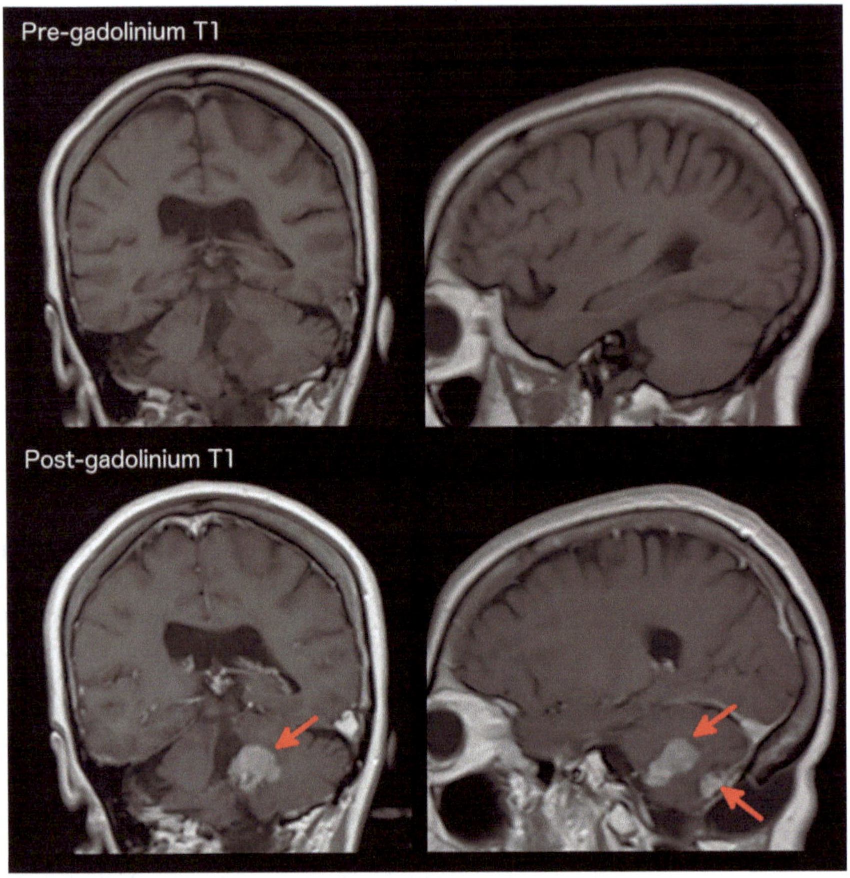

Fig. 5 Brain tumor. Pre- and post-gadolinium T1 sequences with coronal (left) and sagittal (right) views. Comparing pre- and postcontrast T1 scans distinguishes enhancing portions of the tumor (uptake of gadolinium) from non-enhancing portions. Arrows point to an enhancing mass in post-gadolinium scans

see Chapters 14 and 15). Cortical recording from implanted subdural grids can also be used to map areas of activation associated with behaviors, including language.

6 Surgical Resection of Lesions

Patients may undergo surgical resection of lesions for a variety of reasons. For example, patients may undergo surgery to remove part or all of a tumor. Also, for example, about 6% of patients with epilepsy in the USA received a full or partial lobectomy in the early 2000s [82], and this percentage is increasing over time. Participants who have had lesions resected are often not ideal candidates for brain mapping research because even though the resected area can typically be well visualized, their brains have often undergone significant reorganization of structure-function relationships prior to surgery. However, these individuals do provide the opportunity to test language before and after the resection to evaluate the effect of the resection itself. Surgical patients also sometimes have comorbid conditions, such as epilepsy, which further contribute to functional reorganization and can vary from patient to patient. For example, Griffin and Tranel [83] found that presurgical patients with early-onset epilepsy were more likely than those with late-onset epilepsy to have nontraditional cerebral functional organization for language, verbal memory, and visual memory. The early-onset patients also had better postsurgical outcomes following anterior temporal lobectomy, which was attributed to presurgical differences in functional organization between the groups. The brain is expected to undergo functional reorganization after surgery, so it would be best to provide behavioral testing soon after surgery to acquire the best brain mapping results. Pre- and postsurgical testing can also be compared. Additionally, functional imaging techniques can be used to track functional reorganization changes by comparing pre- and postsurgical activation, or by longitudinally tracking changes to activation during recovery [84–86].

If given the opportunity, researchers can identify the association between changes in language (pre- to postsurgical behavioral measures) and extent of resection (pre- and postsurgical volumes) [84, 87–89]. Lesions can be traced after resection on high-resolution scans, such as a T1 MPRAGE or T2 FLAIR scans, and the tracings can include both the resected region and areas of edema surrounding the lesion (Fig. 6). However, it is common to see a shift of neuronal structures and cerebral edema in acute postsurgical patients, which can make it difficult to accurately delineate the lesion location. Even in later recovery stages, it can be difficult to trace a resected region because healthy neuronal

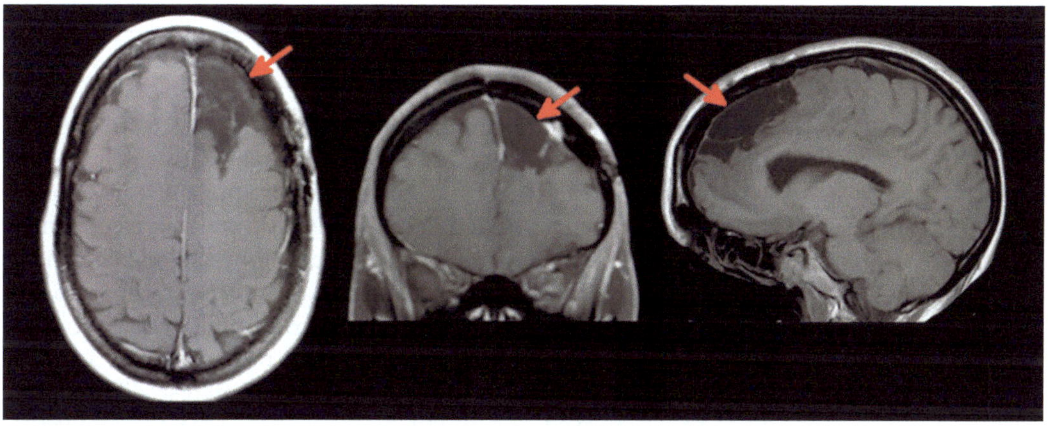

Fig. 6 Surgically resected lesion. T1-weighted images of a patient who has undergone a left frontal craniotomy. Arrow points to the resected region in the left frontal lobe in axial (left), coronal (center), and sagittal (right) views

tissue often shifts into the space created by resection. However, it is possible to trace the resection region using specialized image analysis. For example, the preoperative images can be aligned onto a standard space template, and then the postoperative images can be co-registered to the preoperative image while using a surgical ROI created from the postoperative image to de-weight the lesioned area [90]. Wallerian degeneration (degeneration of white matter tracts) following the surgical resection of cortex/lesion can be detected with DTI. Reduced fractional anisotropy on DTI has been linked to poorer postsurgical clinical outcomes [91].

7 Brain Abscesses

Brain abscesses are loci of infection in the brain and are caused by bacterial, fungi, or parasitic infection. Abscesses can result from a primary infection in another part of the body (e.g., middle ear infection, dental infection, or pulmonary infection) or by trauma (e.g., traumatic brain injury caused by bullets or other foreign bodies). Immunocompromised patients, such as those with HIV, are at greater risk for developing abscesses. CT with contrast, as well as MRI T1 and T2 images, can be used to image abscesses (Fig. 7). However, MRI ADC and DWI imaging and FDG-PET are the best options for differentiation of abscess from tumors [92, 93]. These scans also allow for the assessment of lesion size and location. Abscesses can cause focal deficits; however, because these patients are often very ill, they are not ideal candidates for lesion-symptom mapping research. Herpes encephalitis is one infection that can lead to focal damage in unilateral or bilateral temporal poles and surrounding areas if untreated; but now most cases are effectively treated with acyclovir before significant long-term damage occurs.

8 Transient Lesions

Transcranial magnetic stimulation (TMS) is another tool used by brain mapping researchers. It allows researchers to safely induce temporary cortical lesions and study the subsequent effects. A high-magnitude current is passed in brief pulses through an electromagnetic coil held close to the scalp. The changing electric current in the coil generates a magnetic field that passes through the scalp and induces an electric current in the brain. Localized depolarization of superficial cortical and subcortical neurons then generates a propagating action potential in targeted regions of interest. TMS can be used to either excite or inhibit brain activity. TMS is noninvasive and typically has few side effects, although repetitive TMS (rTMS), where multiple pulses of the same intensity are delivered to a

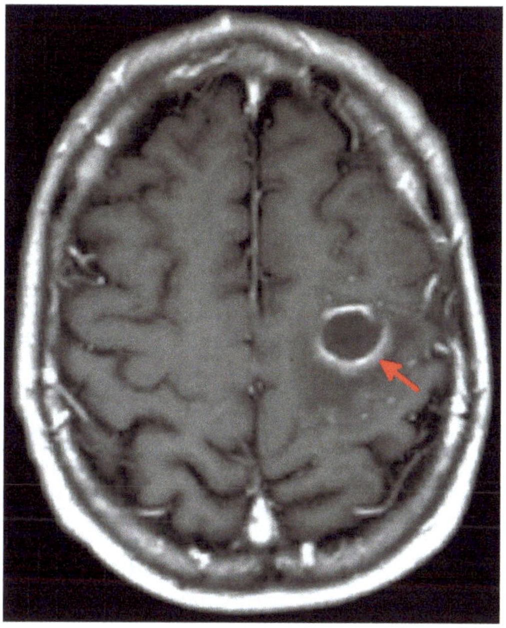

Fig. 7 Brain abscess. Arrow points to a brain abscess on a T1 image

targeted region, has been associated with seizures and syncope [94].

TMS has been successfully used for mapping the cortex through exciting brain activity [95, 96] or through the generation of transient lesions to specific cortical areas [97, 98]. Neuro-navigated TMS allows for the creation of highly focal temporary lesions that can be used for functional mapping of cognitive functions. Sakreida and colleagues [99] used neuro-navigated TMS to induce transient lesions in 12 healthy controls in order to functionally map language functioning in Broca's area, which corresponds to Brodmann areas 44 and 45. They were able to determine that the dorsal portion of area 44 was most susceptible to TMS during an object naming task. Inducing transient lesions via TMS has the advantage of being able to map brain functioning in healthy participants. This allows researchers to enroll large number of subjects and to conduct complex behavioral tasks. Another advantage is that it can be used to identify the effects of a very focal lesion, before opportunity for reorganization of structure-function relationships. The mapping of transient lesions is discussed in greater detail in the chapters on direct cortical stimulation (Chapter 14) and noninvasive brain stimulation (Chapter 15).

9 Deciding Which Lesions Are Appropriate for Brain Behavior Mapping

Several considerations must be taken into account when deciding which populations to include in brain mapping research. These considerations include whether functional reorganization will impact study findings, whether the lesion type is likely to induce focal versus diffuse deficits, or whether patients will be able to engage in behavioral testing, as well as the age of the participants.

As discussed in the sections above, brain lesions are often followed by functional reorganization of cognitive processes. Functional reorganization is particularly important to consider when testing stroke patients in the chronic phase of recovery, patients with slow-growing brain tumors, and patients who have undergone surgical resection of lesions. In these populations, it is likely that cognitive functions that normally depend on the lesioned area may have recovered if different areas in the brain have assumed these functions. Researchers should keep this in mind when interpreting results from these patient populations. Another important factor is whether the lesion type will cause focal or diffuse deficits. Stroke, brain tumors (without the presence of cerebral edema), surgical resection, and TMS can all cause focal deficits that are ideal for brain behavior mapping. Neurodegenerative disorders can also result in relatively focal deficits, particularly in the earlier stages of disease. Any lesion that is accompanied by cerebral edema (i.e., acute TBI, some brain tumors, acute postsurgical resection) will not provide ideal data for specific mapping of function to brain regions because cerebral edema itself causes diffuse deficits. Many chronic patients with TBI or subarachnoid hemorrhage have diffuse deficits that can be difficult to localize with brain mapping methods.

Additionally, researchers must consider whether patients will be able to complete behavioral testing. Patients with brain abscesses and some patients with brain tumors are often too sick to be able to participate in testing. Also, patients in the later stages of neurodegenerative diseases may not be able to understand behavioral tasks. One final point to consider is the age of the patient. Even in normal aging, the brain undergoes changes that can impact both behavioral functioning and functional imaging. For example, it is common to see white matter changes as people age, which are not necessarily tied to reduced cognitive functioning [100]. White matter changes can be visualized using T2 or FLAIR imaging. White matter changes are often tied to vascular changes, such as small vessel disease [101]. Many types of functional imaging depend on tracking changes to hemodynamic functioning, so age-related vascular changes are an important consideration. For example, the BOLD signal relies on neurovascular coupling, the close relationship between brain activation and subsequent changes in cerebral blood flow. Neurovascular coupling can change through the course of normal aging [102] and can also result from brain tumors [74], neurodegenerative diseases, stroke, and hypertension [103]. Investigators who include elderly participants with lesions should ideally use age-matched controls rather than younger controls to help control for these vascular differences between young and elderly participants.

10 Conclusions

There are many different types of brain lesions that can be studied in lesion-symptom mapping research. It is ideal to include participants who are capable of participating in behavioral testing and who have focal rather than diffuse deficits. Different questions can be addressed in patients who have versus have not gone through reorganization of structure-function relationships. Lesion-symptom mapping research will continue to advance with new neuroimaging techniques that allow us to better understand the many structural factors that can impact functioning in the brain. As we continue to examine language and cognitive impairments after lesions and throughout recovery, and in neurotypical populations using TMS, we will gain a better understanding of how the human brain adapts to a lesion in order to restore function.

Acknowledgments

Shannon M. Sheppard is supported by R01 DC015466; Argye E. Hillis is supported by R01 DC05375, R01DC 015466, and P50 DC 014664.

References

1. Saur D, Ronneberger O, Kümmerer D, Mader I, Weiller C, Klöppel S (2010) Early functional magnetic resonance imaging activations predict language outcome after stroke. Brain 133(4):1252–1264. https://doi.org/10.1093/brain/awq021
2. Saur D, Lange R, Baumgaertner A et al (2006) Dynamics of language reorganization after stroke. Brain 129(6):1371–1384. https://doi.org/10.1093/brain/awl090
3. Motta M, Ramadan A, Hillis AE, Gottesman RF, Leigh R (2015) Diffusion–perfusion mismatch: an opportunity for improvement in cortical function. Front Neurol 5. https://doi.org/10.3389/fneur.2014.00280
4. Shahid H, Sebastian R, Schnur T et al (2017) Important considerations in lesion-symptom mapping: illustrations from studies of word comprehension. Hum Brain Mapp 38:2990–3000
5. Hillis AE, Wityk RJ, Tuffiash E et al (2001) Hypoperfusion of Wernicke's area predicts severity of semantic deficit in acute stroke. Ann Neurol 50(5):561–566. https://doi.org/10.1002/ana.1265
6. Olivier Z-W, Walter M-H, Wolf-Dieter H, Jan S (2010) MRI perfusion maps in acute stroke validated with 15O-water positron emission tomography. Stroke 41(3):443–449. https://doi.org/10.1161/STROKEAHA.109.569889
7. Astrup J, Symon L, Branston NM, Lassen NA (1977) Cortical evoked potential and extracellular K+ and H+ at critical levels of brain

ischemia. Stroke 8(1):51–57. https://doi.org/10.1161/01.STR.8.1.51

8. Lutsep HL, Albers GW, Decrespigny A, Kamat GN, Marks MP, Moseley ME (1997) Clinical utility of diffusion-weighted magnetic resonance imaging in the assessment of ischemic stroke. Ann Neurol 41(5):574–580. https://doi.org/10.1002/ana.410410505
9. Sorensen AG, Buonanno FS, Gonzalez RG et al (1996) Hyperacute stroke: evaluation with combined multisection diffusion-weighted and hemodynamically weighted echo-planar MR imaging. Radiology 199(2): 391–401. https://doi.org/10.1148/radiology.199.2.8668784
10. Altamura C, Reinhard M, Vry M-S et al (2009) The longitudinal changes of BOLD response and cerebral hemodynamics from acute to subacute stroke. A fMRI and TCD study. BMC Neurosci 10(1):151. https://doi.org/10.1186/1471-2202-10-151
11. Prabhakaran V, Raman SP, Grunwald MR et al (2007) Neural substrates of word generation during stroke recovery: the influence of cortical hypoperfusion. Behav Neurol 18(1): 45–52
12. Xu F, Ge Y, Lu H (2009) Noninvasive quantification of whole-brain cerebral metabolic rate of oxygen (CMRO2) by MRI. Magn Reson Med 62(1):141–148. https://doi.org/10.1002/mrm.21994
13. Fridriksson J, Rorden C, Morgan PS, Morrow KL, Baylis GC (2006) Measuring the hemodynamic response in chronic hypoperfusion. Neurocase 12(3):146–150. https://doi.org/10.1080/13554790600598816
14. Bonakdarpour B, Parrish TB, Thompson CK (2007) Hemodynamic response function in patients with stroke-induced aphasia: implications for fMRI data analysis. NeuroImage 36(2):322–331. https://doi.org/10.1016/j.neuroimage.2007.02.035
15. Zimmerman RD, Maldjian JA, Brun NC, Horvath B, Skolnick BE (2006) Radiologic estimation of hematoma volume in intracerebral hemorrhage trial by CT scan. Am J Neuroradiol 27(3):666–670
16. Macellari F, Paciaroni M, Agnelli G, Caso V (2014) Neuroimaging in intracerebral hemorrhage. Stroke. https://doi.org/10.1161/STROKEAHA.113.003701
17. Bousser M-G, Ferro JM (2007) Cerebral venous thrombosis: an update. Lancet Neurol 6(2):162–170. https://doi.org/10.1016/S1474-4422(07)70029-7
18. Poon CS, Chang J-K, Swarnkar A, Johnson MH, Wasenko J (2007) Radiologic diagnosis of cerebral venous thrombosis: pictorial review. Am J Roentgenol 189-(6_supplement):S64–S75. https://doi.org/10.2214/AJR.07.7015
19. Bahrainwala ZS, Hillis AE, Dearborn J, Gottesman RF (2014) Neglect performance in acute stroke is related to severity of white matter hyperintensities. Cerebrovasc Dis 37(3):223–230. https://doi.org/10.1159/000357661
20. Basilakos A, Fillmore PT, Rorden C, Guo D, Bonilha L, Fridriksson J (2014) Regional white matter damage predicts speech fluency in chronic post-stroke aphasia. Front Hum Neurosci 8. https://doi.org/10.3389/fnhum.2014.00845
21. Wright A, Tippett DC, Saxena S et al (2018) Leukoaraiosis is independently associated with naming outcome in poststroke aphasia. Neurology 91(6):526–532
22. Werring DJ, Toosy AT, Clark CA et al (2000) Diffusion tensor imaging can detect and quantify corticospinal tract degeneration after stroke. J Neurol Neurosurg Psychiatry 69(2):269–272. https://doi.org/10.1136/jnnp.69.2.269
23. Yu C, Zhu C, Zhang Y et al (2009) A longitudinal diffusion tensor imaging study on Wallerian degeneration of corticospinal tract after motor pathway stroke. NeuroImage 47(2): 451–458. https://doi.org/10.1016/j.neuroimage.2009.04.066
24. Ivanova MV, Isaev DY, Dragoy OV et al (2016) Diffusion-tensor imaging of major white matter tracts and their role in language processing in aphasia. Cortex 85:165–181. https://doi.org/10.1016/j.cortex.2016.04.019
25. Yoshihiro M, Kaoru S, Tatsuya H et al (2006) Effects of cerebral ischemia on evoked cerebral blood oxygenation responses and BOLD contrast functional MRI in stroke patients. Stroke 37(10):2514–2520. https://doi.org/10.1161/01.STR.0000239698.50656.3b
26. Thompson CK, den Ouden D-B, Bonakdarpour B, Garibaldi K, Parrish TB (2010) Neural plasticity and treatment-induced recovery of sentence processing in agrammatism. Neuropsychologia 48(11): 3211–3227. https://doi.org/10.1016/j.neuropsychologia.2010.06.036
27. DeLeon J, Gottesman RF, Kleinman JT et al (2007) Neural regions essential for distinct cognitive processes underlying picture naming. Brain 130(5):1408–1422. https://doi.org/10.1093/brain/awm011

28. Blockley NP, Griffeth VEM, Simon AB, Buxton RB (2013) A review of calibrated blood oxygenation level-dependent (BOLD) methods for the measurement of task-induced changes in brain oxygen metabolism: a review of calibrated BOLD methods. NMR Biomed 26(8):987–1003. https://doi.org/10.1002/nbm.2847
29. Leontiev O, Buxton RB (2007) Reproducibility of BOLD, perfusion, and CMRO2 measurements with calibrated-BOLD fMRI. NeuroImage 35(1):175–184. https://doi.org/10.1016/j.neuroimage.2006.10.044
30. Fair DA, Snyder AZ, Connor LT, Nardos B, Corbetta M (2009) Task-evoked BOLD responses are normal in areas of diaschisis after stroke. Neurorehabil Neural Repair 23(1):52–57. https://doi.org/10.1177/1545968308317699
31. Seghier ML, Ramsden S, Lim L, Leff AP, Price CJ (2014) Gradual lesion expansion and brain shrinkage years after stroke. Stroke 45(3):877–879. https://doi.org/10.1161/STROKEAHA.113.003587
32. Hope TMH, Seghier ML, Leff AP, Price CJ (2013) Predicting outcome and recovery after stroke with lesions extracted from MRI images. NeuroImage Clin 2:424–433. https://doi.org/10.1016/j.nicl.2013.03.005
33. Taylor CA, Bell JM, Breiding MJ, Xu L (2017) Traumatic brain injury-related emergency department visits, hospitalizations, and deaths – United States, 2007 and 2013. MMWR Surveill Summ 66(9):1–16. https://doi.org/10.15585/mmwr.ss6609a1
34. Rigon A, Voss MW, Turkstra LS, Mutlu B, Duff MC (2018) White matter correlates of different aspects of facial affect recognition impairment following traumatic brain injury. Soc Neurosci 14(4):1–15. https://doi.org/10.1080/17470919.2018.1489302
35. Raikes AC, Bajaj S, Dailey NS et al (2018) Diffusion tensor imaging (DTI) correlates of self-reported sleep quality and depression following mild traumatic brain injury. Front Neurol 9. https://doi.org/10.3389/fneur.2018.00468
36. Hanks R, Millis S, Scott S et al (2019) The relation between cognitive dysfunction and diffusion tensor imaging parameters in traumatic brain injury. Brain Inj 33(3):355–363. https://doi.org/10.1080/02699052.2018.1553073
37. Thompson PM, Hayashi KM, de Zubicaray G et al (2003) Dynamics of gray matter loss in Alzheimer's disease. J Neurosci 23(3):994–1005. https://doi.org/10.1523/JNEUROSCI.23-03-00994.2003
38. Bodanapally UK, Sours C, Zhuo J, Shanmuganathan K (2015) Imaging of traumatic brain injury. Radiol Clin 53(4):695–715. https://doi.org/10.1016/j.rcl.2015.02.011
39. Noguchi K, Ogawa T, Inugami A et al (1995) Acute subarachnoid hemorrhage: MR imaging with fluid-attenuated inversion recovery pulse sequences. Radiology 196(3):773–777. https://doi.org/10.1148/radiology.196.3.7644642
40. Narayan RK, Maas AIR, Servadei F, Skolnick BE, Tillinger MN, Marshall LF (2008) Progression of traumatic intracerebral hemorrhage: a prospective observational study. J Neurotrauma 25(6):629–639. https://doi.org/10.1089/neu.2007.0385
41. Lee H, Wintermark M, Gean AD, Ghajar J, Manley GT, Mukherjee P (2008) Focal lesions in acute mild traumatic brain injury and neurocognitive outcome: CT versus 3T MRI. J Neurotrauma 25(9):1049–1056. https://doi.org/10.1089/neu.2008.0566
42. Abu Hamdeh S, Marklund N, Lannsjö M et al (2016) Extended anatomical grading in diffuse axonal injury using MRI: hemorrhagic lesions in the substantia nigra and mesencephalic tegmentum indicate poor long-term outcome. J Neurotrauma 34(2):341–352. https://doi.org/10.1089/neu.2016.4426
43. Adams J, Doyle D, Ford I, Gennarelli T, Graham D, Mclellan D (1989) Diffuse axonal injury in head injury: definition, diagnosis and grading. Histopathology 15(1):49–59
44. Ljungqvist J, Nilsson D, Ljungberg M, Esbjörnsson E, Eriksson-Ritzén C, Skoglund T (2017) Longitudinal changes in diffusion tensor imaging parameters of the corpus callosum between 6 and 12 months after diffuse axonal injury. Brain Inj 31(3):344–350. https://doi.org/10.1080/02699052.2016.1256500
45. Maas AI, Stocchetti N, Bullock R (2008) Moderate and severe traumatic brain injury in adults. Lancet Neurol 7(8):728–741. https://doi.org/10.1016/S1474-4422(08)70164-9
46. Mutch CA, Talbott JF, Gean A (2016) Imaging evaluation of acute traumatic brain injury. Neurosurg Clin 27(4):409–439. https://doi.org/10.1016/j.nec.2016.05.011
47. Currie S, Saleem N, Straiton JA, Macmullen-Price J, Warren DJ, Craven IJ (2016) Imaging assessment of traumatic brain injury. Postgrad Med J 92(1083):41–50. https://doi.org/10.1136/postgradmedj-2014-133211

48. Neary D, Snowden JS, Gustafson L et al (1998) Frontotemporal lobar degeneration. Neurology 51(6):1546–1554
49. Mummery CJ, Patterson K, Price CJ, Ashburner J, Frackowiak RSJ, Hodges JR (2000) A voxel-based morphometry study of semantic dementia: relationship between temporal lobe atrophy and semantic memory. Ann Neurol 47(1):36–45. https://doi.org/10.1002/1531-8249(200001)47:1<36::AID-ANA8>3.0.CO;2-L
50. Chételat G, Landeau B, Eustache F et al (2005) Using voxel-based morphometry to map the structural changes associated with rapid conversion in MCI: a longitudinal MRI study. NeuroImage 27(4):934–946. https://doi.org/10.1016/j.neuroimage.2005.05.015
51. Beyer MK, Janvin CC, Larsen JP, Aarsland D (2007) A magnetic resonance imaging study of patients with Parkinson's disease with mild cognitive impairment and dementia using voxel-based morphometry. J Neurol Neurosurg Psychiatry 78(3):254–259. https://doi.org/10.1136/jnnp.2006.093849
52. Burton EJ, McKeith IG, Burn DJ, Williams ED, O'Brien JT (2004) Cerebral atrophy in Parkinson's disease with and without dementia: a comparison with Alzheimer's disease, dementia with Lewy bodies and controls. Brain 127(4):791–800. https://doi.org/10.1093/brain/awh088
53. Long Z, Irish M, Piguet O, Kiernan MC, Hodges JR, Burrell JR (2019) Clinical and neuroimaging investigations of language disturbance in frontotemporal dementia–motor neuron disease patients. J Neurol 266(4):921–933. https://doi.org/10.1007/s00415-019-09216-0
54. Hardy CJD, Marshall CR, Bond RL et al (2018) Retained capacity for perceptual learning of degraded speech in primary progressive aphasia and Alzheimer's disease. Alzheimers Res Ther 10(1):70. https://doi.org/10.1186/s13195-018-0399-2
55. Lee SE, Sias AC, Mandelli ML et al (2017) Network degeneration and dysfunction in presymptomatic C9ORF72 expansion carriers. NeuroImage Clin 14:286–297. https://doi.org/10.1016/j.nicl.2016.12.006
56. Mesulam M-M, Wieneke C, Hurley R et al (2013) Words and objects at the tip of the left temporal lobe in primary progressive aphasia. Brain 136(2):601–618. https://doi.org/10.1093/brain/aws336
57. Thompson CK, Cho S, Price C et al (2012) Semantic interference during object naming in agrammatic and logopenic primary progressive aphasia (PPA). Brain Lang 120(3):237–250. https://doi.org/10.1016/j.bandl.2011.11.003
58. Mesulam M, Rogalski E, Wieneke C et al (2009) Neurology of anomia in the semantic variant of primary progressive aphasia. Brain 132(9):2553–2565. https://doi.org/10.1093/brain/awp138
59. Meier E, Crinion J, Cebron S, et al (2019) Labeled cortical distance mapping reveals temporal lobe morphometry in primary progressive aphasia. In: Rome, Italy
60. Domínguez JF, Stout JC, Poudel G et al (2016) Multimodal imaging biomarkers in premanifest and early Huntington's disease: 30-month IMAGE-HD data. Br J Psychiatry 208(6):571–578. https://doi.org/10.1192/bjp.bp.114.156588
61. Faria AV, Sebastian R, Newhart M, Mori S, Hillis AE (2014) Longitudinal imaging and deterioration in word comprehension in primary progressive aphasia: potential clinical significance. Aphasiology 28(8–9):948–963. https://doi.org/10.1080/02687038.2014.911241
62. Basser PJ, Pierpaoli C (2011) Microstructural and physiological features of tissues elucidated by quantitative-diffusion-tensor MRI. J Magn Reson 213(2):560–570. https://doi.org/10.1016/j.jmr.2011.09.022
63. Mahoney CJ, Ridgway GR, Malone IB et al (2014) Profiles of white matter tract pathology in frontotemporal dementia. Hum Brain Mapp 35(8):4163–4179. https://doi.org/10.1002/hbm.22468
64. Dani M, Brooks DJ, Edison P (2016) Tau imaging in neurodegenerative diseases. Eur J Nucl Med Mol Imaging 43(6):1139–1150. https://doi.org/10.1007/s00259-015-3231-2
65. Filippi M, Agosta F, Scola E et al (2013) Functional network connectivity in the behavioral variant of frontotemporal dementia. Cortex 49(9):2389–2401. https://doi.org/10.1016/j.cortex.2012.09.017
66. Contreras JA, Avena-Koenigsberger A, Risacher SL et al (2019) Resting state network modularity along the prodromal late onset Alzheimer's disease continuum. NeuroImage Clin 22:101687. https://doi.org/10.1016/j.nicl.2019.101687
67. Amboni M, Tessitore A, Esposito F et al (2015) Resting-state functional connectivity associated with mild cognitive impairment in Parkinson's disease. J Neurol 262(2):

425–434. https://doi.org/10.1007/s00415-014-7591-5
68. Gili T, Cercignani M, Serra L et al (2011) Regional brain atrophy and functional disconnection across Alzheimer's disease evolution. J Neurol Neurosurg Psychiatry 82(1):58–66. https://doi.org/10.1136/jnnp.2009.199935
69. Chen JJ (2019) Functional MRI of brain physiology in aging and neurodegenerative diseases. NeuroImage 187:209–225. https://doi.org/10.1016/j.neuroimage.2018.05.050
70. Mabray MC, Barajas RF, Cha S (2015) Modern brain tumor imaging. Brain Tumor Res Treat 3(1):8–23. https://doi.org/10.14791/btrt.2015.3.1.8
71. Haacke EM, Mittal S, Wu Z, Neelavalli J, Cheng Y-CN (2009) Susceptibility-weighted imaging: technical aspects and clinical applications, part 1. Am J Neuroradiol 30(1):19–30. https://doi.org/10.3174/ajnr.A1400
72. Bulakbasi N, Guvenc I, Onguru O, Edrogan E, Tayfun C, Ucoz T (2004) The added value of the apparent diffusion coefficient calculation to magnetic resonance imaging in the differentiation and grading of malignant brain tumors. J Comput Assist Tomogr 28(6):735–746
73. Haris M, Kumar S, Raj MK et al (2008) Serial diffusion tensor imaging to characterize radiation-induced changes in normal-appearing white matter following radiotherapy in patients with adult low-grade gliomas. Radiat Med 26(3):140. https://doi.org/10.1007/s11604-007-0209-4
74. Metwali H, Raemaekers M, Kniese K, Kardavani B, Fahlbusch R, Samii A (2019) Reliability of functional magnetic resonance imaging in patients with brain tumors: a critical review and meta-analysis. World Neurosurg 125:183–190. https://doi.org/10.1016/j.wneu.2019.01.194
75. Duffau H (2005) Lessons from brain mapping in surgery for low-grade glioma: insights into associations between tumour and brain plasticity. Lancet Neurol 4(8):476–486. https://doi.org/10.1016/S1474-4422(05)70140-X
76. Duffau H, Gatignol P, Mandonnet E, Capelle L, Taillandier L (2008) Intraoperative subcortical stimulation mapping of language pathways in a consecutive series of 115 patients with Grade II glioma in the left dominant hemisphere. J Neurosurg 109(3):461–471. https://doi.org/10.3171/JNS/2008/109/9/0461
77. Lehéricy S, Duffau H, Cornu P et al (2000) Correspondence between functional magnetic resonance imaging somatotopy and individual brain anatomy of the central region: comparison with intraoperative stimulation in patients with brain tumors. J Neurosurg 92(4):589–598. https://doi.org/10.3171/jns.2000.92.4.0589
78. Boatman DF, Miglioretti DL (2005) Cortical sites critical for speech discrimination in normal and impaired listeners. J Neurosci 25(23):5475–5480. https://doi.org/10.1523/JNEUROSCI.0936-05.2005
79. Cervenka MC, Corines J, Boatman-Reich DF et al (2013) Electrocorticographic functional mapping identifies human cortex critical for auditory and visual naming. NeuroImage 69:267–276. https://doi.org/10.1016/j.neuroimage.2012.12.037
80. Lesser RP, Lueders H, Dinner DS, Hahn J, Cohen L (1984) The location of speech and writing functions in the frontal language area: results of extraoperative cortical stimulation. Brain 107(1):275–291. https://doi.org/10.1093/brain/107.1.275
81. Lesser RP, Luders H, Morris H et al (1986) Electrical stimulation of Wernicke's area interferes with comprehension. Neurology 36(5):658–658
82. Englot DJ, Ouyang D, Wang DD, Rolston JD, Garcia PA, Chang EF (2013) Relationship between hospital surgical volume, lobectomy rates, and adverse perioperative events at US epilepsy centers: clinical article. J Neurosurg 118(1):169–174. https://doi.org/10.3171/2012.9.JNS12776
83. Griffin S, Tranel D (2007) Age of seizure onset, functional reorganization, and neuropsychological outcome in temporal lobectomy. J Clin Exp Neuropsychol 29(1):13–24. https://doi.org/10.1080/13803390500263568
84. Limotai C, McLachlan RS, Hayman-Abello S et al (2018) Memory loss and memory reorganization patterns in temporal lobe epilepsy patients undergoing anterior temporal lobe resection, as demonstrated by pre-versus post-operative functional MRI. J Clin Neurosci 55:38–44. https://doi.org/10.1016/j.jocn.2018.06.020
85. Chivukula S, Pikul BK, Black KL, Pouratian N, Bookheimer SY (2018) Contralateral functional reorganization of the speech supplementary motor area following neurosurgical tumor resection. Brain Lang 183:41–46. https://doi.org/10.1016/j.bandl.2018.05.006

86. Sidhu MK, Stretton J, Winston GP et al (2016) Memory network plasticity after temporal lobe resection: a longitudinal functional imaging study. Brain 139(2):415–430. https://doi.org/10.1093/brain/awv365
87. Mattavelli G, Pisoni A, Casarotti A et al (2019) Consequences of brain tumour resection on emotion recognition. J Neuropsychol 13(1):1–21
88. Smith JS, Chang EF, Lamborn KR et al (2008) Role of extent of resection in the long-term outcome of low-grade hemispheric gliomas. J Clin Oncol 26(8):1338–1345. https://doi.org/10.1200/JCO.2007.13.9337
89. Drane DL, Ojemann GA, Aylward E et al (2008) Category-specific naming and recognition deficits in temporal lobe epilepsy surgical patients. Neuropsychologia 46(5):1242–1255. https://doi.org/10.1016/j.neuropsychologia.2007.11.034
90. Yogarajah M, Focke NK, Bonelli SB et al (2010) The structural plasticity of white matter networks following anterior temporal lobe resection. Brain 133(8):2348–2364. https://doi.org/10.1093/brain/awq175
91. Wieshmann UC, Symms MR, Clark CA et al (1999) Wallerian degeneration in the optic radiation after temporal lobectomy demonstrated in vivo with diffusion tensor imaging. Epilepsia 40(8):1155–1158. https://doi.org/10.1111/j.1528-1157.1999.tb00834.x
92. Brouwer M, van de Beek D (2017) Epidemiology, diagnosis, and treatment of brain abscesses. Curr Opin Infect Dis 30(1):129–134
93. Villanueva-Meyer JE, Cha S (2015) From shades of gray to microbiologic imaging: a historical review of brain abscess imaging: RSNA centennial article. Radiographics 35(5):1555–1562. https://doi.org/10.1148/rg.2015140297
94. Rossi S, Hallett M, Rossini PM, Pascual-Leone A (2009) Safety, ethical considerations, and application guidelines for the use of transcranial magnetic stimulation in clinical practice and research. Clin Neurophysiol 120(12):2008–2039. https://doi.org/10.1016/j.clinph.2009.08.016
95. Malcolm MP, Triggs WJ, Light KE, Shechtman O, Khandekar G, Gonzalez Rothi LJ (2006) Reliability of motor cortex transcranial magnetic stimulation in four muscle representations. Clin Neurophysiol 117(5):1037–1046. https://doi.org/10.1016/j.clinph.2006.02.005
96. Weiss C, Nettekoven C, Rehme AK et al (2013) Mapping the hand, foot and face representations in the primary motor cortex—retest reliability of neuronavigated TMS versus functional MRI. NeuroImage 66:531–542. https://doi.org/10.1016/j.neuroimage.2012.10.046
97. Pobric G, Jefferies E, Lambon Ralph MA (2010) Category-specific versus category-general semantic impairment induced by transcranial magnetic stimulation. Curr Biol 20(10):964–968. https://doi.org/10.1016/j.cub.2010.03.070
98. Oliveri M, Finocchiaro C, Shapiro K, Gangitano M, Caramazza A, Pascual-Leone A (2004) All talk and no action: a transcranial magnetic stimulation study of motor cortex activation during action word production. J Cogn Neurosci 16(3):374–381. https://doi.org/10.1162/089892904322926719
99. Sakreida K, Lange I, Willmes K et al (2018) High-resolution language mapping of Broca's region with transcranial magnetic stimulation. Brain Struct Funct 223(3):1297–1312. https://doi.org/10.1007/s00429-017-1550-8
100. Forbes K (2018) MRI brain white matter change: spectrum of change – how can we grade? J R Coll Physicians Edinb 47(3):271–275. https://doi.org/10.4997/JRCPE.2017.313
101. Poggesi A, Pantoni L, Inzitari D et al (2011) 2001–2011: a decade of the LADIS (Leukoaraiosis and DISability) study: what have we learned about white matter changes and small-vessel disease? Cerebrovasc Dis 32(6):577–588. https://doi.org/10.1159/000334498
102. Fabiani M, Gordon BA, Maclin EL et al (2014) Neurovascular coupling in normal aging: a combined optical, ERP and fMRI study. NeuroImage 85:592–607. https://doi.org/10.1016/j.neuroimage.2013.04.113
103. Girouard H, Iadecola C (2006) Neurovascular coupling in the normal brain and in hypertension, stroke, and Alzheimer disease. J Appl Physiol 100(1):328–335. https://doi.org/10.1152/japplphysiol.00966.2005

Chapter 2

Manual Lesion Segmentation

Casey Ferrara, Branch Coslett, and Laurel Buxbaum

Abstract

The reliable identification of healthy versus damaged brain tissue is of primary importance in studies of brain-behavior relationships in patient populations. However, lesion appearance and boundaries are extremely inconsistent and can often vary based on the location in the brain, the appearance of surrounding tissue and structures, the scan type, the patient population, and the individual. Because of the complex nature of this task, manual lesion tracing is considered the "gold standard" in lesion segmentation, as it allows the lesion tracer to call on domain expertise in making these discriminations. In this chapter, we provide an overview of manual lesion segmentation methods, as well as a general template for training in lesion segmentation with the goal of minimizing inter- and intra-observer variability.

Key words Manual segmentation, Lesion analysis, Structural MRI, Reproducibility, Observer agreement

1 What Is Lesion Segmentation?

Lesion segmentation, or the identification and labelling of voxels in a scan based on their relevant properties (e.g., the voxel contains damaged or "lesioned" tissue) [1], aims to capture the extent of a given lesion, distinguishing it from the surrounding tissue. Delineating damaged from non-damaged tissue via segmentation allows for quantitative analysis of lesion volumes and areas of damage and comparisons of lesion size and distribution across patients. For example, to investigate post-stroke brain structure and to relate brain changes to behavioral outcomes, researchers use techniques such as voxel-based lesion-symptom mapping (VLSM) and support vector regression lesion-symptom mapping (SVR-LSM). However, if these analyses are to yield valid clinical inferences, accurate and precise lesion segmentation is of utmost importance [2].

Dorian Pustina and Daniel Mirman (eds.), *Lesion-to-Symptom Mapping: Principles and Tools*, Neuromethods, vol. 180, https://doi.org/10.1007/978-1-0716-2225-4_2,

2 Manual Lesion Segmentation

Considered to be the "gold standard" of reference, manually tracing the lesion is a very commonly used method of performing this segmentation. This involves outlining the lesion periphery, typically done via computer display using a mouse or pen to control the cursor [3, 4]. Manually tracing lesions has been the primary method of lesion segmentation since CT scans became widely available in the 1970s.

The individual who carries out this segmentation – here referred to as a "lesion tracer" but also potentially called "lesion drawer" in the literature – must possess expertise in neuroanatomy in order to incorporate knowledge of the surrounding neuroanatomical features into the process of discriminating between healthy and lesioned tissue. The ability to incorporate this knowledge is often an advantage of manual segmentation over automated methods as the identification of lesion boundaries is complicated by overlapping voxel intensity values between lesion and non-lesioned voxels, which are best disambiguated by considering their position relative to other structural features.

However, the reliance of manual segmentation on subjective judgments raises concerns about ensuring "intra-tracer" or more commonly called *intra-observer reliability* – that is, whether a given lesion is rendered with similar boundaries and lesion volume by the lesion tracer at different time points. Metrics like this may be termed "intra-rater," "intra-observer," or "intra-operator" in the literature. Additionally, the time- and labor-intensive nature of this process may require the involvement of more than one lesion tracer, especially in cases of large sample sizes. This requires additional consideration of inter-observer (or "inter-rater"/"inter-operator") reliability (whether a given lesion is rendered with similar boundaries and lesion volume across different lesion tracers) [5, 6]. We include a brief discussion of the inter-lab variation that exists in the manual lesion segmentation and the implications of those differences later in this chapter. For manual segmentation, consistently high inter-observer reliability is essential and can only be achieved if the segmentation is performed by an individual with considerable experience [2–4, 6, 7].

In order to ensure accurate and consistent lesion segmentation, a detailed training protocol is essential in establishing and maintaining reliability between different lesion tracers over time and across a diversity of potential lesions. In this chapter, we outline our procedures for training lesion tracers as well as our methods for establishing inter-rater reliability. As these methods were developed for the identification and segmentation of stroke lesions (both ischemic and hemorrhagic), we will discuss the distinctions between lesioned tissue and the distant effects of stroke (e.g.,

Wallerian degeneration). For labs studying individuals with other types of pathology, there will likely be different considerations to be included in the training and reliability procedures. To illustrate a variety of issues relevant to lesion segmentation, we will provide figures throughout this chapter from multiple scans of different individuals. In addition to these, we will also be showing a standard stroke case segmentation selected from the publicly available ATLAS dataset, which is used as a standard example for the entire book (Native space ID, c0005s0007t01; lesion size, 56 ml; lesion type, first left hemispheric MCA embolic stroke). These images will be noted as referring to the "ATLAS example" or the "ATLAS case."

3 Manual Segmentation Training: First Steps

When viewing or manually segmenting ("tracing") a scan, a lesion tracer will be employing three different views of the brain, either switching between them in the display or viewing all three simultaneously. One of these views is along the *axial* plan, also referred to as the *transverse, lateral,* or *horizontal* plane. This plane divides the head into superior/inferior (top and bottom) portions, and within a picture from this plane, one will see the left/right and anterior/posterior axes. The second is the *coronal* plane (also known as the *frontal* or *vertical* plane), which divides the head into back and front or posterior/anterior portions. A picture from this perspective allows one to see the left/right and superior/inferior axes. The third is the *sagittal* plane, (also called the *median, longitudinal,* or *anteroposterior* plane), which divides the head into left and right or medial/lateral portions and from which one can see the superior/inferior and anterior/posterior axes. See Fig. 1 for visualizations of these three planes as they extend along the x, y, and z axes, alongside example slices from the ATLAS example.

In programs designed for lesion segmentation, such as MRIcron and ITK-SNAP, the lesion tracer is able to view the scan image from all three perspectives and uses a pen/paint tool, typically controlled by the mouse, to trace on the displayed brain image (see Fig. 2a; note Figs. 2a–c are taken from MRIcron). By outlining and filling in any tissue the lesion tracer considers damaged, they may create a drawn 2D image of the lesion that can be distinguished by virtue of opacity or color. After tracing the extent of the lesion in a given plane, the lesion tracer then moves to the next image along one of the three dimensions (i.e., moving one slice inferior/superior if drawing in the axial view, moving one slice anterior/posterior if drawing in the coronal view, etc.) by adjusting the coordinate values displayed in the toolbar (see Fig. 2b). In the new slice, all lesioned tissue in the display is to be identified once again. In a given "slice" of the brain, if one were to manually trace the lesioned

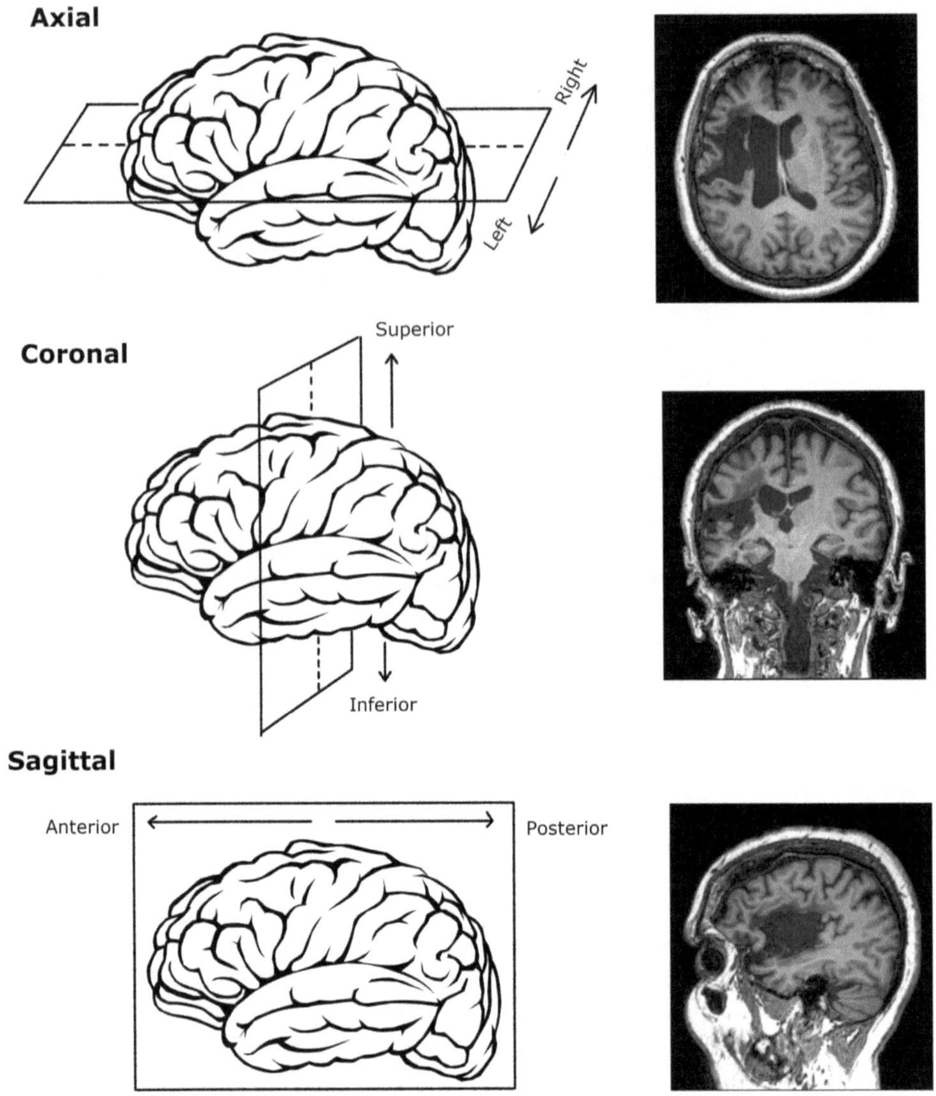

Fig. 1 Visualization of the axial, coronal, and sagittal plane as they extend along the x, y, and z axes. Examples of each view taken from the ATLAS case are included on the right

tissue in the axial plane, corresponding voxels in the coronal and sagittal planes also become marked, as shown in Fig. 3, taken from the ATLAS case. The lesion will ultimately be drawn in each slice of the brain in which there is damaged tissue, creating a 3D rendering of the lesion. General recommendations regarding which plane to segment in and when to switch planes are discussed more in depth later in this chapter.

A crosshair tool (see Fig. 2c) can be used to identify any voxel in all three views (axial, coronal, sagittal) simultaneously; for example, placing the crosshair on a voxel in the lateral thalamus will automatically cause the same voxel to be displayed in all three

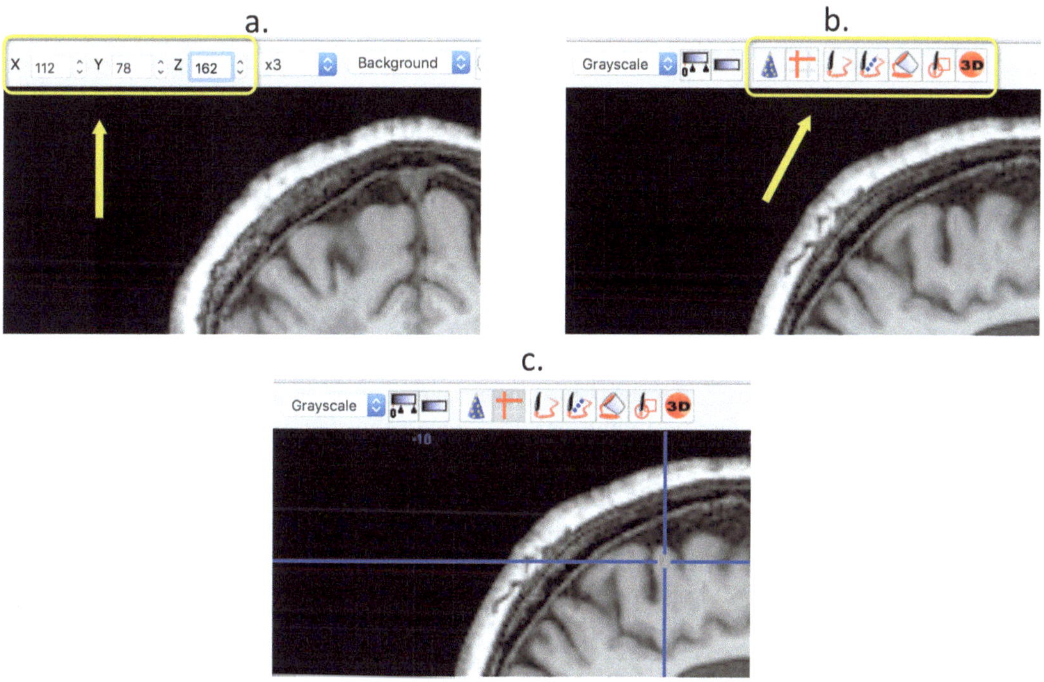

Fig. 2 (**a–c**) MRIcron toolbar showing (**a**) display coordinates, (**b**) paint tools, and (**c**) the crosshair tool

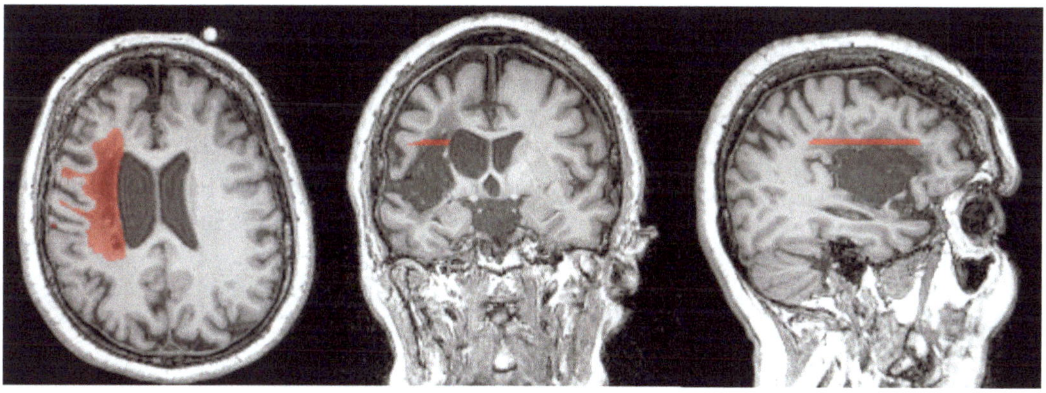

Fig. 3 A single slice drawn in the axial plane, represented in all three views

dimensions. The crosshair tool, therefore, is very useful for determining if a voxel in the lesion should be consistent across all views; that is, a voxel that is lesioned on the axial view should be marked as lesioned on the coronal and sagittal views. By "tracing" the lesion in the axial, coronal, and sagittal view – or the X, Y, and Z axes – we essentially create a 3D model of the area of damage in the brain. Learning how to draw (and think) in three dimensions is often unintuitive and requires a good deal of practice, as will be discussed further below.

4 What Is the Goal of Manual Lesion Segmentation?

What are we actually trying to represent with lesion tracings? While in theory we might imagine the desired outcome to be a three-dimensional representation of dysfunctional vs normally functioning brain tissue, brain function is not interrogated by CT or MRI. Given this limitation, the decision regarding the status of a voxel cannot be based on the functionality of the voxel, which is unknown, but rather the appearance of the voxel. That is, the distinction between healthy and damaged tissue is usually based on differences in the shade of the tissue relative to the surrounding tissue. Because brain tissue that appears structurally anomalous may in fact be fully functional, using the structural appearance of brain tissue as a proxy in order to estimate what is and is not lesioned ultimately yields a level of uncertainty that is inherent to manual tracing.

Moreover, discriminations between gray matter, white matter, cerebral spinal fluid, and lesioned tissue can be extremely difficult to make, because these decisions are tied to the intensity of the voxels as they appear in the scan, and the intensity values for these different types of tissue frequently overlap. In T1-weighted MRI scans, such as the sample scan shown throughout this volume, the appearance (voxel intensity) of lesioned voxels may be difficult to distinguish from healthy gray matter, making voxels that fall within this ambiguous range challenging to assign [5]. These judgment calls require contextual knowledge, such as the voxels' position relative to other structures (e.g., do they appear along the edge of a gyrus where we might expect to see gray matter? Or are they in an area that should be largely white matter?).

The difficulty in identifying damaged tissue based solely on the variations in voxel intensity in any given MRI or CT slice is one reason lesion segmentation algorithms have been challenging to develop. Manual segmentation exploits the lesion drawer's knowledge of context and anatomy. It is therefore important for the lesion tracer to establish a working knowledge of relevant neuroanatomical structures, as well as experience with neuroimaging so that the investigator has an understanding of typical brain variability, the effects of atrophy, and the potential confounding effects of commonly encountered pathology such as "small vessel ischemic disease" that is not directly relevant to the stroke in question. For this reason, our training protocol focuses initially on establishing relevant neuroanatomical knowledge and then emphasizes frequent exposure to typical and atypical brain scans via repeated practice.

Familiarity with neuroanatomy is a prerequisite to the process of lesion tracing. It requires knowing both the location of relevant structures and how those structures appear when represented in two dimensions in an MRI or a CT. To achieve this, our labs at

MRRI and LCNS typically make use of online resources, such as http://da.si.washington.edu/da.html, which provide neuroanatomical reference examples using 2D and 3D views of the brain from cadaver sections, MRI scans, and computer reconstructions.

5 Scan Types

As noted earlier, the appearance of healthy and lesioned tissue on the scan being traced depends on whether the scan is an MRI or a CT and, for MRIs, the type of imaging sequence. Some very commonly used sequences for MRI are T1- and T2-weighted scans. Distinguishing between the two is easily done by looking at the CSF, where CSF is dark in the T1-weighted imaging but bright on the T2-weighted imaging. The brain tissue also appears different in the two scans, with white matter appearing light in the T1w and a dark gray in the T2w and cortex appearing gray in the T1w and light gray in the T2w. The axial slices in Fig. 4 illustrate these differences in appearance between the two sequences. Another common sequence is fluid-attenuated inversion recovery (FLAIR). Although similar to the T2-weighted scan in the appearance of tissue, in FLAIR images, the abnormalities are bright but normal CSF fluid is dark, as in the T1-weighted images. It is worth noting that a given lesion will likely appear different in size as a function of the scan type, with T1-weighted images tending to demonstrate a smaller area of damage than T2 or FLAIR.

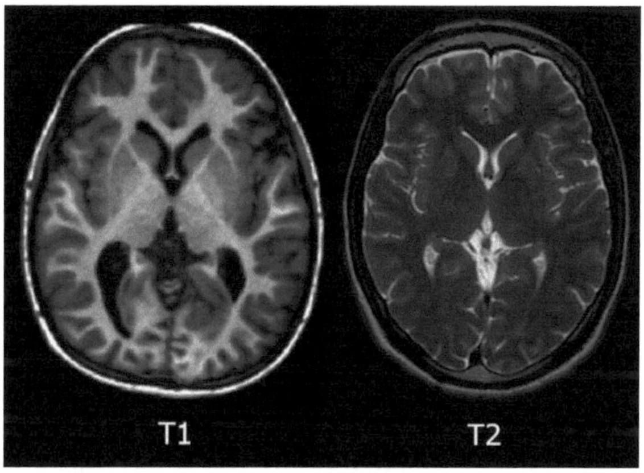

Fig. 4 An axial slice taken from T1-weighted (left) and T2-weighted (right) imaging sequences

6 Template Brains

To gain an understanding of how these tissue types appear in an MRI scan of a neurotypical brain, MRIcron provides an example brain, Montreal Neurological Institute space "Colin27" volume (Fig. 5a). This template MRI volume represents the brain of a single individual (scanned multiple times for stable results) that have been transformed into the Talairach stereotaxic space. The trainee can use the Colin27 brain and overlay various templates to see how area maps such as Brodmann (Fig. 5b) or automated anatomical labelling (AAL; Fig. 5c) correspond to a given scan. Familiarity with a "typical" brain will allow the lesion tracer to quickly note abnormalities in scans of atypical brains.

7 Notes: Scan Artifacts and Brain Abnormalities

Several types of scan artifacts and brain abnormalities may impact the approach to segmentation. The most common scan artifact is *motion artifact*, in which an individual's movement while inside the scanner compromises image quality. These can be caused by voluntary movements by the subject, as well as minor involuntary movements including respiration, swallowing, and eye movements. The result of these movements is a blurring or "ghosting" effect in the scan image, where axial slices from two individuals with or without motion artifacts are compared side by side. Although the relatively high signal-to-noise ratio of structural imaging boosts the ability to visually differentiate between brain anatomy and artifact, the degraded image quality can interfere with the interpretation of healthy vs lesioned tissue, especially in instances of smaller lesions.

To mitigate motion artifacts, preventative measures are preferable over post hoc correction when possible (i.e., if the motion does not occur in the first place, the effects of these artifacts are avoided altogether instead of more complex correction methods being necessary). Essential in reducing subject motion is designing a protocol with the needs of one's population in mind. This includes considerations such as shorter protocols in acute settings or for populations known to have difficulty in scans settings, such as children, or patients with pain, claustrophobia, or discomfort. Taking steps to minimize subject discomfort with comfortable head positioning and support can also reduce head movement in the scanner. This, along with ensuring subjects are familiar and comfortable with the scanner through clear instruction and reminders during the protocol, can be invaluable to data collection from many populations (clinical and nonclinical) [8, 9]. For children in particular, anxiety in the MRI scanner can be a large source of motion. For labs with the funding and access to a mock scanner, practice

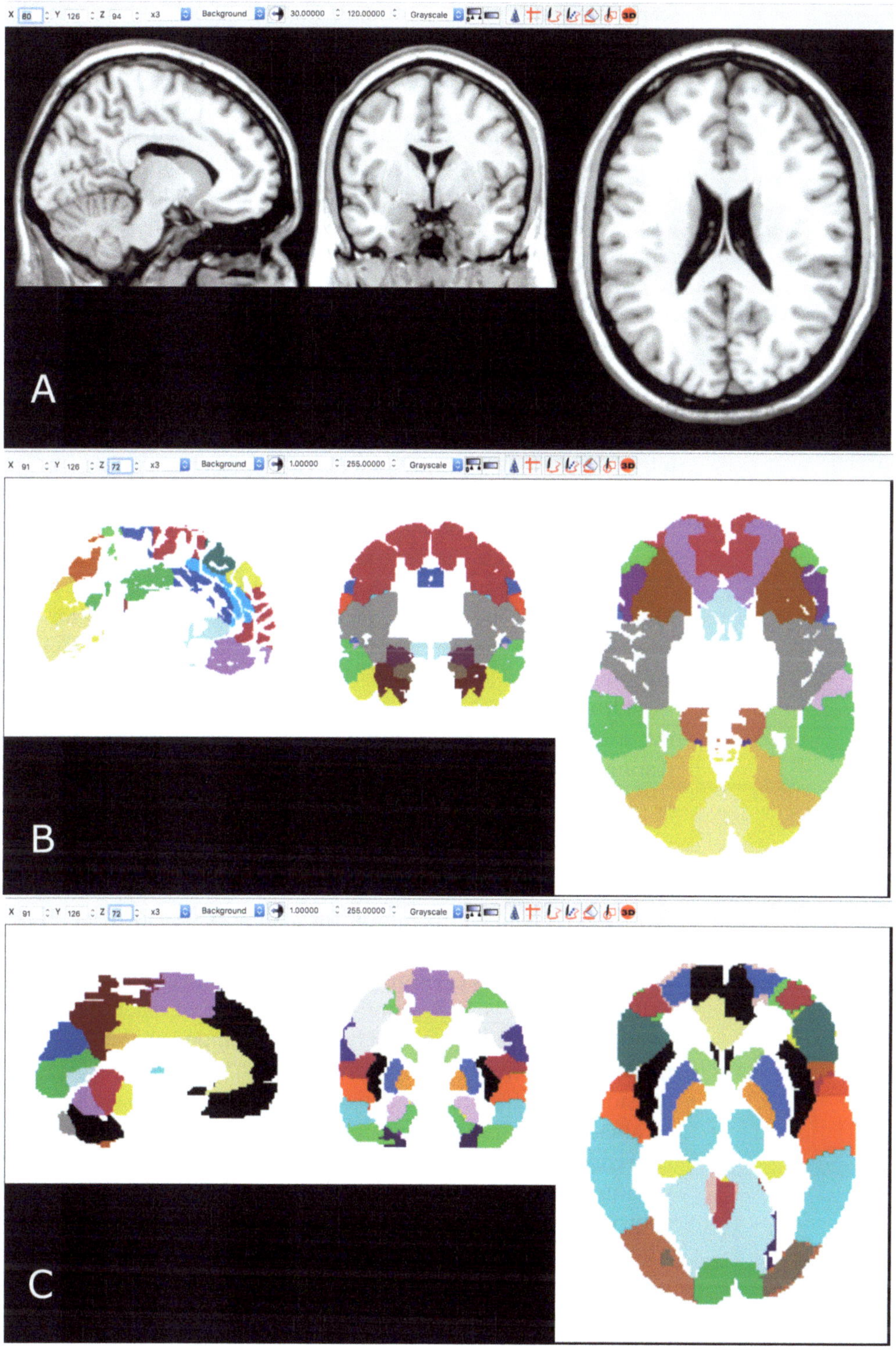

Fig. 5 (**a–c**) Tri-view of templates: (**a**) the Colin27 brain, (**b**) Brodmann areas, and (**c**) AAL regions

scans in a mock scanner to familiarize the child with the machine and procedure can reduce this anxiety [10].

If motion artifacts appear localized to a section of the brain that does not overlap with the majority of the lesion, manual segmentation can typically proceed without issue. It is helpful in such cases to segment the "cleanest" slices first, where the boundaries are most clear, and then use these as a guide when filling in the more ambiguous slices to hopefully create smooth and coherent 3D representation. If, however, the artifact is significant and pervasive, or concentrated around the lesion site, and identifying the appropriate lesion boundaries becomes difficult or impossible for a significant portion of the lesion, the scan may be unusable in its current state. Excluding from analyses those subjects that show high levels of motion degradation may be unavoidable but is generally not ideal, both because of the loss of data and the potential selection bias one may introduce to the dataset as a result. To avoid this, techniques such as the volumetric navigator (vNav) prospective motion correction system [11] and PROPELLER [12] have been found to significantly reduce the extent of motion artifacts in brain images, including motion-induced bias and variance in morphometry, and improve lesion characterization. See [9] for a discussion of some available approaches and their appropriateness for a given task.

Other than motion-related artifacts, another relevant feature of the scan images is the positioning of the head. When in an MRI scanner, subjects are typically laying down, with their head positioned with the help of pillows or foam blocks. This generally produces scans in which the brain is angled in a similar way across patients, with the coronal plane approximately horizontal and the left and right hemispheres symmetrical across the sagittal plane. However, it is not uncommon for a subject's head to be "tilted" in the scanner, either through minor shifting over the course of the scan or due to physical limitations in positioning. This results in an MR image that is "off-center" in the three views. While manual segmentation is not impeded by this, it is important to be aware of this feature when drawing a tilted scan. This is because when considering whether some tissue appears abnormal, it is often invaluable to assess to the opposite hemisphere, which provides a reference point of what the damaged hemisphere might look like without the lesion. This is useful when gauging, for example, the approximate size of the ventricles at a given slice to determine where to draw the boundary of a periventricular lesion. However, in a scan that is tilted, the two hemispheres in a given slice do not always correspond to the same position in a given direction, making cross-hemisphere inferences less straightforward.

For example, in Fig. 6, we can see the individual had a slightly tilted position in the scanner (note the division between the left and right hemispheres do not form a perfectly vertical line). When we

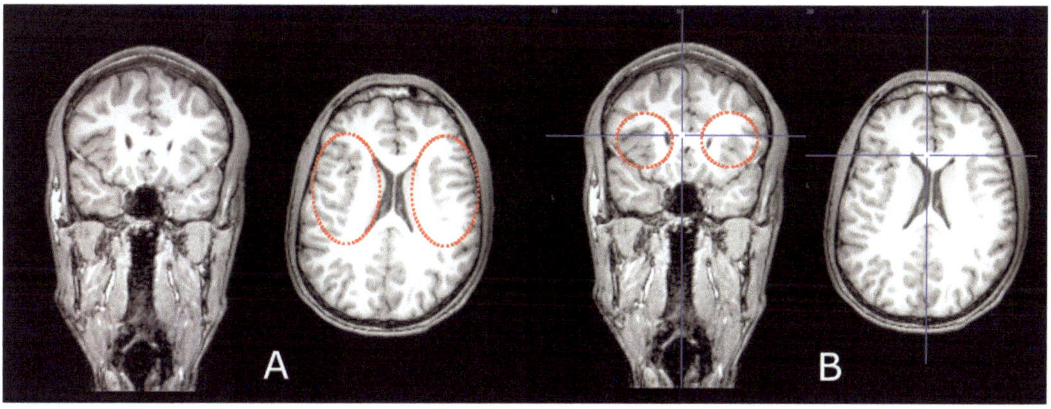

Fig. 6 (**a, b**) A neurotypical brain with a slight tilt. This results in the different appearance of the Sylvian fissure between the two hemispheres in the axial view

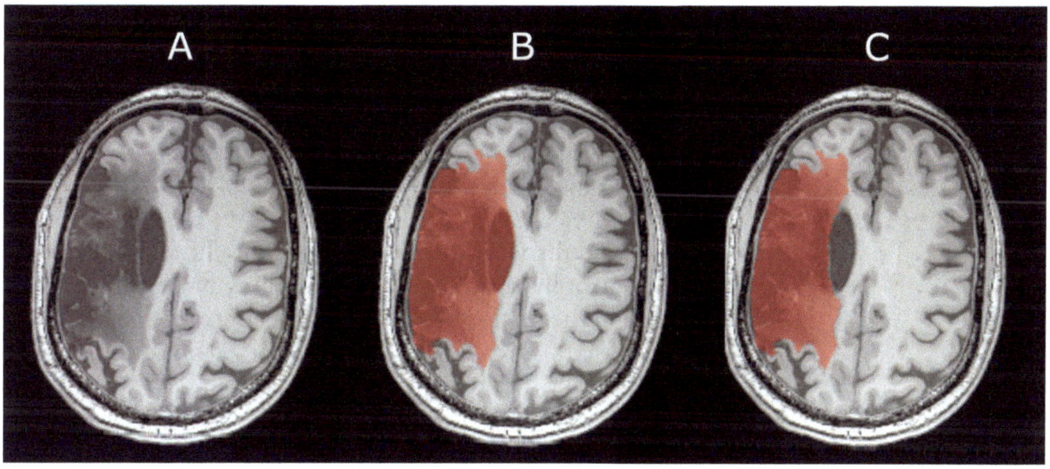

Fig. 7 (**a–c**) Axial view of a moderately tilted brain, shown with an incorrect lesion drawing in (**b**) and correct drawing in (**c**)

look at the axial slice, the Sylvian fissure – circled in red – appears somewhat different between the two hemispheres, with the left hemisphere showing significantly more sulci and gray matter. However, this difference is not due to any structural differences but rather the different positionings the left and right hemispheres have in the z axis due to this tilt. In the coronal view in Fig. 6b, the horizontal blue line shows how these coordinates place us much further into the fissure in the left hemisphere than on the right.

In lesioned brains, this may lead to over-interpreting interhemispheric differences as attributable to the lesion. For example, the axial slice in Fig. 7a shows a sizeable left hemisphere lesion. When determining where to trace the boundaries of this lesion, a novice lesion tracer may be tempted to trace the lesion as in Fig. 7b. While a more experienced tracer would likely recognize the structure that is lesioned in Fig. 7b but spared in Fig. 7c to be the left ventricle, it may be more challenging to recognize for a newly trained tracer without its counterpart in the right hemisphere. ITK-SNAP offers a rotation tool for manually correcting the orientation of the brain, which might be useful to apply prior to lesion segmentation.

In addition to artifacts, features of brain anatomy, such as those related to aging, trauma, or neurodegeneration, can complicate lesion segmentation and should be considered when analyzing scans.

Cerebral Atrophy In studies of aging or conditions that primarily affect older populations, such as stroke, cerebral atrophy is frequently encountered. Atrophy refers to a loss of neurons and neuronal connections, resulting in a decrease in tissue volume. The brain appears to be "shrunken," pulling away from the skull and increasing intergyral space. Cerebrospinal fluid (CSF) volume increases to fill the extra space. This often is a feature of normal aging – MRI studies consistently show an age-related decrease in gray matter and white matter volume along with an increase in CSF volume [13, 14]. However, it is also caused by brain injury, stroke, or diseases and disorders of the brain.

For older adults with stroke, there is typically some degree of whole-brain atrophy related to age, as well as more pronounced focal atrophy in the hemisphere affected by the stroke. The reduced gyri and increased CSF can often be difficult to distinguish from the lesion itself. If the brain appears to be affected by atrophy to comparable degrees in both hemispheres, comparison against the non-lesioned hemisphere is valuable in determining how much tissue loss should be attributed to atrophy as opposed to lesion. However, in cases of pronounced focal atrophy in the damaged hemisphere, comparison with the non-lesioned hemisphere is not as accurate for estimating atrophy-related tissue loss.

Enlarged Ventricles Cerebral ventricular enlargement or *ventriculomegaly* can occur as a result of normal aging but is often associated with and exacerbated by neurodegenerative disorders, traumatic brain injury, stroke, Alzheimer's disease, and schizophrenia, among others. For periventricular lesions, our lab generally attempts to spare the entirety of the ventricle (to the extent that this is discernable), which means that while the expanded boundaries of the ventricle and the increased CSF may be caused by the lesion, we opt to not consider it a "part" of the lesion. Instead we attempt to limit the boundaries of our segmentation to whatever evidence of the structural boundary of the ventricle remains. Other labs may instead choose to include in their segmentation any portions of the ventricle that extend beyond the expected size, likely

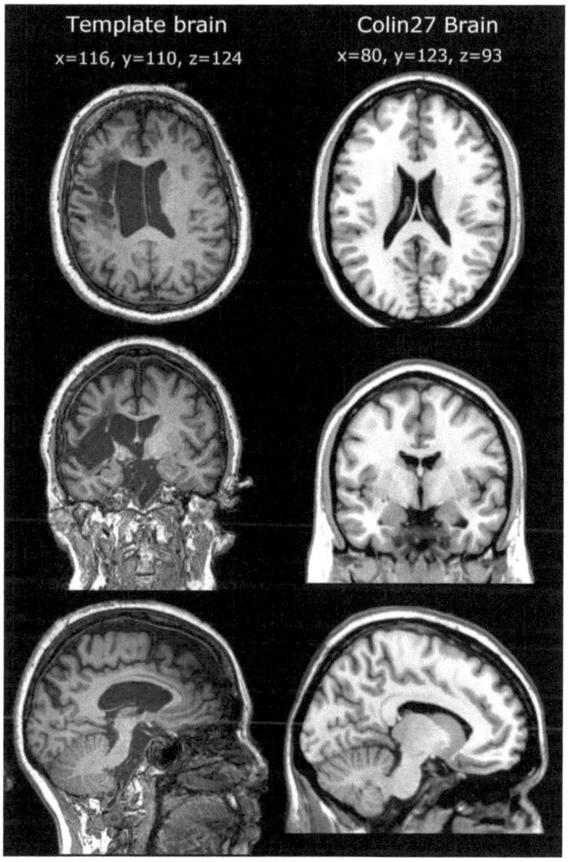

Fig. 8 An example of bilateral enlarged ventricles from the ATLAS brain on the left, alongside slices from a healthy brain (Colin27) on the right

judged by the corresponding hemisphere. For labs studying populations where this will likely be a common occurrence, it is important to implement a consistent procedure regarding sparing or segmenting such ventricles. See Fig. 8 for an example of ventriculomegaly from the ATLAS case alongside comparable slices from the Colin27 brain.

When considering how to segment around the ventricles, it is important to bear in mind that the drawing around the ventricles needs to be consistent from slice to slice to form a coherent 3D shape. In cases of ambiguous lesions where there is no clear border between ventricle and lesion, a rater may suddenly change their mind about where the appropriate boundary should be. This leads to manual segmentations where the lesion suddenly shifts from one slice to the next. These sudden shifts create problems for the registration in template space. Since all the lesions need to be registered in a common space before analyses, this step needs to be accurate. However, if the lesion has a sudden slice that goes too

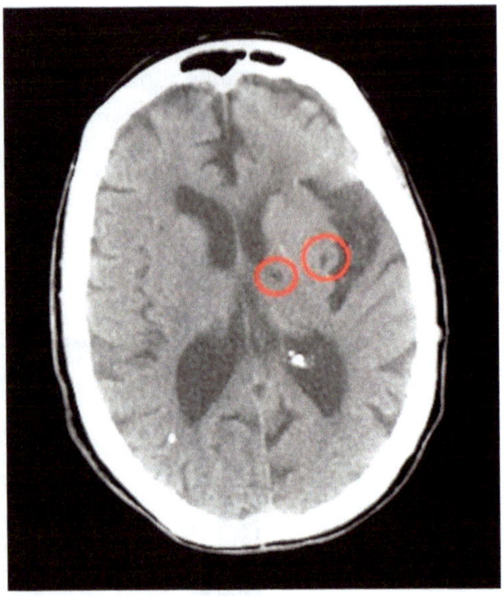

Fig. 9 CT scan of two lacunar strokes. (Image by Mikael Häggström, retrieved from https:/commons.wikimedia.org/wiki/File:CT_of_lacunar_strokes.jpg, CC BY-SA 3.0)

much into the ventricles, it creates a jagged appearance in another plane, which together make the nonlinear deformable registration harder to achieve. During registration, the lesion is "masked out," leaving only non-lesioned tissue for the registration algorithm to "see." The algorithm tries to pull or push the healthy tissue around to match the template. If the tissue visible to the registration is not very uniform, the registration algorithm may have difficulties to properly register the area. Thus, the rater must be careful in avoiding unnecessary sudden changes in lesion shape.[1]

Lacunar Infarcts A very common type of ischemic stroke, lacunar infarcts (LIs) are small infarcts in the white matter, generally thought to arise out of a blockage in a minor artery supplying the brain's subcortical areas [15]. The prevalence of LI increases with aging in the young and adults, and there are many age-related risk factors for LIs, including hypertension, diabetes, and cerebral infarct [16]. LIs appear as small, often isolated areas of damage in a structural MRI, as noted in Fig. 9. Our lab does not include these in our lesion segmentation, as these strokes frequently go unnoticed and are typically not the cerebrovascular event that has given rise to the pathology of interest in our patients.

[1] The problem of jagged border is partially avoided by smoothing the manual lesion and binarizing it again, a strategy used by several groups, which adds another layer of processing on top of what should be the gold standard to start with.

8 General Techniques

To summarize, the first steps in training to become a reliable lesion tracer are to develop a solid background in the relevant neuroanatomy, an understanding of the way in which those structures translate to two-dimensional representations, an awareness of the common artifacts and features of the scans they'll be analyzing, and a basic familiarity with navigating the lesion tracing program. Following from this, these skills can be synthesized and put to use by observing a trainer segmenting lesions and by practicing lesion segmentation.

As mentioned earlier, it is difficult to determine if a given voxel or group of voxels is damaged by its appearance on a single slice or single axis in isolation. What appears to be a damaged tissue from one perspective may actually be an artifact or a sign of atrophy or an individual difference in brain architecture. The determination of the status of a brain region is enhanced by viewing the region on axial, coronal, and sagittal views.

A brief example of this kind of ambiguity can be seen in Figs. 10, 11, and 12, all taken from the ATLAS case and labelled with their z coordinate. When a hypothetical lesion tracer encounters this superior axial slice shown in Fig. 10, his/her eye might first be drawn to the three darker areas of cortex, labelled with yellow arrows. From this slice alone, it is likely hard to determine which shade of darker gray indicates damage, and which does not, and a novice lesion tracer may be equally likely to include any or all of them in their drawing.

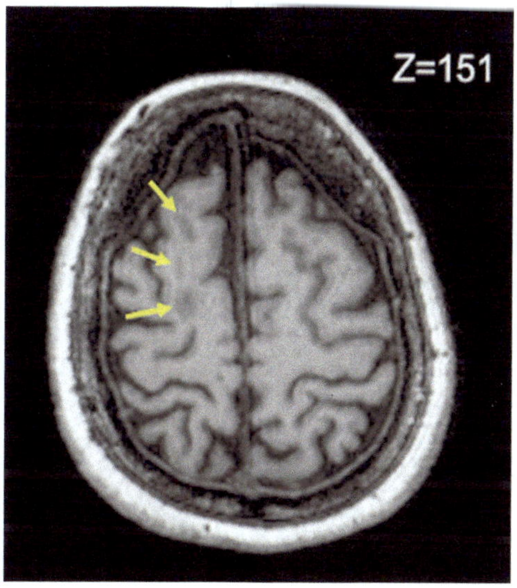

Fig. 10 Axial slice demonstrating three regions of gliosis in white matter

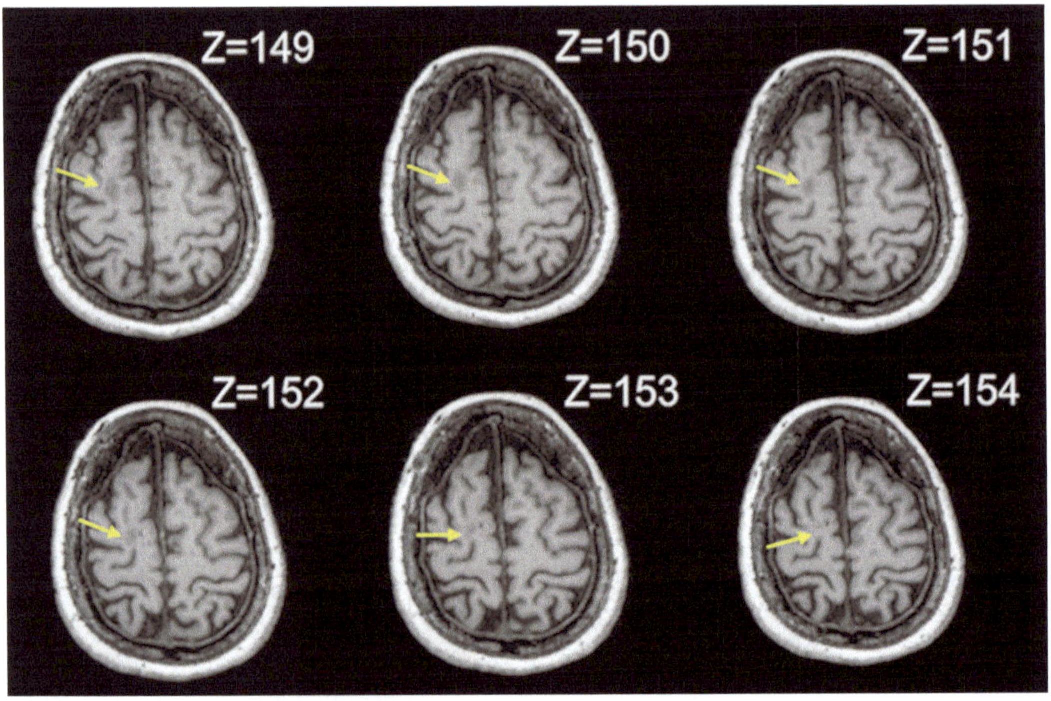

Fig. 11 Axial slices 154 through 149, with only the lesioned voxels noted

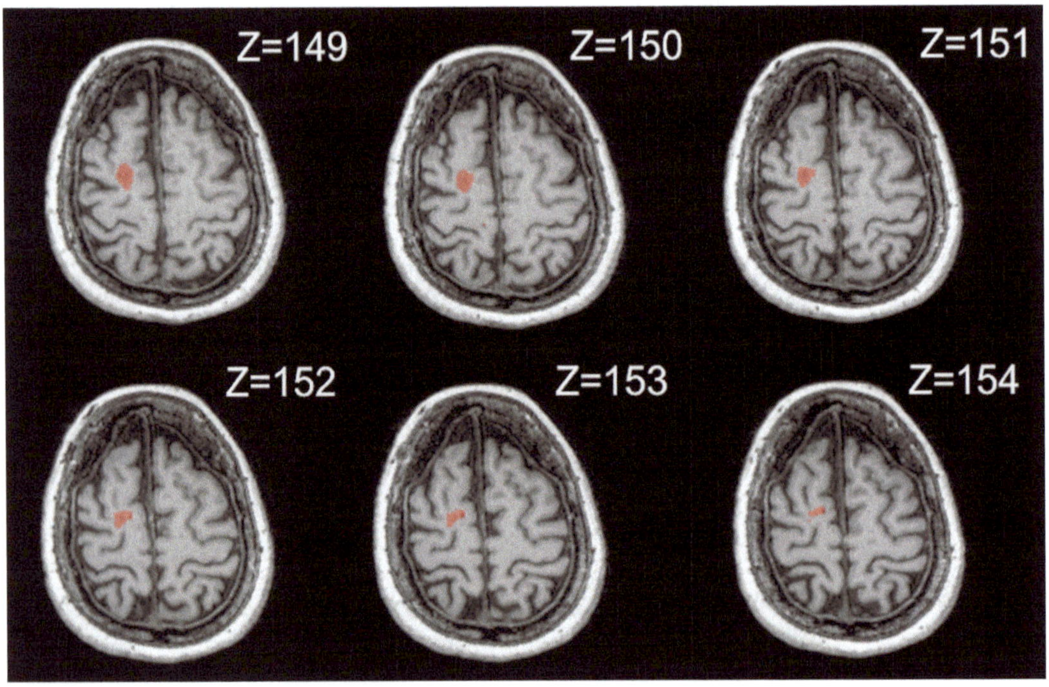

Fig. 12 Final lesion segmentation for slices 154 through 149

However, when we contextualize this slice by examining the slices immediately superior and inferior, our intuition about these voxels will change (*see* Fig. 11). Of the three gray patches of cortex shown in Fig. 10, only one remains a lesion-like darker patch, labelled with a yellow arrow in Fig. 11.

The two relatively anterior patches reveal themselves to be sulci in slices z = 152 and z = 153. For reference, the final lesion segmentation for these slices is shown in Fig. 12.

Only after contextualizing the slice did the nature of those voxel clusters become clear. The same can be said for contextualizing voxel clusters in all three axes. When the crosshairs are placed on each of the three gray patches seen in the axial view, their positions in the coronal and sagittal views are identified as well. When we compare the coronal and sagittal views of the three patches of gray from Fig. 10, their status as sulcus or lesion is also made evident. We can see in Fig. 13a, b that the first two gray patches of voxels are located right at the "beginning" of two sulci, while in Fig. 13c, the third patch is shown to be contiguous with the body of the lesion.

As intuitions about a given group of voxels are subject to change depending on the axis in which we view it, it is perhaps unsurprising that a lesion drawn using only the axial view might look quite different from that same lesion drawn in the coronal view. That means, for example, after finishing the first-pass drawing in the axial view, upon viewing that drawing in the coronal view, there will be adjustments needed, usually along the boundaries of the lesion, where the correct border of the lesion may not have been obvious in one view but becomes clear in another.

As an alternative approach, rather than completing the entire detailed "first-pass" drawing in a single view, the lesion tracer may effectively "down-sample" the slices, drawing the lesion only every five or ten slices in a single view. This (unsurprisingly) is completed in a fraction of the time and creates a skeleton framework of the lesion in the other views, as seen in Fig. 14. At this point, the boundaries of the lesion as determined in the initial view serve as a guidepost in the other views, making the process of drawing the full lesion (i.e., every slice) significantly simpler.

9 Practice, Practice, Practice

Subsequent to observation and supervised practice, the trainee should begin a phase of semi-independent practice, working their way through a set of pre-identified lesions and consulting with their trainer during and after each drawing is completed. We have found it useful to set aside a set of scans for which we already have drawings that are finished, checked, and approved by an expert that will be used for practice by the trainee. Having complete

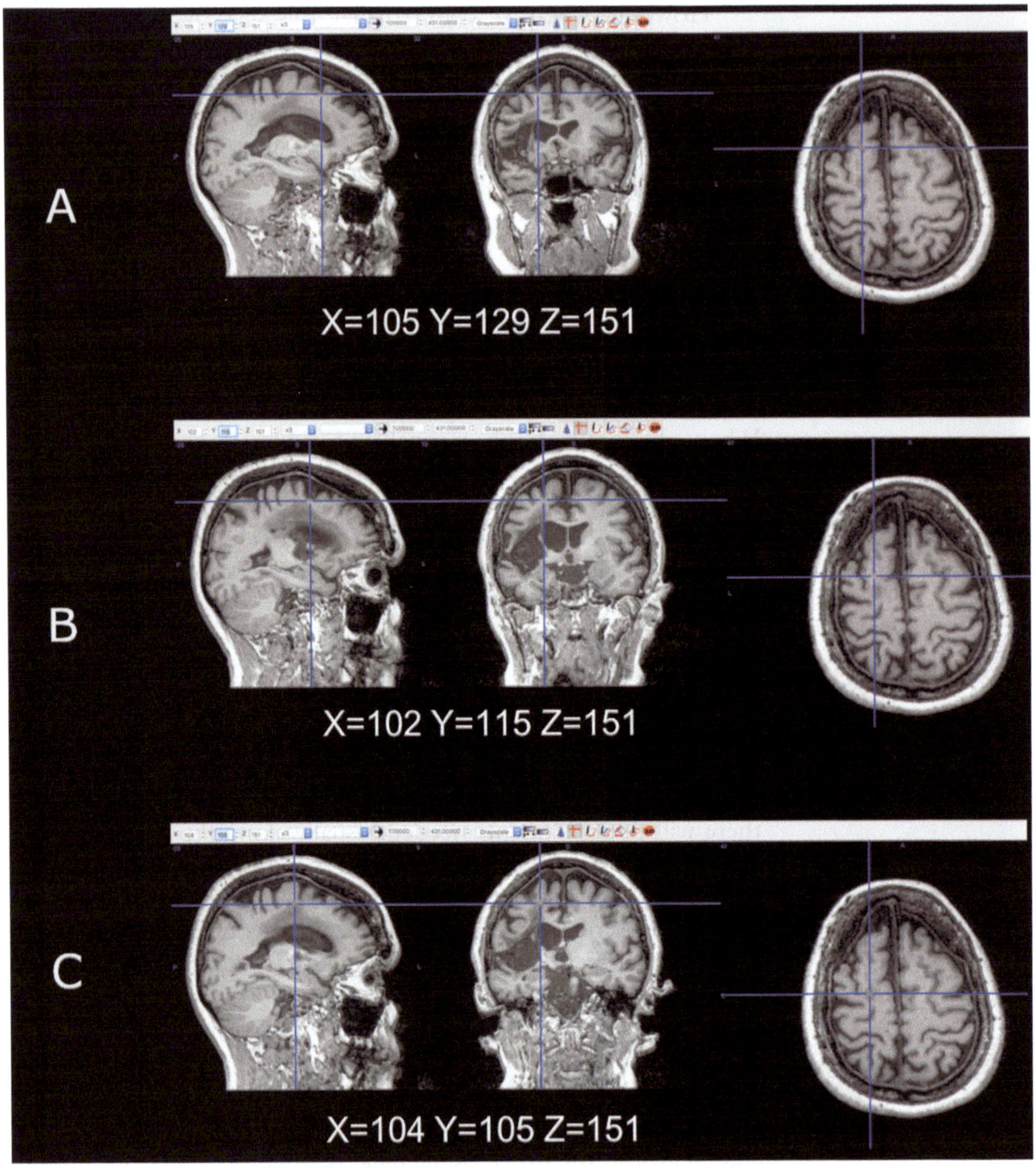

Fig. 13 (**a–c**) Sagittal, coronal, and axial views of the regions of gliosis identified in Fig. 12

drawings already at hand is important to serve as a "correct" version against which the practice drawings can be compared. We have typically set aside six to ten lesions for this purpose, grouped into "easy" and "hard" categories (or "easy," "medium," "hard") to give the trainee practice drawing a varied group of lesions.

What makes a lesion "easy" or "hard" to draw depends on a number of factors, including abnormal brain architecture, lesions that have a large number of "ambiguous" areas (where tissue might appear healthy or damaged depending on the view it is displayed

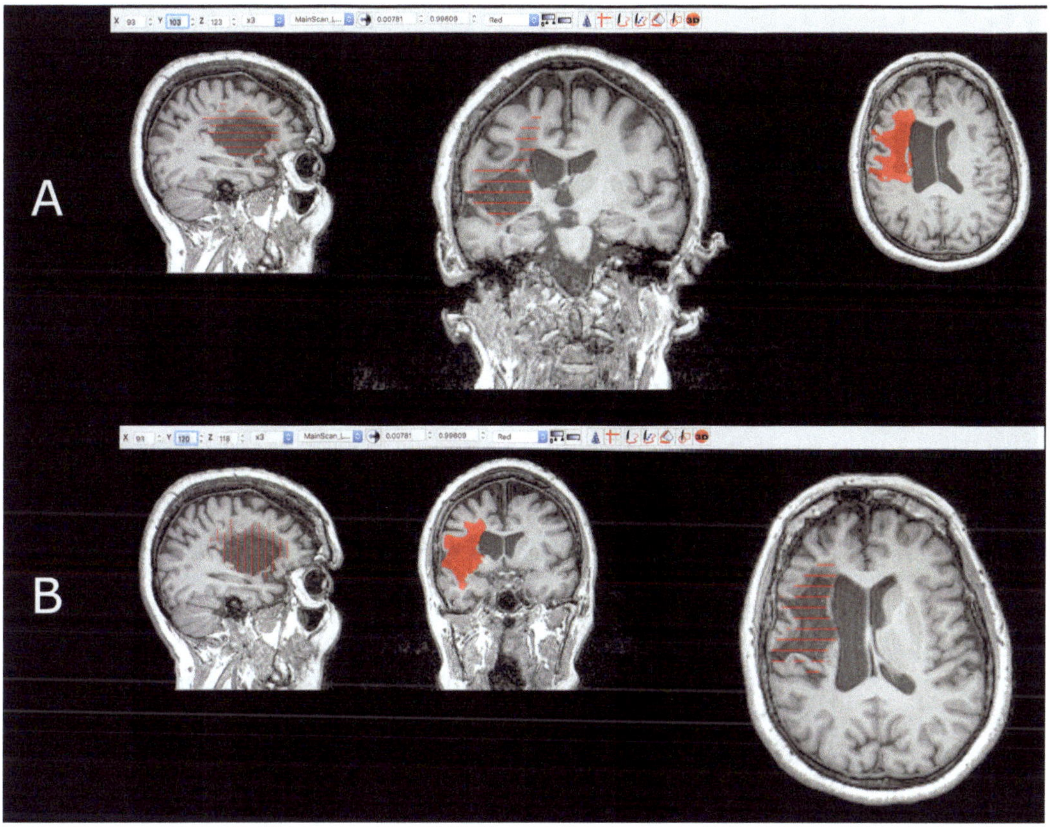

Fig. 14 (**a, b**) Lesion skeleton approach. Sagittal, coronal, and axial views demonstrating the appearance of the segmentation when (**a**) down-sampling in the axial view and (**b**) when down-sampling in the coronal view

in), or lesions in which the boundaries are largely composed of various shades of gray (resulting in more ambiguity) rather than clear edges delineating damaged tissue that appears significantly different than the healthy surrounding tissue. See Fig. 15 for some examples of these different boundaries. Additionally, areas of tissue that appear functional based on their saturation (i.e., they appear light/white in a T1-weighted MRI) but that seem to be essentially isolated from any other functional tissue can make judgment calls more difficult.

The trainee, beginning with the easier scans and making their way to the difficult ones, may complete all practice scans in the designated set, either receiving feedback through the drawing process or upon completion of each lesion. Practice drawings can be compared against the final correct drawing in MRIcron by overlaying both files on the scan image and adjusting the opacity to easily identify areas of discrepancy. The number of practice lesions included in the set may vary by the individual needs of the trainee, but we have found that between 8 and 13 practice lesions have allowed us to include a variety of lesion types in the training. The

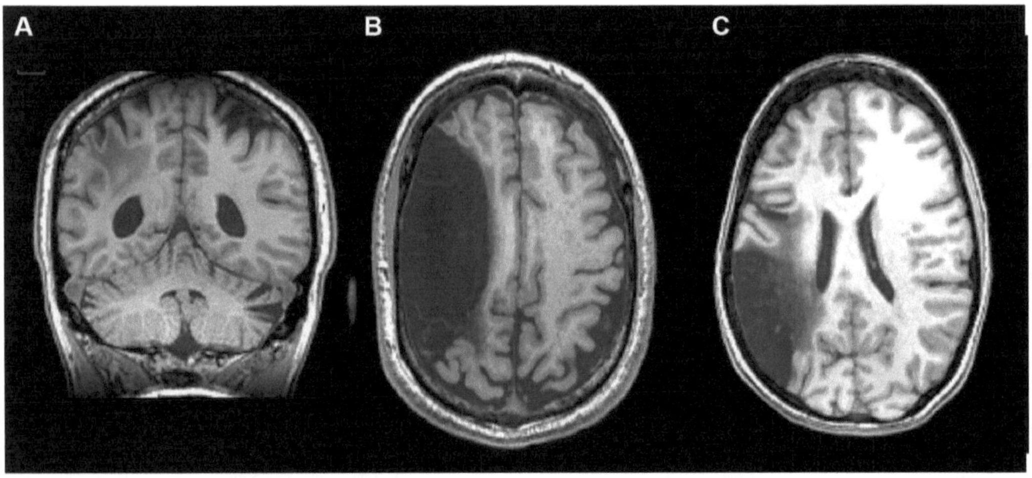

Fig. 15 (**a–c**) Examples of different types of "boundaries" found on lesions. Note the subtler contrast between the shades of gray in the left hemisphere lesion in image (**a**), while in (**b**) and (**c**) the gradient from lesion and CSF to white matter is more stark

amount of time required to manually segment a lesion from start to finish varies greatly by individual and by the lesion difficulty and size, but we have estimated that it may take between 7 and 10 hours per scan, with a novice lesion tracer likely requiring more time early on. This task is (unsurprisingly) best spread out over several days to avoid fatigue and to allow the tracer to approach the lesion with "fresh eyes" multiple times. This means that if a trainee completed a scan roughly every 3 days (~3 hours per day), the practice phase of training might take approximately 18 days. This time may be greatly extended depending on the number of hours a trainee is able to devote to tracing per day, as well as the number of practice lesions to be completed. For this practice phase of training, having frequent access to experienced individuals who have reached reliability in lesion tracing will be essential. Ideally, an expert in neuroanatomy will also be available to the trainee for consultation throughout.

10 Reliability

After completing all designated practice lesions, if the trainee (and the trainer) feels that their drawings have been consistent with the correct drawings and that they are ready to proceed, the trainee can be tested for reliability. Our labs at MRRI and LCNS typically set aside two sets of three scans, unseen during training or practice, to be used for reliability testing. As with the practice lesions, all reliability lesions were the result of a left cerebral vascular accident (CVA) and were imaged using MRI. As with the practice lesions, all lesion boundaries were identified and traced using MRIcron and were visually compared to that of an expert, at which point they were included in the reliability test set. The reliability test requires the trainee to manually draw three lesions. This task is generally meant to mirror how lesion tracing might proceed in reality, so we typically do not impose time limits or other constraints on this task. Unlike typical lesion tracing, we specify that the trainee cannot seek advice from their trainer or other lesion tracers. Upon completion, the trainee's lesion drawings are statistically compared against the finalized "correct" drawings. This is done by calculating the voxel overlap between the two files, typically via measures such as the dice overlap (or "dice coefficient"), which gives a measure of the proportion of the total number of voxels identified as lesioned in the two drawings that are lesioned in both. This metric allows us to examine how similar the volumes are, as well as whether the same voxels appear in each lesion mask or not. This is calculated by the following formula (where X and Y represent voxels from each lesion):

$$DC = \frac{2\,|X \cap Y|}{|X| + |Y|}$$

To be considered "reliable," the overlap percentage must meet a certain threshold. The reliability measures we have used as a threshold are a dice overlap of 0.8 or greater and a low mean percentage discrepant voxels[2, 3] (approximately 4%), which are consistent with reliability measures previously reported in [17] and [18]. For comparison, the manual segmentation for ATLAS dataset used 11 individuals who were trained ("Training consisted of a detailed protocol and instructional video, and all tracers were guided through the training process by an expert tracer") on a set of five brains, which they traced twice (at two different time points). Inter-rater reliability was first calculated using the lesion volumes, yielding an inter-rater reliability of 0.76 ± 0.14.[4] Reliability was then calculated using dice coefficient (DC). The average inter-rater DC for first segmentation was 0.745 ± 0.19 and for the secondary segmentations was 0.766 ± 0.16.[5]

[2] Measures such as non-overlapping voxels are also useful in assessing reliability. The percentage of nonoverlapping voxels in a volume pair is the difference between the number of voxels in the total volume and the number of voxels shared by both volumes, divided by the number of voxels in the total volume.

[3] From [5]: "Relatively small differences in the placement of lesion boundaries can lead to large cumulative effects over an entire volume. . . .To give additional perspective upon the nonoverlap between two volumes, nonoverlapping voxels that were located more than 2 voxels from one or both volume surfaces were classified as discrepant voxels (the 2-voxel criterion was based upon results from the inter-surface analyses)., the percentage of discrepant voxels in the total volume was computed (number of discrepant voxels divided by the number of voxels in the total volume)."

[4] Intra-rater reliability using lesion volume was 0.84 ± 0.09.

[5] Intra-rater DC scores (calculated by comparing the initial segmentation to the secondary segmentation for each tracer) were averaged to obtain a final intra-rater DC score (0.831 ± 0.13).

If the threshold selected by our lab is not met after averaging across these values for the three test lesions, the trainer and trainee review the drawings together and discuss the areas of discrepancy, with an eye toward any overall trends in the trainee's drawings that hurt their reliability. These may be things such as being overly liberal or conservative with what is considered lesion or failing to adequately "clean up" the drawing in one of the views. The trainee is then given a second opportunity to practice additional unseen lesions and compare their drawings to the completed ones alongside the trainer to rectify any remaining confusion or refine their abilities further. We do not have a set amount of time for this secondary practice phase, largely leaving it to the discretion of the trainee and trainer to determine when to test for reliability again. This second reliability test will make use of the second set of test scans, previously set aside and unseen by the trainee. The reliability task proceeds as before, and once the three lesions have been traced, overlap and discrepancy are calculated again. We have not had any individual trainees fail to meet our reliability threshold after a second round of training and testing. If a trainee fails to meet reliability a second time, it is up to the discretion of the lab how they choose to proceed.

When considering the lesions to be included in the reliability test, it is advisable to choose lesions representative of the population that will be tested across a range of expected difficulty and size. It is also important to consider the final drawings against which the trainee's will be compared. If, for example, a lab hired a series of three RAs over the course of a few years, and each one trained the next one hired on lesion tracing, it is possible that idiosyncrasies in an individual RA's "style" of tracing might be passed on to the next and so forth, creating "drift" in the judgments of the lesion tracers over time. This can be mitigated by using a single "gold standard" drawing of a lesion that each new RA is tested against, rather than testing each trained RA against their trainer's drawing. Moreover, for labs where multiple individuals are responsible for segmentation, it is advisable to compute reliability of each trainee against the "gold standard" as well as against each other and/or other trained lesion tracers. This is a useful tool because, while each individual may be above the minimum threshold against the standard drawing, they may differ in their areas of discrepancy with that standard, such that they no longer meet "reliability" against each other. This knowledge is useful in identifying any systematic trends in the way lesions are drawn that differ among the different drawers, which can then be reconciled.

This type of discrepancy or drift can also be mitigated by oversight on the part of an expert, such as a neurologist, who can assess lesion segmentations upon completion for acceptability. In our setting, all lesion drawings are sent to a neurologist who makes any necessary adjustments before approving them to be registered onto a common template and included in analyses. Aside from ensuring validity in our data, this allows any consistent or significant mistakes or oversights on the part of the lesion tracer to be identified and discussed as they come up.

The reliability we have found using this procedure has been high relative to the inter-rater reliability reported in the literature. However, research groups vary in their methods as well as in the reported metric. Below, we have included a table briefly summarizing a few papers that have reported inter- and intra-rater reliability or agreement values for comparison [3, 5–7, 19–21]. Inter-rater reliability is especially important in labs where the individual responsible for lesion tracing may change every few years. In labs where a single individual will perform the lesion tracing over a long period of time, additional intra-rater reliability measures should be implemented as well, as described in [22], where trainees traced a given lesion at two different time points, and lesion volume and dice coefficient were used to assess reliability within an individual (Table 1).

In this chapter, we have outlined a method for manual lesion segmentation that emphasizes the importance of an in-depth, hands-on period of training for prospective lesion tracers and a high level of inter-rater reliability across lesion tracers (both past and present as well as across multiple contemporaries). This helps ensure consistency in how lesions are represented across a dataset, which is essential to any valid scientific inference based on that data.

Table 1
Summary of reliability/agreement measures from a selection of relevant publications

Authors	Raters	Lesions	Variable	Inter-rater	Intra-rater
Zhu et al. (2010)	2	10	Reliability (R^2)	0.92	
Ito et al. (2018)	11	5	Dice similarity coefficient (DC)	0.75 ± 0.18	0.83 ± 0.13
Paty et al. (1994)	?	?	Reproducibility of tracing	?	6%
Grimaud et al. (1996)	3	8	Coefficient of variability (CV) of precision	$11.0 \pm 5.8\%$	$9.0 \pm 5.2\%$
Filippi et al. (1995)	?	?	Variability	14%	6.5%
Filippi et al. (1995)	3	20	Median agreement	93.4% (77.3–98.3%)	95% (85.1–99.4%)
			Variability	14.1%	
Fiez et al. (2000)	2	10	Reliability (intraclass correlation coefficient)	0.88	0.86, 0.95 (mean = 0.905)

The training and reliability protocol outlined here have hopefully illustrated what the implementation of this "gold standard" method – manual segmentation– might look like and can serve as a template to be modified to fit the needs of a given lab.

References

1. Crainiceanu C, Sweeney EM, Eloyan A, Shinohara RT (2016) A tutorial for multisequence clinical structural brain MRI. In: Ombao H, Lindquist M, Thompson W, Aston J (eds) Handbook of neuroimaging data analysis. CRC Press/Taylor & Francis Group, Boca Raton/London/New York, pp 109–133
2. Ito KL, Kim H, Liew SL (2019) A comparison of automated lesion segmentation approaches for chronic stroke T1-weighted MRI data. Hum Brain Mapp 40:4669–4685. https://doi.org/10.1002/hbm.24729
3. Filippi M, Horsfield MA, Bressi S et al (1995) Intra- and inter-observer agreement of brain MRI lesion volume measurements in multiple sclerosis. Brain 118:1593–1600. https://doi.org/10.1093/brain/118.6.1593
4. Paty DW, Li DKB (1993) Interferon beta-1b is effective in relapsing-remitting multiple sclerosis: II. MRI analysis results of a multicenter, randomized, double-blind, placebo-controlled trial. Neurology 43:662–667. https://doi.org/10.1212/wnl.43.4.662
5. Fiez JA, Damasio H, Grabowski TJ (2000) Lesion segmentation and manual warping to a reference brain: intra- and interobserver reliability. Hum Brain Mapp 9:192–211. https://doi.org/10.1002/(SICI)1097-0193(200004)9:4<192::AID-HBM2>3.0.CO;2-Y
6. Filippi M, Horsfield MA, Tofts PS et al (1995) Quantitative assessment of MRI lesion load in monitoring the evolution of multiple sclerosis. Brain 118:1601–1612. https://doi.org/10.1093/brain/118.6.1601
7. Paty DW, Li DKB, Oger JJ-F et al (1994) Magnetic resonance imaging in the evaluation of clinical trials in multiple sclerosis. Ann Neurol 36:S95–S96. https://doi.org/10.1002/ana.410360721
8. Havsteen I, Ohlhues A, Madsen KH et al (2017) Are movement artifacts in magnetic resonance imaging a real problem? – a narrative review. Front Neurol 8:1–8. https://doi.org/10.3389/fneur.2017.00232
9. Zaitsev M, Maclaren J, Herbst M (2015) Motion artifacts in MRI: a complex problem with many partial solutions. J Magn Reson Imaging 42:887–901. https://doi.org/10.1002/jmri.24850
10. De Bie HMA, Boersma M, Wattjes MP et al (2010) Preparing children with a mock scanner training protocol results in high quality structural and functional MRI scans. Eur J Pediatr 169:1079–1085. https://doi.org/10.1007/s00431-010-1181-z
11. Tisdall MD, Reuter M, Qureshi A et al (2016) Prospective motion correction with volumetric navigators (vNavs) reduces the bias and variance in brain morphometry induced by subject motion. NeuroImage 127:11–22
12. Nyberg E, Sandhu GS, Jesberger J et al (2012) Comparison of brain MR images at 1.5T using BLADE and rectilinear techniques for patients who move during data acquisition. Am J Neuroradiol 33:77–82. https://doi.org/10.3174/ajnr.A2737
13. Courchesne E, Chisum HJ, Townsend J et al (2000) Normal brain development and aging: quantitative analysis at in vivo MR imaging in healthy volunteers. Radiology 216:672–682. https://doi.org/10.1148/radiology.216.3.r00au37672
14. Good CD, Johnsrude IS, Ashburner J et al (2001) A voxel-based morphometric study of ageing in 465 normal adult human brains. NeuroImage 14:21–36. https://doi.org/10.1006/nimg.2001.0786
15. Wardlaw JM (2005) What causes lacunar stroke? J Neurol Neurosurg Psychiatry 76:617–619
16. Cai Z, He W, Peng CY et al (2016) The prevalence of lacunar infarct decreases with aging in the elderly: a case-controlled analysis. Clin Interv Aging 11:733–738. https://doi.org/10.2147/CIA.S108166
17. Rorden C, Brett M (2000) Stereotaxic display of brain lesions. Behav Neurol 12:191–200
18. Schwartz MF, Kimberg DY, Walker GM et al (2009) Anterior temporal involvement in semantic word retrieval: voxel-based lesion-symptom mapping evidence from aphasia. Brain 132:3411–3427. https://doi.org/10.1093/brain/awp284
19. Grimaud J, Lai M, Thorpe J et al (1996) Quantification of MRI lesion load in multiple sclerosis: a comparison of three computer-assisted techniques. Magn Reson Imaging 14:

495–505. https://doi.org/10.1016/0730-725X(96)00018-5

20. Ito KL, Kim H (2018) A comparison of automated lesion segmentation approaches for chronic stroke T1- weighted MRI data. bioRxiv
21. Zhu LL, Lindenberg R, Alexander MP, Schlaug G (2010) Lesion load of the corticospinal tract predicts motor impairment in chronic stroke. Stroke 41:910–915. https://doi.org/10.1161/STROKEAHA.109.577023
22. Liew S-L, Anglin JM, Banks NW et al (2018) A large, open source dataset of stroke anatomical brain images and manual lesion segmentations. Sci Data 5:1–11. https://doi.org/10.1038/sdata.2018.11

Chapter 3

Automated Lesion Segmentation

Joseph C. Griffis and Dorian Pustina

Abstract

Studies investigating the effects of brain lesions on cognition and behavior have played a key role in shaping modern theories of brain function. While early studies in this domain typically involved post-mortem dissections of selected patients who presented with cognitive and/or behavioral deficits following brain injuries or neurological disease, modern lesion studies typically involve analyzing noninvasive neuroimaging data acquired from large numbers of patients using techniques such as computed tomography (CT) or magnetic resonance imaging (MRI). Modern studies therefore depend critically on the accurate segmentation of the lesion from healthy brain tissue in neuroimaging scans. Lesion segmentation has traditionally been performed manually by expert raters, which is a time-intensive procedure that poses obstacles to reproducibility. There is growing interest in automated lesion segmentation techniques that can be applied to large datasets collected from patients with brain lesions. Here, we provide an overview of the state of the art with respect to semiautomated and fully automated lesion segmentation methods that have been implemented in publicly available software packages. We discuss considerations for selecting an automated lesion segmentation method and highlight the potential advantages and disadvantages of different approaches. The second appendix at book's end provides hands-on tutorials for using some of the selected approaches.

Key words Automated lesion segmentation, Lesion-symptom mapping, Stroke, Tumor, Brain lesion

1 Introduction to Automated Lesion Segmentation

Before the proliferation of noninvasive neuroimaging techniques such as computerized tomography (CT) and magnetic resonance imaging (MRI), brain lesion identification involved detailed post-mortem examinations of deceased patients. Modern lesion identification, in contrast, typically involves the creation of 3D lesion maps based on CT or MRI scans. This is usually accomplished by having an expert neuroanatomist visually identify and manually trace the lesion(s) (*see* Chapter 2). This manual approach is currently considered the gold standard for lesion identification.

Despite being the current gold standard, the manual approach has several limitations. For example, it requires that researchers have access to someone (typically a board-certified neurologist or

Dorian Pustina and Daniel Mirman (eds.), *Lesion-to-Symptom Mapping: Principles and Tools*, Neuromethods, vol. 180, https://doi.org/10.1007/978-1-0716-2225-4_3, © Springer Science+Business Media, LLC, part of Springer Nature 2022

radiologist) with the expertise necessary to identify and differentiate brain lesions by visually inspecting scans from the relevant imaging modalities. Further, manual lesion delineation can be very time-consuming and may require between 20 min and several hours to completely segment a single lesion [1, 2]. This approach may therefore be impractical for analyses of extremely large lesion databases [3]. Another drawback of this approach is that it depends on the subjective impressions of the expert rater. This may limit the reproducibility of subsequent lesion analyses, as even though inter-rater reliability may often be quite good (e.g., intraclass correlations >0.9) [1, 4], similar segmentations can differ in meaningful ways [5, 6].

These and other limitations have motivated researchers to develop a diverse array of automated methods for lesion identification. Automated lesion segmentation (ALS) aims at generating lesion maps automatically through computer algorithms. While all ALS methods aim to enable objective and efficient lesion identification, they may differ in important ways. For example, some methods require user interaction to help the algorithm make useful decisions (i.e., semiautomated methods), while others are fully automated.

The imaging characteristics of brain lesions depend on factors such as the type of pathology (e.g., ischemic stroke, hemorrhagic stroke, tumor), the imaging method (e.g., CT, T1w MRI), and the age of the lesion (e.g., acute, subacute, chronic). For example, an acute lesion may not be visible on a T1w MRI scan, but a chronic lesion appears clearly to be dark, indicating tissue death. The acute lesion can instead be more promptly identified when measuring the diffusion of water molecules using diffusion-weighted MRI; yet, an acute lesion exhibits an initial restricted diffusion and looks bright in apparent diffusion coefficient (ADC) maps, while at later stages, a chronic lesion exhibits unrestricted diffusion and looks dark in ADC maps. Given the variety of imaging characteristics, many methods are specifically designed to identify a particular type of pathological lesion using data from a specific imaging modality—or set of modalities—collected at a particular temporal stage of lesion development. If applied on a different type of lesion or different modality, the results would most likely be suboptimal. ALS tools can vary also with regard to whether they require data from a control group, the statistical or machine learning algorithm that is employed, and the computational resources required to run, all of which are discussed throughout this chapter.

The above shows that there are many important factors for researchers to consider when selecting a method to use in their study. The aim of this chapter is therefore to guide researchers into the world of ALS tools by summarizing the metrics typically used to evaluate their performance and by providing a summary of the methods currently available. We highlight key considerations when selecting a method and briefly discuss future potential directions of ALS methods.

2 Concepts and Terminology

The terminology used for ALS is a blend of concepts from clinical neurology, neuroimaging, and bioengineering. It might, therefore, be useful to introduce some of the terms to familiarize the reader with their meaning in subsequent paragraphs.

The term "segmentation" typically defines the process through which a lesion map is obtained. When performed manually, lesion segmentation consists simply of drawing the lesion borders on a screen (*see* Chapter 2); thus, the process might be called lesion "drawing." In image processing jargon, the lesion map is sometimes referred to as a "mask," but it can be called also a "label" image. Independently on which term is used, the lesion map is almost invariably a binary image with just two values, 0 and 1 (but note that the MRIcron software uses the value 255 instead of 1).

The temporal stage of a stroke lesion is referred to with terms such as "acute" or "chronic." Acute stroke lesions are new, dynamic, and not settled in their shape and form to a definitive area. Chronic stroke lesions are several months old and are not expected to change much in the coming months. This said, there is evidence that lesions change even in the chronic stage, most likely by expanding [7]. But longitudinal change of chronic lesions is quite slow and is not expected to affect dramatically the initially drawn lesion maps. It is also unclear whether the apparent expansion of a lesion is related to degradation of more tissue at the lesion borders or rather a natural expansion occurring due to less mechanical pressure following natural aging atrophy. A more detailed explanation of the type and stages of stroke lesions can be found in the first two chapters of this book.

The algorithms used to automatically draw lesion maps usually fall in one of the two categories: "supervised" or "unsupervised." These bioengineering terms describe how an algorithm learns. Supervised methods require a training set of ground truths—i.e., manually segmented lesion maps—to learn what should be considered a lesion. In contrast, unsupervised methods do not need manually segmented lesions and learn to identify lesions based on inherent anomalies of the data—e.g., low-intensity voxels in an area where voxels should normally have high intensity. It should be noted that both types of algorithm detect the anomalies by knowing more on how a normal brain looks like, but the difference is whether the algorithm requires or not examples of ground-truth lesion maps.

Two other terms frequently mentioned in the literature are "univariate" and "multivariate." These terms come from statistics and refer to the method used to fit the data or learn the lesion anomaly. Univariate methods consider the relationship between a single input and a single output; e.g., when the voxel is dark, it might be lesioned. Multivariate methods consider the full relationship between multiple inputs and a single output. An input in a multivariate model may be limited value because it is considered in the context of all the other inputs; e.g., the T1w value of a single voxel can be considered in the context of values from other modalities (ADC) or values of neighboring voxels to determine if the voxel of interest should be considered lesioned.

Finally, a word of note on the binary nature of lesion maps. By design, each voxel is considered either lesioned or healthy, 0 or 1. Both manually and automatically segmented lesions follow this binary decision process. However, a binarized conceptualization of lesions is more of an operational artifact than a biologic truth corresponding to reality. There is no reasonable way to be sure to what degree the underlying tissue functions normally, particularly since lesion maps are drawn on structural MRI scans that do not carry any functional information. It is reasonable to assume that neural tissue sometimes is partially functional or has limited blood perfusion or improper vascular reactivity, etc. Making a binary decision on whether a voxel is lesioned or healthy is sometimes hard even for experienced raters. The variability of manual segmentations performed by different raters can be attributed to some degree to this uncertainty. A more realistic approach would be to attribute a graded level of tissue functionality to each voxel, but these approaches are neither easy nor well explored. Computer-generated maps can produce different degrees of uncertainty of the algorithm, and, more importantly, do not carry the inter-rater variability. But these probabilities, or algorithm confidence values, are not based on any functional information. In this chapter, we do not discuss the potential use of graded maps, but young researchers should note that graded maps can be subject of future research. The binary classification of voxels to produce lesion maps, either manually or automatically, is only a representation of the biological reality, which is likely more complex.

3 Performance Evaluation Metrics

The accuracy of lesion segmentation, whether performed manually or through automated methods, is of critical importance; all subsequent statistical analyses depend critically on the assumption that the lesion data consist in faithful representations of the true lesion topographies that are present in the patient sample. Before discussing specific methods, it is important to consider how

Table 1
Common performance evaluation measures

Measure	Definition	Description
True positive rate (recall, sensitivity)	TP/(TP + FN)	Ratio of TPs to the total number of TPs and FNs
False positive rate	FP/(FP + TN)	Ratio of FPs to the total number of FPs and TNs
True negative rate (specificity)	TN/(TN + FP)	Ratio of TNs to the total number of TNs and FPs
False negative rate	FN/(FN + TP)	Ratio of FNs to the total number of FNs and TPs
Positive predictive value (precision)	TP/(TP + FP)	Ratio of TPs to the total number of TPs and FPs
Negative predictive value	TN/(TN + FN)	Ratio of TNs to the total number of TNs and FNs
Dice similarity coefficient (DSC)	$2 \mid A \cap M \mid /\lvert A\rvert + \mid M\rvert$; 2JC/(1 + JC); 2TP/(2TP + FP + FN)	Ratio of twice the number of intersecting elements in A and M to the sum of all elements in A and M
Jaccard coefficient (JC)	$\lvert A \cap M \mid /\lvert A \cap M\rvert$; DSC/(2-DSC); TP/(TP + FP + FN)	Ratio of the number of intersecting elements in A and M to the number of elements in the union of A and M
Hausdorff distance (HD)	$\max\{d(M, A), d(A, M)\}$	Maximum distance between all surface points on A and M
Average symmetric surface distance (ASSD)	$\Sigma_{m \in M}\min_{a \in A} d(m, a)/ \mid M\rvert$	Average distance between surface points in A and M
Lesion volume correlation	$\mathrm{Corr}(M_{vol}, A_{vol})$	Correlation between volumes of A and M
Percent volume difference	$100 *(abs(M_{vol} - A_{vol})/M_{vol})$	Difference in volumes between A and M expressed as a percentage of M

TP true positive, *FP* false positive, *TN* true negative, *FN* false negative, *M* manual, *A* automated, *vol* volume

performance is evaluated. Prior to selecting a method for use in their own work, researchers should understand the strengths and weaknesses of the original validation study and any independent validations that have been subsequently published. Validation studies often report multiple metrics that can provide different perspectives on the performance of a given method, and such metrics may be sometimes complex to understand in the limited space dedicated to a research article. This section summarizes that most commonly used performance metrics reported in new ALS publications. Table 1 contains the list of these metrics and the corresponding definitions. Figure 1 contains a conceptual depiction of the areas.

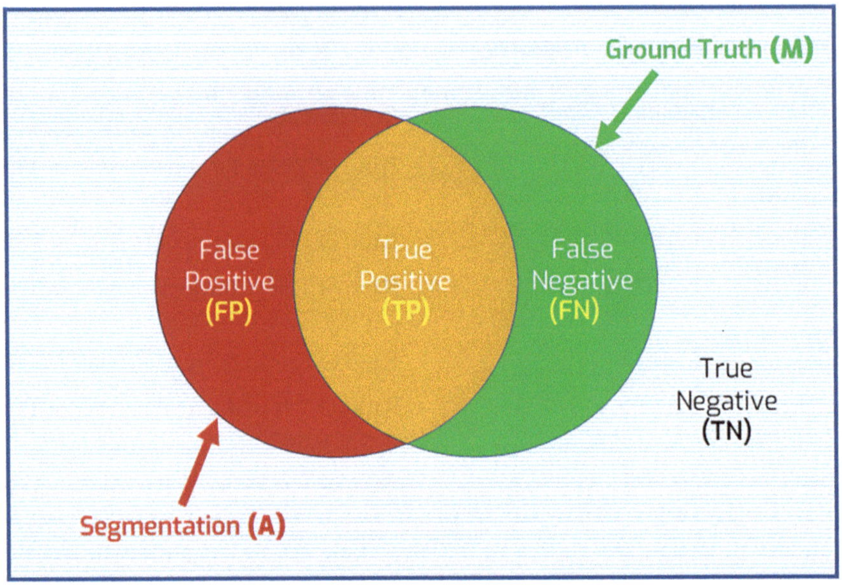

Fig. 1 Illustration of key measures of segmentation accuracy

Because automated lesion identification methods aim to separate lesioned tissue from healthy tissue in neuroimaging datasets, it is important for a method to demonstrate both high sensitivity and high specificity for lesioned tissue with respect to "ground-truth" lesion segmentations. Ground truths typically correspond to manual lesion segmentations, but sometimes they may correspond to simulated lesions during algorithm development [8, 9]. **Sensitivity** refers to the true positive rate—the proportion of true lesioned voxels that are correctly identified as lesioned. **Specificity** refers to the true negative rate—the proportion of true non-lesioned voxels that are correctly identified as non-lesioned. Related measures such as the positive predictive value (also referred to as precision), negative predictive value, false positive rate, and false negative rate are also commonly used to evaluate segmentation performance (*see* Table 1 for definitions). Unlike sensitivity and specificity, positive and negative predictive values depend on the prevalence of the outcome of interest (in this case, the ratio of lesioned vs. healthy voxels). Note that because manual segmentation by an expert rater is the accepted gold standard, these and other performance measures are typically computed by comparing the lesion segmentations produced by an automated method to the manual segmentations by an expert rater.

Segmentation agreement along shape and location is often measured using spatial overlap measures. The Dice similarity coefficient or DSC [10, 11] is perhaps the most popular measure for evaluating the performance of automated lesion segmentation methods, and it is likely to be reported in most validation studies.

The DSC is a measure of the spatial similarity between the binarized lesion segmentations produced by an automated method and the ground truths produced by an expert rater. The **DSC** between two objects (i.e., segmentations) A and M is equal to twice the number of overlapping elements (i.e., voxels) between A and M divided by the total number of elements in A and M (Table 1). DSC can more simply be thought of as the percentage overlap of true and predicted lesion maps with respect to the total volume occupied by the two. Higher DSC values indicate higher spatial similarity between a lesion segmentation and its corresponding ground truth, with DSC = 1 indicating perfect agreement and values greater than 0.6 typically being considered "good" in the literature [2, 9, 12]. The DSC can also be expressed in terms of true positives, false positives, and false negatives (Table 1). The DSC is particularly useful because it uses a standardized scale from 0 to 1, which is often used as a scale of comparison of better vs. worse algorithms. However, DSC has a major drawback: it is prevalence (i.e., lesion size) dependent. Similar shifts in space lead to a stronger drop of DSC values for smaller lesions than for larger lesions. This happens because small lesions have less voxels, and their misclassification leads to a larger proportion of the lesion to be wrong [11]. DSC values should, therefore, be interpreted cautiously and in the context of other relevant measures such as lesion size (*see* next paragraph). The datasets used in the LSM literature vary wildly in the average lesion size of the participants; these differences can, therefore, create a confounding regarding the perceived accuracy of ALS methods because datasets with larger lesions appear to produce more "accurate" segmentations (higher DSC in average). It follows that DSC is not well comparable between methods tested on different datasets despite using a standardized 0–1 value range. Importantly, DSC cannot distinguish between differences that result from different types of overlap errors (e.g., lesion borders that extend into adjacent tissue vs. completely wrong prediction of lesion clusters in areas clearly healthy).

The Jaccard coefficient (Table 1) is a measure closely related to DSC, often referred to as "intersection over union." The Jaccard coefficient can be converted to the DSC by the equation shown in Table 1.

Distance-based measures (*see* Table 1) address some of the limitations of DSC/Jaccard [8, 13–16]. These measures are sometimes referred to as "metric" measures because they measure pure distance, independently of lesion size or lesion overlap. A commonly used distance measure is the **Hausdorff distance**, which measures the maximal displacement between the contours of two objects [17]. For objects A and M, it is defined as the maximum of all distances from each contour point in A and the nearest point in M (Table 1). Hausdorff distance is sensitive to outliers, i.e., single voxels that are erroneously predicted as lesion far from the lesion contour. For this reason, Hausdorff distance can be difficult to interpret without other metrics of reference [16]. Another common measure is the **average symmetric surface distance** (ASSD), which is defined as the average distance between the contours of two objects, computed in both directions (i.e., $A \rightarrow M$, $M \rightarrow A$). Usually, for metric measures, smaller values indicate better performance because they reflect smaller distances between the relevant points in the automatically generated segmentation and analogous points in the ground truth. ASSD can provide a good, balanced understanding of the quality of the segmentation because it takes into account both outliers and the lack of overlap, without being affected by lesion size. For this reason, comparing ASSD between ASL tools can yield a more unbiased understanding of the ALS tools.

Another dimension that is particularly relevant to lesion segmentation performance is the predicted lesion volume. There is plenty of evidence that lesion volume is one of the major predictors of the severity of cognitive deficits, and many lesion studies incorporate lesion volume as a covariate. It is therefore of critical importance that ALS produces lesion maps of maximal volume accuracy. The agreement of the predicted vs. manual lesion volume is sometimes measured with the percent volume difference (*see* Table 1). The correlation between volume estimates obtained from sets of automated and manual segmentations is also reported in some publications [8, 12, 15]. This correlation is important because, independently of how far off are the lesion volumes for each individual subject, a close linear relationship between lesion volumes is important when lesion volume is used as a variable of interest or a nuisance covariate in statistical analyses. For volume difference measurements, smaller values indicate better performance as they reflect smaller differences in segmentation volumes. For volume correlation measurements, larger values indicate better performance as they reflect stronger relationships between volume estimates for automated and manual segmentations [12, 15].

More detailed descriptions of the above (and other) measures can be found in dedicated journal articles. For a recent review on this topic, *see* [18].

4 Other Considerations for Selecting an Automated Lesion Segmentation Tool

As explained at the beginning of the chapter, ALS methods are typically designed to detect lesions with a specific etiology, at a specific stage, and using data from a specific imaging modality (or set of modalities) [19–23]. The range of data accepted by the method can become important when conducting retrospective studies that use clinical scans with various modalities. A method built for a single modality and a single type of lesion will be useful only in that research context.

An important aspect to consider when choosing a method is the need for additional scans from a healthy control group. Most methods use healthy control data during building and testing, but some methods—i.e., those that rely on anomaly detection—require control data also during regular use. The viability of these methods will therefore depend on the availability and quality of data from the reference control group. When opting for methods that require healthy control data, it is important to ensure that the control group is demographically comparable to the patient population of interest, as anomaly detection techniques are sensitive to systematic deviations from the reference group data. If the lesion data are primarily collected from elderly patients, then it is important to ensure that the reference group also consists of elderly subjects. If data from healthy young adults are used as the reference, the method might produce many false positives driven by aging-related structural changes such as sulcal widening and ventricular enlargement.

When identifying candidate methods for future use, it is important to ensure that they have been implemented as publicly available software. Many published methods are developed and validated in proof-of-concept studies, without being translated into publicly available tools. These methods are not likely to be immediately useful to researchers, unless the user has the knowledge and the will to build the tool based on the publication description. Methods implemented as freely available software packages can be readily used and tested by the broader community, they are more frequently adopted, thus increasing the chances that the method is vetted by multiple groups using independent data (e.g., [24]). Besides the immediate availability of the software, new users may want to consider the long-term support of the tool or software. Segmentation methods are often initiated as ad hoc projects by one or few researchers, who make the tool available for public use. There is often no funding or incentive for the researchers to maintain and continue improving the method. However, sometimes there might be dedicated funding to continue the development, which is a good indication that bugs and features will be updated, and more research groups will adopt the tool. When selecting a method for long-term use, it may be a good practice to ask the authors if they intend to support, maintain, and further develop the tool.

ALS tools are built within a computer platform and require other software to work properly. Considering the dependencies and requirements is an important step. For example, some publicly available methods run in the MATLAB environment (e.g., Clusterize, ALI, lesion_gnb), which requires a paid license. Other methods rely on open-source free software (e.g., LINDA). Additional toolboxes and packages may be required that can be free or may require payment but in all cases will need dedicated attention for installation. For example, Clusterize, ALI, and lesion_gnb require the Statistical Parametric Mapping (SPM) MATLAB toolbox, while LINDA requires the Advanced Normalization Tools for R (ANTsR) package. Sometimes a segmentation tool requires a specific version of the software or package it depends on; e.g., the lesion_gnb method is specifically built to work with SPM version 12 and also requires access to the Statistics and Machine Learning Toolbox as implemented in MATLAB versions R2014b and later to run correctly. Some tools (i.e., LINDA) cannot run natively on Windows operative systems and require Linux or macOS. A lab that has access only to Windows computers may need to spend extra efforts to find a solution for using Linux software within Windows (i.e., by adopting Windows Linux Subsystem).

An aspect of increased concern in neuroimaging research is the reproducibility of the results. The degree to which a result is reproducible when run on the same computer or different computers depends on several factors. Subtle changes can occur because the machine learning models rely on random initialization of parameters, or because other intermediate steps (e.g., registration in template space) are not fully reproducible without additional parameters. Besides the reproducibility issues that depend on the software tool, upgrades to system libraries may change the way a software processes the data, further impairing the ability to reproduce the same lesion map in the future. To resolve the reproducibility issues, researchers should consider using containerized versions of the tool (i.e., Docker or Singularity containers). The container includes a small operating system in miniature, with all the necessary libraries and software, which can be moved easily between computers and archived with the data for future replication of results.

5 Overview of Currently Available ALS Tools

Lesion-symptom mapping studies are often performed on data obtained from stroke patients (*see* Chapter 1) [23, 25] at either the acute/subacute or chronic stage of recovery [26]. While some packages may exist for the automated segmentation of other types of lesions (e.g., lesions associated with neurodegenerative diseases such as multiple sclerosis), methods that are designed to detect acute and/or chronic stroke lesions are likely to be most useful for lesion-symptom mapping studies, and they will accordingly be the focus of this section. The methods described in the following sections are associated with figures of their application on the same example case (case c0005s0007t01 of the ATLAS dataset, left embolism in the middle cerebral artery, lesion size 56 ml, [4]).

5.1 Acute and Subacute Lesion Segmentation Methods

Acute lesion segmentation is typically performed using data from clinical imaging investigations acquired in an emergency medical setting using CT or diffusion-weighted imaging (DWI) [6, 14, 22]. Of note, acute lesions are often not visible on typical anatomical MRI scans (e.g., T1w, T2w, etc.). The clinical purpose of acute data poses many challenges for automated lesion identification. For example, data from only a single imaging modality may be available for each patient, and different patients may have data from different lesion modalities (some CT, others DWI). Acute lesions also differ in their appearance on DWI (i.e., hyperintense) vs. CT (i.e., hypointense). Thus, it is ideal for acute/subacute lesion identification methods to (1) only require data from a single imaging modality rather than requiring multiple images and (2) be capable of utilizing data from various potentially relevant imaging modalities. A recently developed approach to acute stroke lesion identification that meets these criteria is summarized below.

5.2 Iterative Region-Growing and Clustering: The Clusterize Toolbox

This is a semiautomated lesion segmentation method that can be applied to imaging data from multiple MRI modalities as well as CT [1]. This method uses the Clusterize algorithm, which is an automated clustering algorithm that uses local intensity maxima and iterative region-growing [27]; the aim is to segment the image volume into clusters from which the final lesion segmentation is manually selected by the user. Besides accepting data from multiple imaging modalities, the method can also be applied to lesions with different etiologies and/or at different temporal stages. Control data are not required to utilize this method, but user interaction is necessary for good performance (*see* Fig. 2). The method is summarized below, and a detailed description of the method can be found in the original validation paper by De Haan and colleagues [1].

Method Summary First, the input image is scaled to an intensity range of 0–1000. Then, a 2 × 2 mm full-width half-maximum (FWHM) Gaussian smoothing kernel is applied separately to each slice in the image volume to reduce irregular edges and improve the detection of lesion clusters. The algorithm then identifies, for each slice, local intensity maxima that contain enough voxels to surpass a user-specified extent threshold (e.g., 100 voxels). The maximum intensity threshold is then iteratively relaxed in successive steps, and each voxel is assigned to a cluster at each step. To ensure that every voxel is only assigned to a single cluster, voxels that are connected to multiple clusters are assigned to the nearest cluster to which they are connected. For each slice and each intensity threshold, the clusters are stored in a 2D matrix, resulting in a single 3D cluster matrix for each axial slice (i.e., with dimensions of threshold, x dimension, y dimension). To avoid assigning to a cluster background voxels (i.e., out-of-brain voxels), a minimum intensity

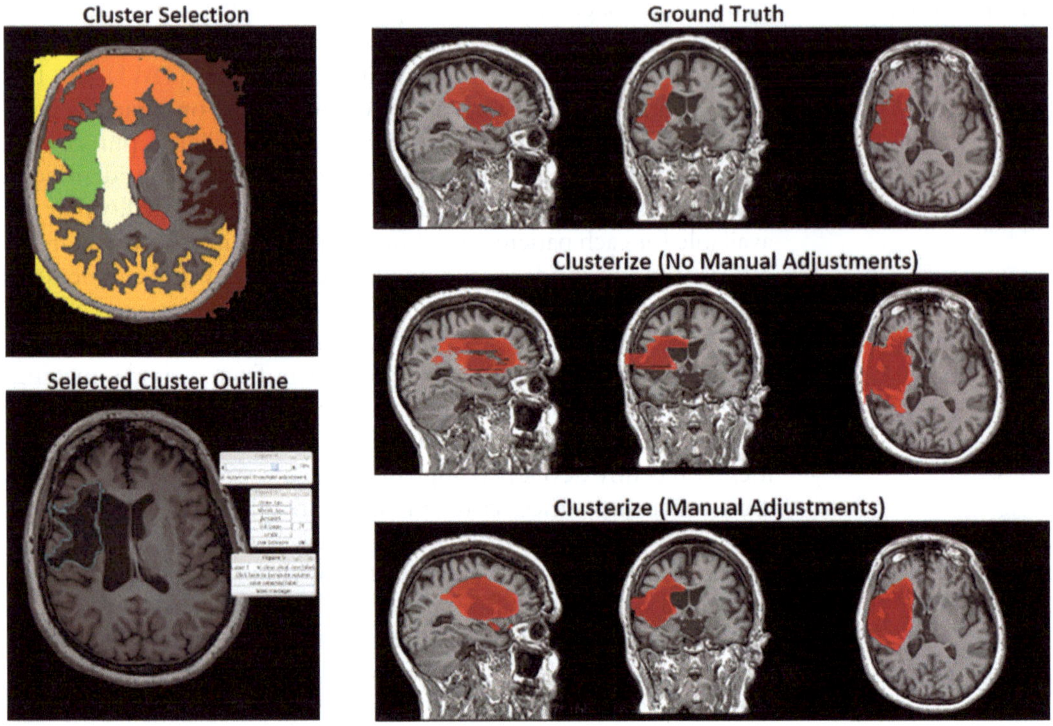

Fig. 2 Example of the Clusterize toolbox. Left, the top image shows the clusters identified using default settings and automatic threshold adjustment. The selected cluster is shown in green. The bottom image shows the outline of only the selected cluster overlaid on the same slice of the T1w image volume. The threshold used to define the clusters can be manually adjusted using the slider bar, and the other menus allow the user to perform additional modifications and/or save the cluster as a new file. Note that while the example is of a chronic lesion, the Clusterize toolbox can be applied to any neuroimaging data modality. Right, the top row of images shows the ground-truth segmentation of the chronic stroke lesion from the T1w MRI scan. The middle row shows the lesion segmentation obtained using Clusterize without performing any cluster modifications. The bottom row shows the lesion segmentation obtained using Clusterize with manual cluster modifications

threshold (e.g., 20% max intensity) can be defined prior to running the algorithm. Finally, the cluster(s) corresponding to the lesion segmentation are then manually selected for each slice in a final step that requires user interaction (*see* Figure). In this step, the user manually selects the lesion cluster(s) in at least one slice (successive slices are updated once a cluster is selected) and applies modifications (e.g., smoothing, growing, shrinking, etc.) as needed. An example of the cluster selection is shown in Fig. 2. It is ideal for the user to step through the entire image and modify the clusters identified for each slice as necessary. The user can also remove voxels from clusters or draw their own cluster boundaries if none of the automatically generated clusters provide a good fit for the lesion on a given slice.

Validation This method was directly validated using data from various modalities: CT data from 13 acute stroke patients, DWI data from 16 acute stroke patients, T2FLAIR MRI data from 15 acute stroke patients, and T1w MRI data from 11 chronic stroke patients. With respect to manual segmentation, the authors reported a mean DSC of 0.85 for the CT dataset, a mean DSC of 0.86 for the DWI dataset, a mean DSC of 0.85 for the T2FLAIR dataset, and a mean DSC of 0.87 for the T1w dataset. The Clusterize method allowed for significantly faster lesion identification compared to manual tracing. However, a subsequent validation in an independent dataset found that without manual correction of the user-selected clusters (which can still be a time-consuming step), the Clusterize method performed relatively poorly [24]. This highlights the importance of manual cluster correction. The impact of manual correction on the final segmentation output is illustrated in Fig. 2.

Advantages The advantages of the method are that it only requires data from a single imaging modality, it can be applied to data from a range of imaging modalities collected at different time points post-stroke, and it does not require comparable data from a reference group.

Disadvantages The method requires substantial user interaction to achieve a final lesion delineation, although it can be argued that all methods require some degree of user interaction as visual inspection and manual correction is an important part of quality control even for fully automated methods.

Availability The Clusterize toolbox is a publicly available software package and can be downloaded at https://www.medizin.uni-tuebingen.de/kinder/en/research/neuroimaging/software/?download=clusterize-toolbox. The toolbox requires access to MATLAB and the Statistical Parametric Mapping (SPM) toolbox (https://www.fil.ion.ucl.ac.uk/spm/).

5.3 Chronic Lesion Identification

LSM studies are more typically performed on chronic stroke patients, that is, when both the lesion and the behavioral deficit are stabilized (*see* Chapters 1 and 2) [26]. Methods that are capable of segmenting chronic stroke lesions from T1w MRI scans are of high interest; they can be used not only for basic mapping of cognitive functions but also for rehabilitation studies [24] and functional MRI studies [9]. Typically, a single high-resolution T1w scan is acquired for lesion identification and/or spatial co-registration and normalization, and the lesions appear as dark in T1w MRI [19] similar to the low intensity of gray matter or CSF [2, 9, 12, 28, 29]. The indirect effects of the lesion, such as abnormally enlarged ventricles and/or widened sulci, are often prominent in scans obtained from chronic patients [2]. Separating the chronic lesion from isointense compartments such as CSF and distinguishing the direct and indirect effects of the lesion are therefore major challenges for automated identification of chronic lesions in T1w MRI (manual tracing presents a similar issue; *see* Chapter 2). This section will provide an overview of the three methods for automatically identifying lesions in T1w MRI that have been implemented in publicly available software packages: ALI, lesion_gnb, and LINDA.

5.4 Unsupervised Fuzzy Clustering with Fixed Prototypes: Automatic Lesion Identification (ALI) Toolbox

Voxel-based morphometry (VBM) was one of the first methods to be explored for automatically identifying chronic stroke lesions in T1w MRI scans [30]. VBM is a method for detecting differences in brain anatomy between groups or between an individual and a reference group by statistically comparing the assignment of voxels in tissue classes (e.g., gray matter, white matter, CSF, etc.) in a common stereotactic space [31, 32]. The rationale behind using VBM for automated lesion identification is that because lesions represent major deviations from the normal brain anatomy, lesioned voxels show statistically significant differences on their normal tissue assignment when compared to the same voxels in non-lesioned brains. The rationale behind VBM is quite reasonable, although traditional VBM has been shown to have poor sensitivity for lesioned voxels [30]. However, some approaches attempt to overcome the limitations of traditional VBM and achieve automatic lesion segmentation [9, 29]. We consider here ALI, the method developed by Seghier and colleagues [9] in a publicly available software package (Automated Lesion Identification toolbox for SPM). The ALI method is an unsupervised anomaly detection method that involves comparing voxels in a patient's scan to their counterparts in a reference set of scans obtained from individuals without brain lesions. It is frequently employed in lesion-symptom mapping studies that include large-scale data-driven studies of stroke recovery, such as PLORAS [3].

Method Summary First, the patient T1w MRI scan(s) are segmented into different tissue classes and spatially normalized to the MNI template space using the unified segmentation-normalization algorithm implemented in the SPM software package [31]. In healthy brains, three tissue classes are segmented and registered in template space. However, attributing lesions to one of the three classes (GM or CSF) would lead to major deformations during registration because the algorithm attempts to find the corresponding location of the lesioned tissue in the healthy brain, which may be mistaken as GM or CSF. To addresses tissue misclassification, prior probability information about an extra (i.e., lesion) tissue class is incorporated into the unified segmentation-

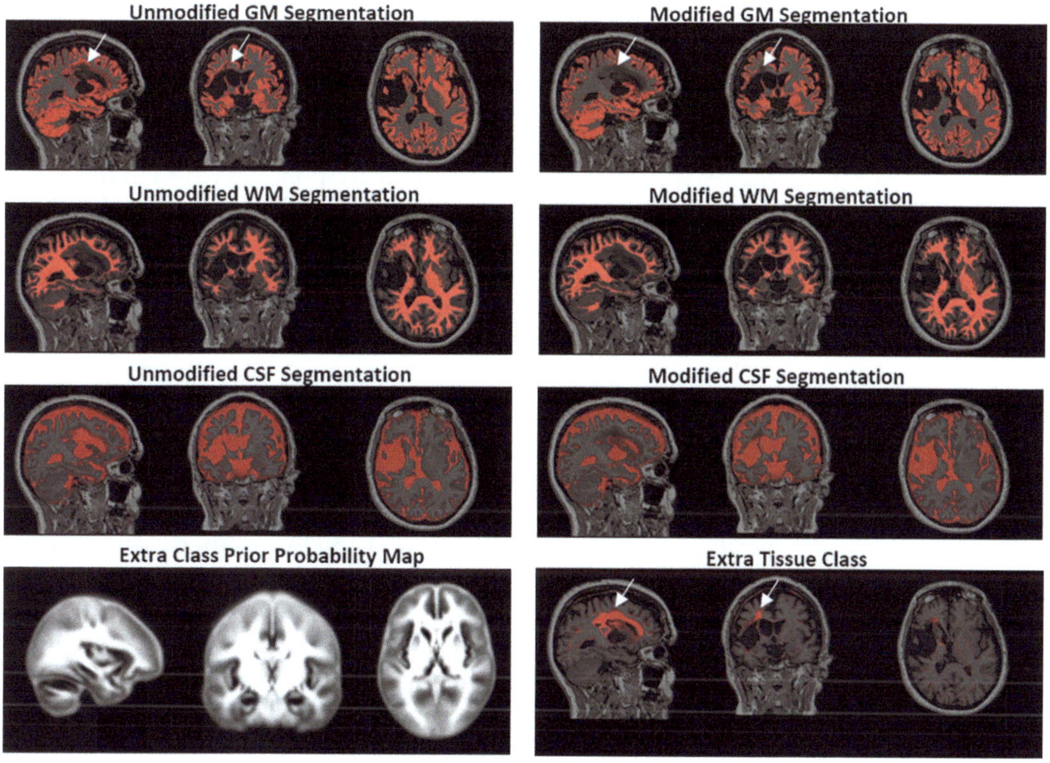

Fig. 3 Effect of ALI extra tissue class on tissue segmentation. The first three rows in the left column show the results of the unmodified tissue segmentation. The white arrows show that there are portions of the lesion that are misclassified as intact GM. The bottom row in the left column shows the extra class prior probability map used by ALI. The first three rows in the right column show the results of the modified segmentation incorporating the extra tissue class, and the last row shows the results for the extra tissue class. As shown by the arrows, the lesion voxels that were previously misclassified as intact GM are now assigned to the extra tissue class. The results from the modified segmentation can now be compared to segmentations from a control group to identify abnormal voxels (not shown due to lack of required control data)

normalization algorithm. This extra tissue class is defined as the mean of the prior probabilities (encoded in spatial prior probability maps) for white matter and CSF tissue classes and is designed to primarily capture lesion voxels that are assigned low gray matter or white matter probabilities (*see* Fig. 3). It is also intentionally biased toward voxels located within the white matter because white matter lesions tend to be disproportionately misclassified as intact gray matter by the unified segmentation-normalization algorithm [9, 29], as illustrated in Fig. 3. The modified unified segmentation-normalization procedure is then iteratively applied to the T1w MRI scan(s). At the first iteration, the prior probabilities for the extra tissue class are defined as the mean of the white matter and CSF probabilities (*see* Fig. 3). For subsequent iterations, the segmentation outputs for the extra tissue class obtained from the preceding iteration are used to update the prior probabilities for the extra tissue class. After the desired/predefined number of iterations is complete, spatial smoothing (8 mm FWHM) is applied to the atlas-normalized gray matter and white matter tissue probability maps obtained from the final iteration. A fuzzy clustering algorithm is then used to identify outlier voxels in the patient tissue based on their statistical distance from the corresponding voxels in the tissue maps obtained from a control population. The continuously valued lesion map is then defined as the union of the outlier voxels from the gray and white matter tissue classes. A user-defined threshold parameter (i.e., U) that ranges from 0 to 1 can be applied to obtain a final binary lesion map.

Validation The original validation of this method used T1w MRI data from eight patients with chronic lesions [9]. For a U threshold of 0.3, the authors reported an average DSC of 0.64. In a comparison study, ALI method was found to generally exhibit poorer performance than lesion_gnb or LINDA (methods described in the following sections) [24].

Advantages ALI has the advantage of combining lesion identification with tissue segmentation and spatial normalization in a single procedure. This allows for the lesion information to be incorporated into the segmentation and normalization procedures, improving their performance. This also allows lesion segmentation to be easily incorporated into standard functional MRI preprocessing pipelines, since tissue segmentation and spatial normalization are often needed during fMRI preprocessing.

Disadvantages The primary limitation of this method is that it requires data from a comparable group of healthy controls, which are not provided with the software. It also requires substantial processing time (i.e., 370 s +248 s per healthy brain in the reference set) [24].

Availability This method is publicly available as a toolbox for the SPM software package (https://www.fil.ion.ucl.ac.uk/spm/ext/#ALI). This method requires access to MATLAB.

5.5 Voxel-Based Gaussian Naïve Bayes Classification: lesion_gnb

Griffis and colleagues [12] recently developed a supervised method based on Bayes classification of voxels. The method, which is implemented in a tool called lesion_gnb, is capable of identifying chronic stroke lesions from single T1w MRI scan without the need for data from a reference set of comparable controls. Whereas the ALI method described in the previous section used a modified tissue segmentation procedure to avoid the misclassification of lesioned voxels as CSF or intact gray matter, lesion_gnb takes advantages of tissue segmentation errors by using them to identify voxels in the lesioned hemisphere that are anomalous relative to their counterparts in the un-lesioned hemisphere.

Method Summary First, the T1w MRI scan is processed using the New segmentation algorithm implemented in the SPM12 software package. The New segmentation algorithm is similar to the unified segmentation-normalization algorithm implemented in the earlier versions of SPM (and used by ALI), but it differs in its treatment of mixing proportions, its use of an improved registration model, and its incorporation of expanded prior probability maps (PPMs) for out-of-brain voxels (SPM12 Manual, FIL Methods Group). The tissue probability maps (TPMs) obtained from the new segmentation procedure are then smoothed using a default 8 mm FWHM Gaussian kernel. Matching smoothed prior probability maps (PPM) are used to obtain tissue classification into gray matter, white matter, and CSF. Two feature maps are created by using image algebra to take advantage of segmentation errors caused by the lesion and to finally identify voxels corresponding to "missing" and "abnormal" tissues (Fig. 4). The "missing" tissue corresponds to the lesion core, while "abnormal tissue" corresponds to the remaining tissue abnormalities affected by, e.g., gliosis. The two feature maps, "missing" and "abnormal," are then used as inputs to a trained Gaussian Naïve Bayes (GNB) classifier to obtain a posterior lesion probability to each voxel. The posterior probability is then converted to a discrete lesion class.

Postprocessing steps such as smoothing, cluster-thresholding, and re-segmentation using the final map as a PPM for a lesion tissue class (i.e., as in the preprocessing for ALI) can be applied to reduce false positives and improve anatomical fidelity of the final lesion mask (Fig. 4). While not included in the assessments reported by the original validation paper, the author recommends that postprocessing includes a re-segmentation step where the final binarized lesion mask is re-smoothed using an 8 mm FWHM Gaussian kernel and used as a PPM for a lesion tissue class in a second iteration of the new segmentation procedure similar to later implementations of ALI [33]. The rationale for this is that it corrects tissue misclassifications, may improve normalization, and produces a continuously valued tissue probability map for the lesion tissue class that can be manually thresholded to achieve greater anatomical fidelity than the predicted lesion segmentation obtained from the GNB classifier [34]. This additional step could be applied using a lesion segmentation obtained from any method (i.e., not just lesion_gnb) and is illustrated in Fig. 4. The code to perform this additional postprocessing step is included in the lesion_gnb toolbox. Example segmentations are shown in Fig. 4.

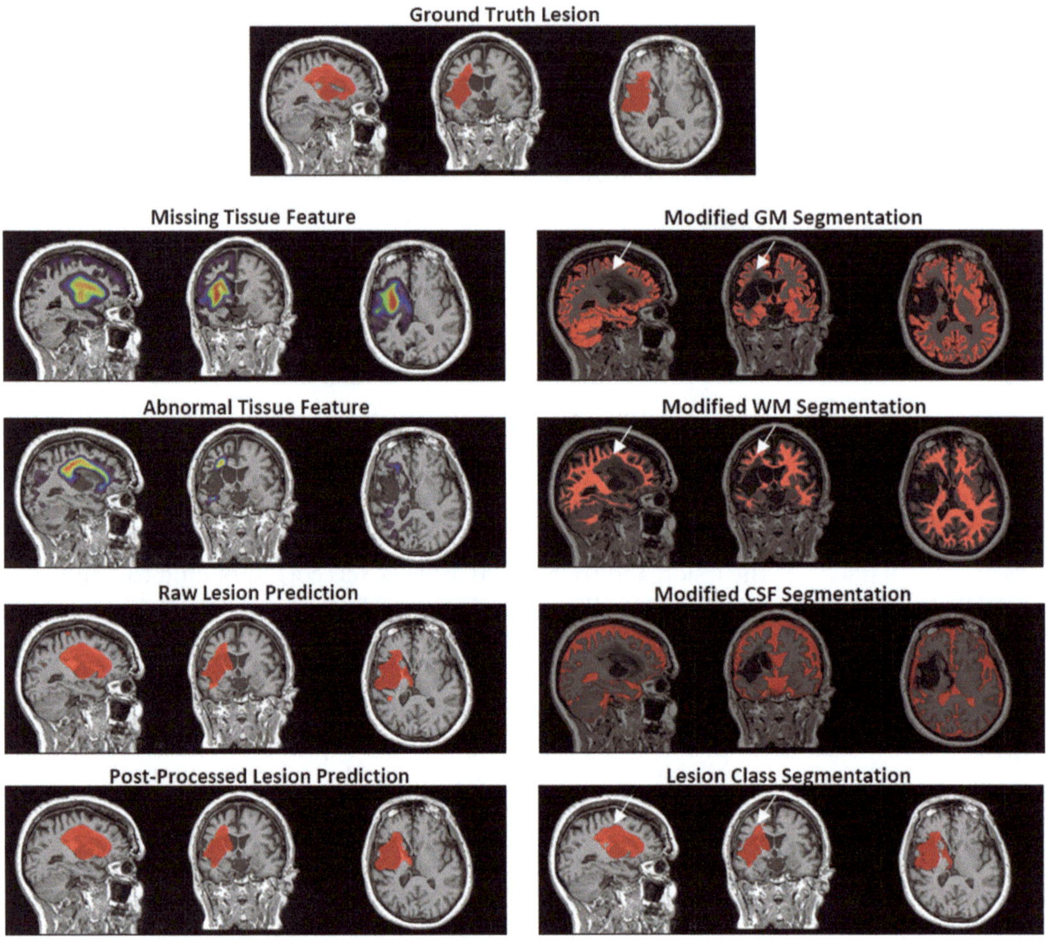

Fig. 4 Lesion_gnb features and results. The ground-truth lesion is shown at the top of the figure. The column on the right shows the features that are computed by lesion_gnb along with the raw and postprocessed lesion predictions. The right column shows the results of re-running the segmentation using the smoothed post-processed lesion prediction as an extra tissue class prior. Analogously to ALI, this prevents the misclassification of lesion voxels (white arrows). It also produces a new lesion tissue class that often has finer detail than the original lesion prediction

Validation The original validation of this method utilized a leave-one-out cross-validation approach where the GNB classifier was iteratively trained on 29/30 of the manual lesion segmentations and T1w MRI scans, and an out-of-sample prediction was obtained for the held-out case. This resulted in out-of-sample predictions for all 30 patients. The mean/median DSC for the predicted lesion segmentations was 0.58/0.66, and the mean/median percent volume difference (PVD) was 30.56/24.73. Simple postprocessing (i.e., smoothing at 8 mm FWHM and cluster-thresholding to remove all voxel clusters with fewer than 100 voxels) resulted in higher mean/median DSCs of 0.66/0.73 and reduced the mean/median PVDs to 28.91/22.36. The lesion volume correlation between automated and manually segmented lesions was high ($r = 0.97$). An independent comparison of lesion_gnb, ALI, Clusterize, and LINDA (*see* next section) reported that lesion_gnb generally produced superior lesion segmentations to ALI and Clusterize and required the least processing time of all methods evaluated (i.e., 246 s) [24].

Advantages A major advantage of this method is that it can be used to segment lesion in single T1w MRI scans without requiring additional data from a reference group. It is also fast compared to other automated approaches. The addition of the extra processing step also makes it useful for aiding in template registration/normalization.

Disadvantages The major disadvantages of this method include the potential for increased false positives, particularly for small lesions. It also lacks an ability to discriminate between stroke lesions and other lesions or abnormalities that may introduce similar errors into the new segmentation procedure [24]. This latter disadvantage is illustrated by the inclusion of the enlarged ventricle in the lesion segmentation shown in Fig. 4.

Availability This method is available as the lesion_gnb toolbox for SPM12 (https://www.fil.ion.ucl.ac.uk/spm/ext/). This method requires access to MATLAB and the Statistics and Machine Learning Toolbox to obtain lesion predictions, although the "missing" and "abnormal" feature maps can be generated without the Statistics and Machine Learning Toolbox.

5.6 Lesion Identification with Neighborhood Data Analysis: LINDA

While the ALI and lesion_gnb approaches differ in many ways, they are both voxel-based approaches—they treat each voxel as an independent observation without considering information about its neighbors. This makes them potentially vulnerable to increased false positives and ignores the fact that lesioned voxels tend to be adjacent to other lesioned voxels [8, 35]. To overcome this limitation, Pustina and colleagues [8] developed a method that incorporates voxel neighborhood information and identifies the lesion using graded hierarchical steps of finer voxel resolution. The method was implemented in a tool called LINDA; it relies on machine learning (random forest) models of anomalies on six imaging features computed from a single T1w scan. This method requires data from a reference group during training, but reference maps and a trained model are included in the software implementation to avoid these needs for the end user.

Method Summary LINDA starts by applying bias correction, denoising, skull-stripping, and intensity scaling on the lesioned

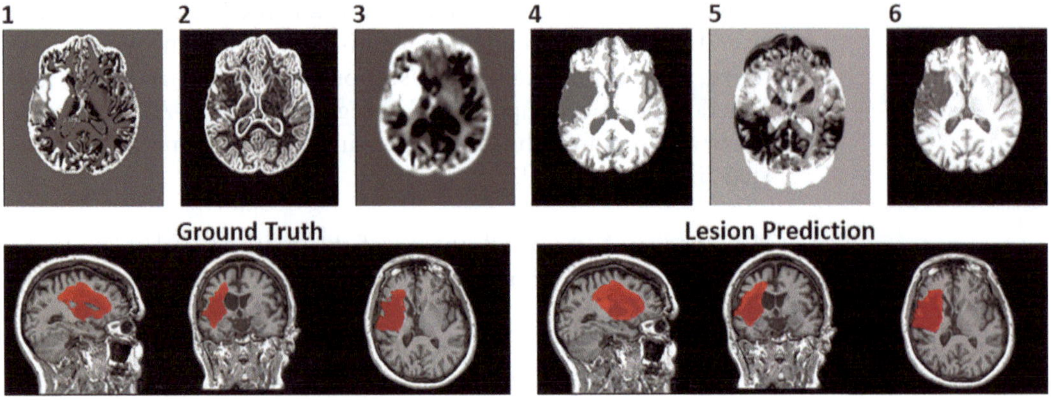

Fig. 5 LINDA features and results. The top row shows the six features used by LINDA for the ATLAS lesion example. Feature numbers correspond to the numbers provided in the main text. Note that the lesion features are in template space. Below, the ground-truth lesion is shown along with the lesion prediction obtained from LINDA

T1w and then registers the resulting brain-only scan to a standard brain-only template and computes six features: (1) the deviation of tissue segmentation results with respect to an average tissue segmentation obtained from healthy controls, (2) a gradient magnitude image that outlines the sudden intensity changes, (3) deviance of T1w image intensities with respect to the reference T1w dataset, (4) the tissue segmentation of the T1w scan, (5) deviance of the T1w image intensity asymmetry from asymmetries of the reference dataset, and (6) the raw bias-corrected T1w scan (Fig. 5). The features are then input into a series of random forest models trained at different resolutions, from lowest to highest, where the previous low-resolution posterior probability constitutes an additional feature for the next high-resolution prediction. These models attempt to predict the tissue class of each voxel (lesion, GM, WM, CSF) using information not only from a single voxel but from a layer of surrounding voxels as well. The current LINDA implementation has three resolution steps, and the entire prediction routine is repeated three times, starting from the registration to template space, to improve the initial template registration and segment the lesion more accurately. An example segmentation is shown in Fig. 5.

Validation This method was validated using data from 60 patients with chronic left hemispheric strokes [8] using k-fold and leave-one-out cross-validation approaches. For the main performance measures, the mean (standard deviation) was 0.7 (standard deviation = 0.16) for the DSC, 17.9 mm (9.8 mm) for Hausdorff distance, 2.54 mm (1.38 mm) for average displacement, 0.72 (0.19) for sensitivity, and 0.73 (0.16) for specificity. The correlation of lesion volumes between the automated and manual segmentations was also high ($r = 0.96$). The original validation also found that LINDA performed similarly using data obtained from a different institution and also showed that LINDA outperformed ALI. An independent evaluation found that LINDA generally produces more accurate segmentations than Clusterize, ALI, and lesion_gnb, but it also requires the most processing time (64 min per case) of the three approaches [24].

Advantages LINDA has several advantages that include the incorporation of information about voxel neighborhoods, the inclusion of control averages needed to compute the image features in the software implementation, and the implementation of the method in a free programming environment (i.e., R) [36].

Disadvantages LINDA requires a long processing time due to the multiple registrations in template space. It is not suitable for small (e.g., 5 ml) or bilateral lesions because it was trained on a dataset with relatively big unilateral lesions. It is possible to train LINDA with new lesions, but the currently proposed features would require a control group. LINDA works natively in Linux or macOS, but not on Windows; however it is possible to use it in the Windows Linux Subsystem.

Availability LINDA is available at http://dorianps.github.io/LINDA/; it requires the ANTsR package and the randomForest package but can be found as a Docker container with all the dependencies already installed (https://github.com/dorianps/docker).

6 Best Practices After Obtaining Automatic Segmentations

Once lesion segmentations have been obtained, they should undergo visual inspection to ensure that segmentations are acceptable and do not represent grossly erroneous labels. What should and should not be labeled as lesion may sometimes depend on the local lab protocols; lesion segmentation is not an exact science, and different raters may reach different conclusions. There is no standard advice we can provide in this chapter of when a segmentation should be considered unacceptable. Some areas of the brain are larger and might be more resilient to segmentation error during LSM analysis (e.g., frontal lobe), while other areas require very precise segmentation for proper LSM analysis (e.g., brainstem). The researcher must always consider that every voxel that is improperly labeled can introduce errors to the findings and interpretations of results and decide to what extent the analysis that will be performed might be affected by the imperfections of the automated segmentations. Visual inspections may also allow to tune segmentation method better in simple steps, e.g., apply a cluster size threshold to remove small false positive clusters, change the number of iterations in LINDA, etc. Manually correcting ALS segmentations is typically easier than full manual tracing from scratch; thus, manual corrections can be part of the quality control process if needed. Visual inspections and manual corrections can be done with software such as MRIcron (https://www.nitrc.org/projects/mricron) or ITK-SNAP (http://www.itksnap.org/).

7 Considerations for Future Approaches

Automated lesion segmentation is an area of ongoing research with potential impact on basic and clinical research and possibly on clinical practice. A disadvantage that is common to many automated lesion segmentation approaches is that they tend to perform most poorly for small lesions. This is likely due, in part, to the fact that small lesions are often difficult to differentiate from normal anatomical variability (i.e., larger sulci, atrophy, etc.). Other times the method is designed and trained for larger lesions and can still be optimized for smaller ones. An important avenue for future work is to improve the detection of small lesions without compromising on the accuracy of large lesion detection. Automated methods also tend to perform poorly on brainstem and cerebellar lesions, where lesions are not only small but poorly represented in the imaging features typically used to detect lesions [24]. An important avenue for future work is to identify features that are sensitive to brainstem lesions to expand the coverage of potential lesions ALS methods can capture. ALS methods are often built and tested on single datasets, while their performance on a different dataset is either unclear or impaired [4].

Currently, there is a lot of interest in using deep learning methods based on convolutional neural networks to perform segmentations. These approaches have already shown promising results in segmenting brain regions or removing the skull. Deep learning methods are seen by many researchers as a breakthrough for achieving highly accurate segmentations in ultrafast algorithms. The reader of this chapter is invited to try new methods once they become available.

References

1. de Haan B, Clas P, Juenger H, Wilke M, Karnath H-O (2015) Fast semi-automated lesion demarcation in stroke. NeuroImage Clin 9: 69–74. https://doi.org/10.1016/j.nicl.2015.06.013
2. Wilke M, de Haan B, Juenger H, Karnath HO (2011) Manual, semi-automated, and automated delineation of chronic brain lesions: a comparison of methods. NeuroImage 56: 2038–2046. https://doi.org/10.1016/j.neuroimage.2011.04.014. Epub 2011 Apr 14
3. Seghier ML, Patel E, Prejawa S, Ramsden S, Selmer A, Lim L, Browne R, Rae J, Haigh Z, Ezekiel D, Hope TM, Leff AP, Price CJ (2015) The PLORAS database: a data repository for predicting language outcome and recovery after stroke. NeuroImage. https://doi.org/10.1016/j.neuroimage.2015.03.083
4. Liew SL, Anglin JM, Banks NW, Sondag M, Ito KL, Kim H, Chan J, Ito J, Jung C, Khoshab N, Lefebvre S, Nakamura W, Saldana D, Schmiesing A, Tran C, Vo D, Ard T, Heydari P, Kim B, Aziz-Zadeh L, Cramer SC, Liu J, Soekadar S, Nordvik JE, Westlye LT, Wang J, Winstein C, Yu C, Ai L, Koo B, Craddock RC, Milham M, Lakich M, Pienta A, Stroud A (2018) A large, open source dataset of stroke anatomical brain images and manual lesion segmentations. Sci data 5:180011. https://doi.org/10.1038/sdata.2018.11
5. Fiez JA, Damasio H, Grabowski TJ (2000) Lesion segmentation and manual warping to a reference brain: intra- and interobserver reliability. Hum Brain Mapp 9:192–211
6. Lansberg MG, Albers GW, Beaulieu C, Marks MP (2000) Comparison of diffusion-weighted MRI and CT in acute stroke. Neurology 54(8): 1557–1561. https://doi.org/10.1212/wnl.54.8.1557
7. Seghier ML, Ramsden S, Lim L, Leff AP, Price CJ (2014) Gradual lesion expansion and brain shrinkage years after stroke. Stroke 45: 877–879. https://doi.org/10.1161/STROKEAHA.113.003587. Epub 2014 Jan 14
8. Pustina D, Coslett HB, Turkeltaub PE, Tustison N, Schwartz MF, Avants B (2016) Automated segmentation of chronic stroke lesions using LINDA: lesion identification with neighborhood data analysis. Hum Brain Mapp 37(4):1405–1421. https://doi.org/10.1002/hbm.23110
9. Seghier ML, Ramlackhansingh A, Crinion J, Leff AP, Price CJ (2008) Lesion identification using unified segmentation-normalisation models and fuzzy clustering. NeuroImage 41: 1253–1266. https://doi.org/10.1016/j.neuroimage.2008.03.028. Epub 2008 Mar 28
10. Dice LR (1945) Measures of the amount of ecologic association between species. Ecology 26(3):297–302. https://doi.org/10.2307/1932409
11. Zou KH, Warfield SK, Bharatha A, Tempany CM, Kaus MR, Haker SJ, Wells WM 3rd, Jolesz FA, Kikinis R (2004) Statistical validation of image segmentation quality based on a spatial overlap index. Acad Radiol 11(2): 178–189. https://doi.org/10.1016/s1076-6332(03)00671-8
12. Griffis JC, Allendorfer JB, Szaflarski JP (2016) Voxel-based Gaussian naive Bayes classification of ischemic stroke lesions in individual T1-weighted MRI scans. J Neurosci Methods 257:97–108. https://doi.org/10.1016/j.jneumeth.2015.09.019
13. Maier O, Wilms M, von der Gablentz J, Kramer UM, Munte TF, Handels H (2015) Extra tree forests for sub-acute ischemic stroke lesion segmentation in MR sequences. J Neurosci Methods 240:89–100. https://doi.org/10.1016/j.jneumeth.2014.11.011. Epub 2014 Nov 21
14. Maier O, Menze BH, von der Gablentz J, Hani L, Heinrich MP, Liebrand M, Winzeck S, Basit A, Bentley P, Chen L, Christiaens D, Dutil F, Egger K, Feng C, Glocker B, Gotz M, Haeck T, Halme HL, Havaei M, Iftekharuddin KM, Jodoin PM, Kamnitsas K, Kellner E, Korvenoja A, Larochelle H, Ledig C, Lee JH, Maes F, Mahmood Q, Maier-Hein KH, McKinley R, Muschelli J, Pal C, Pei L, Rangarajan JR, Reza SMS, Robben D, Rueckert D, Salli E, Suetens P, Wang CW, Wilms M, Kirschke JS, Kr Amer UM, Munte TF, Schramm P, Wiest R, Handels H, Reyes M (2017) ISLES 2015 - a public evaluation benchmark for ischemic stroke lesion segmentation from multispectral MRI. Med Image Anal 35:250–269. https://doi.org/10.1016/j.media.2016.07.009
15. Mitra J, Bourgeat P, Fripp J, Ghose S, Rose S, Salvado O, Connelly A, Campbell B, Palmer S, Sharma G, Christensen S, Carey L (2014) Lesion segmentation from multimodal MRI using random forest following ischemic stroke. NeuroImage 98:324–335. https://doi.org/10.1016/j.neuroimage.2014.04.056. Epub 2014 May 2
16. Gerig G, Jomier M, Chakos M (2001) Valmet: a new validation tool for assessing and improving 3D object segmentation. In: Medical image computing and computer-assisted

intervention – MICCAI 2001. Springer, Berlin/Heidelberg, pp 516–523

17. Hausdorff F (1962) Set theory. Chelsea Pub. Co, New York
18. Taha AA, Hanbury A (2015) Metrics for evaluating 3D medical image segmentation: analysis, selection, and tool. BMC Med Imaging 15: 29. https://doi.org/10.1186/s12880-015-0068-x
19. Allen LM, Hasso AN, Handwerker J, Farid H (2012) Sequence-specific MR imaging findings that are useful in dating ischemic stroke. Radiographics 32 (5):1285–1297; discussion 1297–1289. https://doi.org/10.1148/rg.325115760
20. Forkel SJ, Catani M (2018) Lesion mapping in acute stroke aphasia and its implications for recovery. Neuropsychologia 115:88–100. https://doi.org/10.1016/j.neuropsychologia.2018.03.036
21. Okuda T, Korogi Y, Shigematsu Y, Sugahara T, Hirai T, Ikushima I, Liang L, Takahashi M (1999) Brain lesions: when should fluid-attenuated inversion-recovery sequences be used in MR evaluation? Radiology 212(3): 793–798. https://doi.org/10.1148/radiology.212.3.r99se07793
22. Rekik I, Allassonniere S, Carpenter TK, Wardlaw JM (2012) Medical image analysis methods in MR/CT-imaged acute-subacute ischemic stroke lesion: segmentation, prediction and insights into dynamic evolution simulation models. A critical appraisal. NeuroImage Clin 1:164–178. https://doi.org/10.1016/j.nicl.2012.10.003. eCollection
23. Rorden C, Karnath HO (2004) Using human brain lesions to infer function: a relic from a past era in the fMRI age? Nat Rev Neurosci 5: 813–819. https://doi.org/10.1038/nrn1521
24. Ito KL, Kim H, Liew SL (2019) A comparison of automated lesion segmentation approaches for chronic stroke T1-weighted MRI data. Hum Brain Mapp 40(16):4669–4685. https://doi.org/10.1002/hbm.24729
25. Sperber C, Karnath HO (2018) On the validity of lesion-behaviour mapping methods. Neuropsychologia 115:17–24. https://doi.org/10.1016/j.neuropsychologia.2017.07.035
26. Karnath HO, Rennig J (2017) Investigating structure and function in the healthy human brain: validity of acute versus chronic lesion-symptom mapping. Brain Struct Funct 222(5):2059–2070. https://doi.org/10.1007/s00429-016-1325-7
27. Clas P, Groeschel S, Wilke M (2012) A semi-automatic algorithm for determining the demyelination load in metachromatic leukodystrophy. Acad Radiol 19(1):26–34. https://doi.org/10.1016/j.acra.2011.09.008
28. Crinion J, Holland AL, Copland DA, Thompson CK, Hillis AE (2012) Neuroimaging in aphasia treatment research: quantifying brain lesions after stroke. NeuroImage 73:208–214. https://doi.org/10.1016/j.neuroimage.2012.07.044. Epub 2012 Jul 27
29. Stamatakis EA, Tyler LK (2005) Identifying lesions on structural brain images—validation of the method and application to neuropsychological patients. Brain Lang 94:167–177. https://doi.org/10.1016/j.bandl.2004.12.010
30. Mehta S, Grabowski TJ, Trivedi Y, Damasio H (2003) Evaluation of voxel-based morphometry for focal lesion detection in individuals. NeuroImage 20(3):1438–1454. https://doi.org/10.1016/s1053-8119(03)00377-x
31. Ashburner J, Friston KJ (2005) Unified segmentation. NeuroImage 26(3):839–851. https://doi.org/10.1016/j.neuroimage.2005.02.018
32. Gitelman DR, Ashburner J, Friston KJ, Tyler LK, Price CJ (2001) Voxel-based morphometry of herpes simplex encephalitis. NeuroImage 13(4):623–631. https://doi.org/10.1006/nimg.2000.0734
33. Sanjuán A, Price CJ, Mancini L, Josse G, Grogan A, Yamamoto AK, Geva S, Leff AP, Yousry TA, Seghier ML (2013) Automated identification of brain tumors from single MR images based on segmentation with refined patient-specific priors. Front Neurosci 7:241. https://doi.org/10.3389/fnins.2013.00241
34. Griffis JC, Nenert R, Allendorfer JB, Szaflarski JP (2017) Linking left hemispheric tissue preservation to fMRI language task activation in chronic stroke patients. Cortex 96:1–18. https://doi.org/10.1016/j.cortex.2017.08.031
35. Kimberg DY, Coslett HB, Schwartz MF (2007) Power in voxel-based lesion-symptom mapping. J Cogn Neurosci 19:1067–1080. https://doi.org/10.1162/jocn.2007.19.7.1067
36. Liew SL, Zavaliangos-Petropulu A, Jahanshad N, Lang CE, Hayward KS, Lohse KR, Juliano JM, Assogna F, Baugh LA, Bhattacharya AK, Bigjahan B, Borich MR, Boyd LA, Brodtmann A, Buetefisch CM, Byblow WD, Cassidy JM, Conforto AB, Craddock RC, Dimyan MA, Dula AN, Ermer E, Etherton MR, Fercho KA, Gregory CM, Hadidchi S, Holguin JA, Hwang DH, Jung S, Kautz SA, Khlif MS, Khoshab N, Kim B, Kim H, Kuceyeski A, Lotze M, MacIntosh BJ, Margetis JL, Mohamed FB, Piras F, Ramos-

Murguialday A, Richard G, Roberts P, Robertson AD, Rondina JM, Rost NS, Sanossian N, Schweighofer N, Seo NJ, Shiroishi MS, Soekadar SR, Spalletta G, Stinear CM, Suri A, Tang WKW, Thielman GT, Vecchio D, Villringer A, Ward NS, Werden E, Westlye LT, Winstein C, Wittenberg GF, Wong KA, Yu C, Cramer SC, Thompson PM (2020) The ENIGMA Stroke Recovery Working Group: big data neuroimaging to study brain-behavior relationships after stroke. Hum Brain Mapp. https://doi.org/10.1002/hbm.25015

Chapter 4

Mapping the Spatial Distribution of Lesions in Stroke: Effect of Diffeomorphic Registration Strategy in the ATLAS Dataset

Brian B. Avants and Nicholas J. Tustison

Abstract

Large-scale neuroimaging datasets from chronic stroke patients offer the opportunity to better understand patterns of injury, concomitant behavioral impairments, and long-term patient outcomes. Traditional exploratory approaches to this problem rely on mapping lesioned brains into a common atlas coordinate system via deformable image registration. The majority of these methods assume that one-to-one and invertible transformations exist between the patient brain and the template or reference brain. However, this assumption is violated by the presence of visible lesions in the brain, which is also confounded with the general difficulty of intersubject brain mapping. We review and compare image registration strategies for mapping T1-weighted neuroimaging of stroke patients into a single common coordinate system.

Key words Cost masking, Enantiomorphic, ANTs, ANTsR, ANTsPy, Lesion filling

1 Introduction

Cognitive neuroscientists and neurologists have relied, for over a century, on relating the consequences of localized brain injury to behavior [1, 2]. Recently, large datasets of patients with a common type of brain injury have been assembled and made accessible to a wider audience [3, 4]. These data offer the opportunity to study brain behavior relationships, and their subtleties, within specific cohorts and disease types.

The Anatomical Tracings of Lesions After Stroke (ATLAS) dataset comprises perhaps the largest publicly available collection of labeled T1-weighted neuroimages from stroke patients ($n = 229$) [3]. Such datasets present an opportunity to establish reference standards for technical improvements in the analysis of lesioned brains. ATLAS and related competitive challenge datasets (ISLES 2015 [4]) promote continually improving technical contributions that mark progress with respect to these benchmarks.

Dorian Pustina and Daniel Mirman (eds.), *Lesion-to-Symptom Mapping: Principles and Tools*, Neuromethods, vol. 180, https://doi.org/10.1007/978-1-0716-2225-4_4,

Without such standards, progress languishes within the confines of private datasets and research groups that prevent open comparisons of methodology.

Open-source methodology and freely available reference data are two key ingredients to open science. Scientific software developers link algorithm libraries such as the Insight Toolkit [5] to application libraries such as Advanced Normalization Tools (ANTs) [6] and, ultimately, interface with statistical computing platforms such as R [7] and python. For example, ANTs interfaces with R and python via ANTsR and ANTsPy, respectively [7, 8]. These high-level interfaces provide open-source, comprehensive analytics platforms which support everything from low-level operations (image smoothing) to high-level machine learning and enable rapid development of cutting-edge solutions in specific problem domains [9].

Here, we employ ATLAS data with ANTsR to review and compare image registration strategies for neuroimaging in stroke. The data from ATLAS includes a chronic stroke subject with T1-weighted (T1w) magnetic resonance imaging (MRI) and a manually generated lesion segmentation. While we focus on T1w neuroimaging in stroke, the strategies discussed in this document are not restricted to this modality or this condition.

2 Materials

This report relies on open-source software, the ANTsX ecosystem (https://github.com/ANTsX), as well as a publicly available dataset, ATLAS [3] (http://fcon_1000.projects.nitrc.org/indi/retro/atlas.html).

2.1 Imaging Data

See [3] for a detailed description of the images. We use a specific exemplar subject with generous permission from the ATLAS authorship. The subject experienced ischemic stroke (left MCA embolism), leading to a visible lesion with estimated size of 56 ml. The "native space" ID for this subject is c0005s0007t01. Figure 1 illustrates the subject and the lesion label. Note the left-lateralized nature of the lesion.

2.2 Software

We use R version 3.6.0 ("Planting of a Tree") [10] for basic statistical processing and ANTsR version 0.5.1 for core image analysis, data organization, and early development efforts. ANTsR, itself, links to ITKR, ANTs, and the Insight Toolkit (ITK). For deep learning, we rely on TensorFlow version 2.0 as accessed via ANTsRNet [11] – a deep learning network library extension of ANTsR. ANTs and its derived packages are available at http://github.com/ANTsX, whereas TensorFlow is available through CRAN and elsewhere.

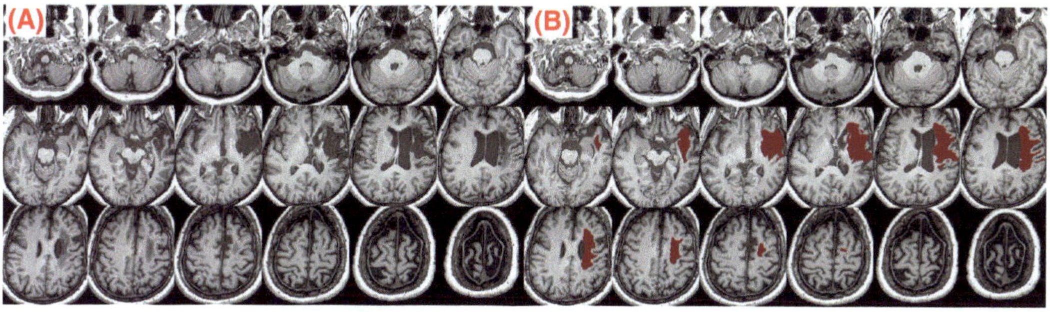

Fig. 1 (**a**) The example ATLAS subject T1 with left-lateralized manually drawn lesion label overlaid in (**b**)

3 Methods

The overall strategy performs the following steps:

1. Perform brain extraction based on deep learning models.
2. Affinely register the extracted brain to the template brain.
3. Deformably register the extracted brain to the template brain.

We focus on step 3 but provide some commentary on step 1, which comprises a relatively recent development.

3.1 Brain Extraction

Image registration methods can accommodate "clutter" within the image domain but generally reach better performance more quickly if irrelevant features are removed from the image. As the goal, here, is to accurately map the anatomical features of cortical and subcortical anatomy, it is prudent to extract the brain from the "background" features of the face, head, and neck. Performance metrics in brain extraction depend, in part, on the context in which the task is being performed [12, 13]. In this protocol, we use brain extraction to allow the registration to focus more specifically on anatomical features of interest, i.e., brain-specific features. As such, the extraction does not need to be "perfect" to serve this purpose. Rather, we prioritize speed and robustness to the presence of lesions, which is why we focus on recently developed deep learning approaches to brain extraction. For the interested reader, traditional approaches are reviewed in [14].

Deep learning-based segmentation provides a computationally fast and robust alternative to traditional brain extraction algorithms provided by ANTs, FreeSurfer, and related methods [15]. While the details of deep learning are beyond the scope of this document, the U-net model [16] is generally considered to be a reliable architecture on which to base domain-specific segmentation methods [17]. Here, we contrast two different U-net models for brain extraction, one provided by Nobrainer [18] and one by ANTsRNet.

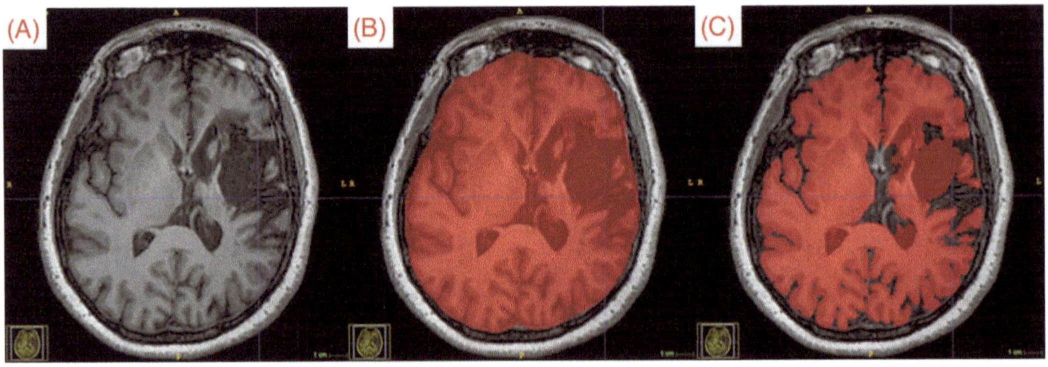

Fig. 2 (**a**) The example ATLAS image slice. (**b**) The *ANTsRNet* result. (**c**) The *Nobrainer* result which leaves out lesioned brain tissue. Undoubtedly, future work could fine-tune the *Nobrainer* network to perform well on this task

Nobrainer Model

The Nobrainer model is based on a large collection ($n = 11{,}480$) of precomputed FreeSurfer results. The model was trained with example images from over 100 sites and, as such, should be relatively insensitive to variability in T1 contrast. At the same time, this model was trained on control data and is sensitive to the presence of lesions. This is likely due to the fact that it is based on the specific needs of FreeSurfer, which seeks a surface-based matching of cortex.

ANTsRNet Model

The ANTsRNet model, on the other hand, provides a brain extraction that is perhaps more similar, in spirit, to the "Brain Extraction Tool (BET)" [15]. Relative to the Nobrainer model, ANTsRNet brain extraction provides an overall smoother interface between cerebral and non-cerebral tissue. That is, the priors built into this model tend more toward the geometric than the intensity-based. This makes the model less sensitive to the presence of lesions. Therefore, we use this network in the remainder of the study. Figure 2 shows a comparison of the two methods. *See* **Note 1** for further discussion of how these models might be used.

3.2 Affine Registration

The following affine and deformable registration sections assume that the input images comprise only normal brain tissue and possible lesions, i.e., little or no extracerebral tissue is present. The process of image registration is used to determine the correspondence between two image-based object representations. This process is typically performed via iterative optimization of a chosen measure quantifying the degree of similarity of the current correspondence between the images within the constraints of a specified transformation [5]. Such transformations span the spectrum of flexibility from linear transforms, such as rigid or affine, to highly deformable, biologically plausible transformations, such as diffeomorphisms.

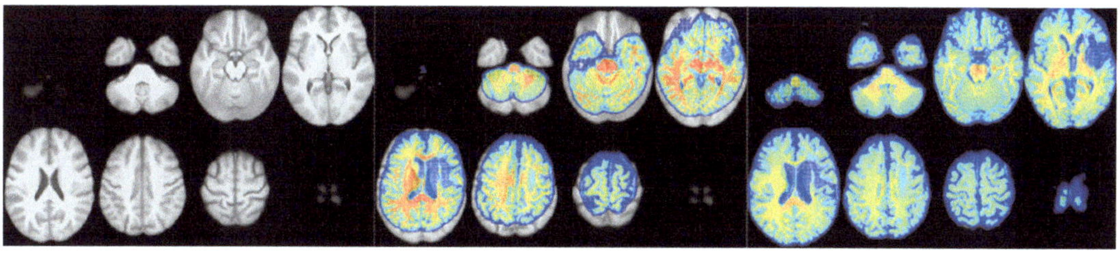

Fig. 3 Affine registration results. The template is depicted on the left, and the initial alignment of the subject to the template is shown in the middle where the subject brain has been overlaid on top of the template. The affine result is illustrated on the right where the subject brain is overlaid on the template

Registration on brain extracted data greatly reduces the number of local minima in which the optimization can be stuck. For the case of affine registration, the majority of the solution space is constrained by the overall size and shape of the brain. In contrast to deformable registration, the fine-scale details are far less important. Consequently, affine registration is less sensitive to the presence of lesions than the subsequent deformable mapping (Fig. 3). In this study, we utilize the default ANTs registration affine parameters as incarnated in the ANTsR toolkit.

3.3 Deformable Registration Methods

Deformable registration approaches are characterized by greater transformation flexibility for establishing finer-scale correspondence between image pairs. From a technical perspective, similar to the linear methods described previously, deformable registration methods consist of a *transformation model* and a *similarity criterion*. Additionally, a *regularization term* is also included within the optimization scheme to modulate the flexibility of the transformation [5]. For the purposes of the data processing below, we use a fixed transformation model and regularization and optimization method, the collection of which is known as symmetric normalization (SyN) [6]. The SyN transformation model belongs to the family of "digital diffeomorphisms" – transformations that are, in the discrete space, verifiably and consistently invertible. These types of transformations have also demonstrated excellent performance in spatial normalization of neuroimaging data [19]. The (default) regularization for SyN is a linear operator that acts as a Gaussian smoothing on the velocity field that is generated by the gradient of the transformation with respect to the similarity criterion. SyN also employs a specific, custom optimization strategy that updates four transformations in total while ensuring that they remain in the space of digital diffeomorphisms. For the data below, we use these core components to investigate three different strategies for normalization between a lesioned brain and that of a "normal" template image – the common space for mapping all images which, for this study, is represented by the UK Biobank group-averaged T1 template [20].

3.3.1 Automated Registration

In order to facilitate usage, we have wrapped commonly used parameter sets within the antsRegistration functionality provided in ANTsR. For the data described below, we use one of these recommended parameter sets. By default, the similarity metric is computed over the complete image domain with no special handling for lesions. The similarity metric is the Insight Toolkit implementation of the neighborhood cross-correlation (NCC) metric as discussed in [6]. NCC quantifies the strength of the linear association of intensities in corresponding neighborhoods and is robust to intensity scale and offset differences characteristic of MRI. This metric and the SyN transformation model combination have performed well in a variety of tasks and remains a top performer in registration evaluation studies [21]. All of the methods utilized below are diffeomorphic in that they are differentiable and invertible transformations with a differentiable inverse. This diffeomorphic constraint is a minimal, yet general, prior assumption for anatomical plausibility of the resulting correspondence for most image registration workflows. The code chunk, below, indicates how to run this specific antsRegistration configuration in ANTsR with an eight core execution time of roughly 90 s on a 2018 MacBook Pro. *See* **Note 2** for a brief discussion on this topic and what it implies for brain mapping. *See* result in Fig. 4.

Automatic Registration Script

```
library( ANTsR )
template = antsImageRead( "data/template_t1_ukbiobank_dyadic.
nii.gz" )
newdir=diag(3); newdir[c(1,5)]=-1; antsSetDirection( tem-
plate, newdir )
templateMask = antsImageRead( "data/template_t1_brain_ukbio-
bank_dyadic.nii.gz" ) %>%
 antsCopyImageInfo2( template )
template = iMath( template * templateMask, "Normalize" )
subjectBrain = antsImageRead( "data/example_image_brain.nii.
gz" ) %>% iMath("Normalize")
autoreg = antsRegistration( template, subjectBrain, 'SyN',
synMetric='CC', synSampling=2)
trans = antsRegistration( template, subjectBrain, 'Transla-
tion', affIterations=c(10,0,0,0))$warpedmovout
affout = antsApplyTransforms( template, subjectBrain, autoreg
$fwdtransforms[19] )
png( 'figures/aff.png', width=1028, height=256 )
layout(matrix(1:3,nrow=1))
plot( template, axis=3, colorbar=F, nslices=8 )
plot( template, trans, axis=3, colorbar=F, nslices=8, al-
pha=0.8 )
plot( template, affout, axis=3, colorbar=F, nslices=8 , al-
```

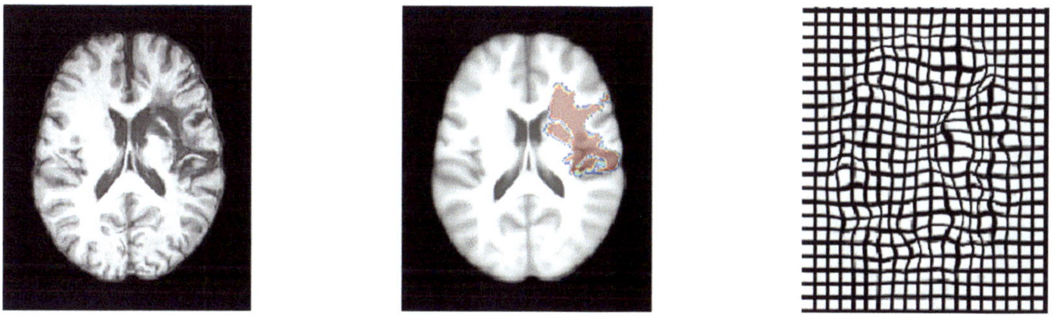

Fig. 4 Automated registration results. At left is the subject brain deformed to the template space. Center shows the lesion overlaid on the template at the corresponding slice. At the right, we see a slice of the deformed grid. The remainder of the registration result figures follows this same layout

```
pha=0.8 )
dev.off()
subjectLesionMask = antsImageRead( "data/example_image_le-
sion_label.nii.gz" ) %>%
 thresholdImage( 1, Inf )
temp = antsApplyTransforms(template,subjectLesionMask, auto-
reg$fwdtransforms )
grid = createWarpedGrid( template, transform = autoreg
$fwdtransforms[14],
 gridDirections=c(T,T,F), fixedReferenceImage=template )
png( 'figures/auto.png', width=1028, height=256 )
layout(matrix(1:3,nrow=1))
plot( autoreg$warpedmovout*255, axis=3, colorbar=F, slice=95 )
plot( template*255, temp, axis=3, colorbar=F, slice=95, al-
pha=0.8 )
# plot( abs(template-iMath(autoreg$warpedmovout,"Normalize"))
*255, axis=3, colorbar=F, slice=95 )
plot( grid*255, axis=3, colorbar=F, slice=95, alpha=0.8 )
dev.off()
```

3.3.2 Cost Function Masking

Cost function masking (CFM) is an approach that changes only the image region over which the similarity metric is evaluated [22–24]. CFM relies on a user-defined lesion mask, which is passed as a parameter to the image registration method. The method ignores the masked regions within the optimization framework and, in theory, is therefore only relying on "healthy" brain tissue to perform the registration, which is essential for lesion-symptom mapping studies. *See* **Note 3** for comments on how one should assess a lesion mask and prepare it for inclusion in CFM. *See* result in Fig. 5.

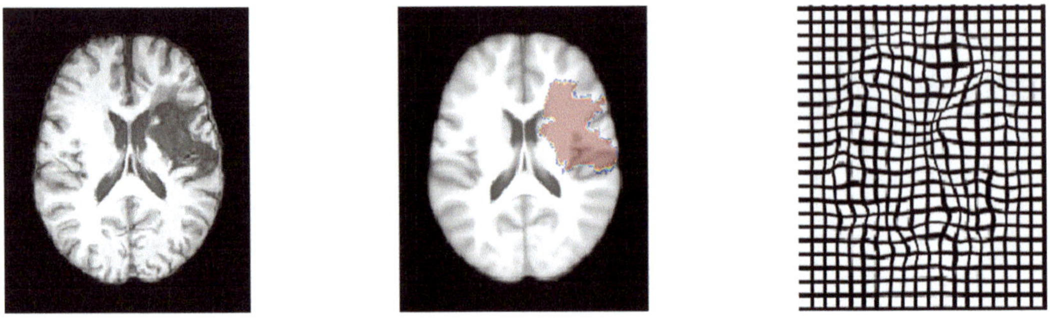

Fig. 5 CFM registration results. The center panel demonstrates that lesion volume is relatively preserved in comparison to the automated approach, which tends to excessively shrink the lesion due to overfitting. The mutual information between the template and deformed subject is 0.769, where higher values indicate greater similarity

CFM Registration Script

```
# Read in the lesion mask
subjectLesionMask = antsImageRead( "data/example_image_le-
sion_label.nii.gz" ) %>%
 thresholdImage( 1, Inf )
# Perform binary dilation on the lesion mask
subjectLesionMaskDil = morphology( subjectLesionMask, "di-
late", 5 )
# Calculate the inverse mask, i.e., foreground/background
switch
subjectLesionMaskNeg = iMath( -subjectLesionMaskDil, "Normal-
ize" )
subjectLesionMaskNegAff = antsApplyTransforms( template, sub-
jectLesionMaskNeg,
 autoreg$fwdtransforms[19] )
# Register the subject brain/mask to the template
cfmreg = antsRegistration( template, subjectBrain, 'SyNOnly',
synMetric='CC',
 synSampling = 2, initialTransform = autoreg$fwdtransforms
[19],
 mask = subjectLesionMaskNegAff, verbose = TRUE )#
temp = antsApplyTransforms(template,subjectLesionMask, cfmreg
$fwdtransforms )
# Create the warped grid describing the subject to template
deformation
grid = createWarpedGrid( template, transform = cfmreg
$fwdtransforms[14],
 gridDirections=c(T,T,F), fixedReferenceImage=template )

# Display the results
png( 'figures/cfm.png', width=1028, height=256 )
layout(matrix(1:3,nrow=1))
```

```
plot( cfmreg$warpedmovout*255, axis=3, colorbar=F, slice=95 )
plot( template*255, temp, axis=3, colorbar=F, slice=95, al-
pha=0.8 )
# plot( abs(template-cfmreg$warpedmovout), axis=3, colorbar=F,
slice=95 )
plot( grid*255, axis=3, colorbar=F, slice=95, alpha=0.8 )
dev.off()
```

3.3.3 Enantiomorphic Registration

An enantiomorph is a mirror image. Enantiomorphic registration [25] uses an existing lesion mask to in-paint the lesioned area with healthy brain tissue from the contralateral side or, potentially, from other subject images. The advantage of this approach, in comparison to CFM, is that the image registration is able to leverage realistic anatomical features in order to guide the mapping. In theory, this may lead to a more accurate proxy spatial mapping of the missing tissue in comparison to CFM, which simply uses the surrounding tissue to interpolate the correspondence. One complexity of this method is that it assumes the contralateral side can be plausibly mapped to the lesioned hemisphere. This assumption, in itself, may be difficult to verify. Nevertheless, after reviewing enantiomorphic inpainting in the ATLAS dataset, we have found few, if any, clear misregistrations. As such, we recommend that researchers at least consider this approach in the future.

Enantiomorphic Filling and Registration Script (*See* Figs. 6 and 7)

```
# Obtain the enantiomorphic image by reflecting across the
sagittal plane
subjectBrainReflect = reflectImage( subjectBrain, axis = 0, tx
= 'SyN',
 verbose = FALSE, totalSigma = 0.5 )$warpedmovout
# Smooth the lesion mask to decrease distortion
slesion = smoothImage( subjectLesionMask, 1 ) %>% iMath("Nor-
malize")
# Perform in-painting
slesionNeg = 1 - slesion
subjectFill = subjectBrain * slesionNeg + subjectBrainReflect
* slesion
# Register enantiomorphic brain
enantreg = antsRegistration( template, subjectFill, 'SyNOnly',
synMetric='CC',
 synSampling = 2, initialTransform = autoreg$fwdtransforms
[19] )
enantmap = antsApplyTransforms( template, subjectBrain, enan-
treg$fwdtransforms )
# layout(matrix(1:2,nrow=1))
```

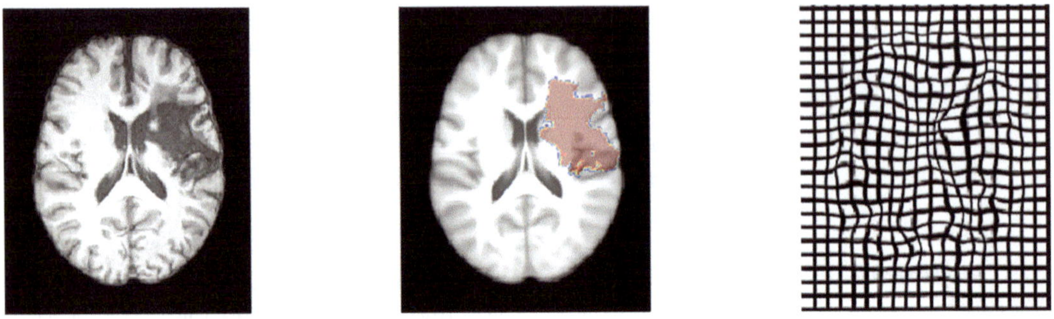

Fig. 6 Enantiomorphic registration results. The deformation grid suggests that the "shape" of the deformation field has additional structure in comparison to the smoothly interpolated deformations in the CFM. Furthermore, lesion volume remains relatively preserved, like the CFM approach. The final similarity value is intermediate to the automatic and CFM results

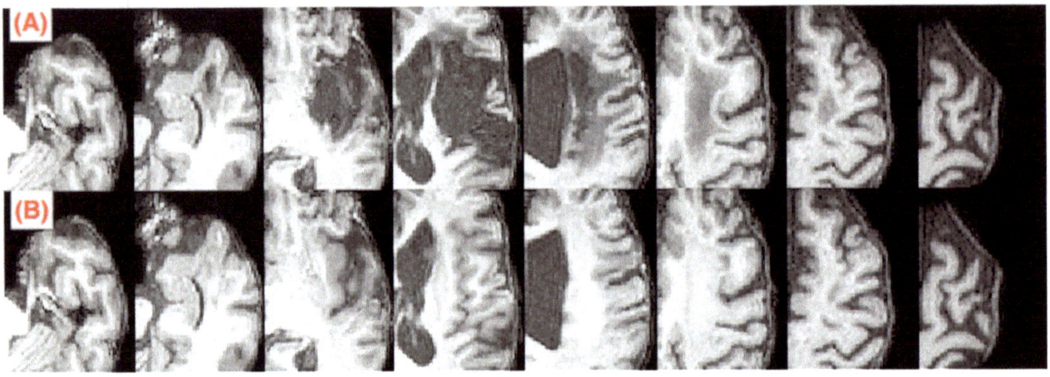

Fig. 7 Enantiomorphic filling of the lesion for the example subject. The original lesioned area is in (**a**), whereas (**b**) shows the lesioned area filled by the healthy tissue from the contralateral hemisphere

```
# plot( cropImage( subjectBrain, subjectLesionMaskDil ),
axis=3, colorbar=F )
# plot( cropImage( subjectFill, subjectLesionMaskDil ),
axis=3, colorbar=F )
temp = antsApplyTransforms(template,subjectLesionMask, enan-
treg$fwdtransforms )
grid = createWarpedGrid( template, transform = enantreg
$fwdtransforms[14],
 gridDirections=c(T,T,F), fixedReferenceImage=template )
png( 'figures/enantreg.png', width=1028, height=256 )
layout(matrix(1:3,nrow=1))
plot( enantmap*255, axis=3, colorbar=F, slice=95 )
plot( template*255, temp, axis=3, colorbar=F, slice=95, al-
pha=0.8 )
# plot( abs(template-enantmap), axis=3, colorbar=F, slice=95,
alpha=0.8 )
plot( grid*255, axis=3, colorbar=F, slice=95, alpha=0.8 )
dev.off()
```

```
png( 'figures/enant2.png', width=1028, height=256 )
layout(matrix(1:2,nrow=2))
plot(255* cropImage( subjectBrain, subjectLesionMaskDil ),
nslices = 8, ncol=8, axis=3 )
plot( 255*cropImage( subjectFill, subjectLesionMaskDil ),
nslices = 8, ncol=8, axis=3 )
dev.off()
```

See **Note 4** for discussion on alternatives to enantiomorphic filling.

3.3.4 Effect of Lesion Size and Registration Strategy on Spatial Mapping Results

In this section, we make suggestions on how to quantify the impact of lesion size on registration results. Automated registration provides a baseline (reference) result against which we can compare other strategies. The basic idea is to measure the degree to which cost function masking or other lesion-specific approaches impact the nature of the computed spatial transformations, relative to the baseline approach. For example, naive deformable registration between a "normal" template and a subject with a left hemisphere infarct of the frontal lobe clearly violates basic assumptions of bijectivity and differentiability (discussed, e.g., in [6]) for diffeomorphic registration and image similarity. In this case, lesion masking should have a strong influence on the result. On the other hand, a single voxel lesion will lead to a minimal difference in registration outcome regardless of whether one uses cost function masking or not.

We may quantify the effects of normalization strategy by computing basic distributional statistics on the lesion volumes, deformation fields, as well as the similarity measurements. Given a pair of registration results with transformations ϕ_1 (e.g., from the fully automated approach) and ϕ_2 (e.g., from CFM), compute:

- The expectation of the mean Euclidean distance in template space:

$$E\left(\frac{1}{|\Omega|}\sum_x \phi_1(x) - \phi_2(x)\right),$$

  where $|\Omega|$ indicates the size of the spatial domain of the template

- The expectation of the maximum Euclidean distance in template space:

$$E\left(\mathrm{argmax}_x \phi_1(x) - \phi_2(x)\right),$$

  that is, the maximum difference in Ω

- The expectation of the mutual information between the transformed patient and the template image (as assessed by antsImageMutualInformation which returns the *negative* mutual information).
- The change in the lesion volume from the original image space.

Linear regression then allows one to infer statistical relationships between registration strategy, lesion volume, and these quantitative metrics.

Computing Lesion Volume for Assessing Registration Effects

```
# lesion volumes
originalVolume = labelStats( subjectLesionMask, subjectLesion-
Mask )$Volume[19]
affMask = antsApplyTransforms( template, subjectLesionMask,
 enantreg$fwdtransforms[19], interpolator = 'nearestNeighbor'
)
affVolume = labelStats( affMask, affMask )$Volume[19]
autoMaskDef = antsApplyTransforms( template, subjectLesion-
Mask,
 autoreg$fwdtransforms, interpolator = 'nearestNeighbor' )
autoMaskDefVolume = labelStats( autoMaskDef, autoMaskDef )
$Volume[19]
deformedMask = antsApplyTransforms( template, subjectLesion-
Mask,
 enantreg$fwdtransforms, interpolator = 'nearestNeighbor' )
deformedVolume = labelStats( deformedMask, deformedMask )$Vo-
lume[19]
```

Brain-behavior associations will be impacted by the details of each choice outlined above. We encourage researchers who are interested in evaluating such choices to contact the ANTsX team. For guidance on opening such discussions, please *see* **Note 5**.

3.3.5 Brain and Behavior Inference

Population-level registration of a large-scale dataset such as ATLAS allows one to estimate the spatial distribution of lesions within the group template space. While such a map may be of value, the motivation for much of the work in the registration of patients with ischemic stroke is to map the relationship between structure and function in the brain. Ideally, a better registration method – or a better method for handling lesioned tissue during registration – would lead to more reproducible, reliable, and biologically plausible inference between brain and behavior. This type of assessment remains a challenge to the field in part because datasets like ATLAS lack associated patient outcomes. Such public data would allow the community to establish more standardized mapping between lesion locations and patient outcomes; we hope that such data becomes available in the near future.

4 Notes

Note 1: Deep learning inference can be improved with ensembling approaches.

That is, small modifications to the input via intensity changes (noise, bias) and/or orientation changes will lead to slightly different segmentation outcomes. These can be averaged in order to determine a consensus segmentation which may improve over any of the individual segmentations. This strategy is particularly effective because inference is, relatively speaking, computationally cheap. Both model architectures can take advantage of this strategy to further refine results. The models are available and documented in ANTsRNet (https://github.com/ANTsX/ANTsRNet). A python variant is also available at https://github.com/ANTsX/ANTsPyNet. The Nobrainer model is available here https://github.com/neuronets/nobrainer.

Note 2: Diffeomorphic maps should be interpreted with care in brain mapping.

The diffeomorphism group is infinite dimensional, in the mathematical sense. Practically speaking, it encompasses a collection of invertible transformations that can model geometrically continuous maps. In the neurobiological setting, we use these maps to model a potentially feasible path of shape transformations from one anatomical configuration to another. The advantage of this model is that it makes very minimal assumptions about the underlying images and transformation processes. As such, software such as ANTs can be applied to an enormous range of image registration problems. This also means that transformation models such as used in ANTs are not brain-specific. Therefore, we caution users not to over-interpret the descriptive capabilities of the derived deformation fields. It is crucial to note that these models are, at best, approximations to one possible geometric transformation. The benefit to these methods is not that they are biologically true. Rather, they are *statistically uncertain* but reasonable models that can be applied at scale with very low failure rate, thus making them appropriate for large-scale data processing.

Note 3: Process the lesion mask for CFM.

Lesion masks are fundamentally imperfect. Depending on the lesion drawing protocol, the user may want to either morphologically erode (if the protocol includes "healthy"-looking tissue) or dilate (if the protocol includes only severely hypointense tissue) the lesion mask before passing to the registration method. Furthermore, some methods may provide a "soft" probabilistic mask. In this case, the user may want to select either a higher or lower probability threshold depending on how these probabilities align with perceptually healthy tissue. Lastly, masks passed to registration software may be *inclusive*. ANTs, in particular, expects the mask to denote which voxels should be included in the registration, not which voxels to exclude.

Note 4: Enantiomorphic filling and inpainting methods.

There are a variety of inpainting methods that may be used for interpolating missing anatomy [26]. Joint intensity fusion (JIF) [27] – a continuous space extension to joint label fusion (JLF) available in the ANTsX ecosystem – provides another alternative to filling lesions. JIF will infill anatomy based on control subjects instead of the contralateral hemisphere. JIF, like JLF, will first require deformable registration of a set of control subjects to the target subject. It will then fill each lesioned voxel with a weighted average of the intensities of the library of normal subjects. This method was previously applied to fill lesions, which appear in both hemispheres of the brain, in multiple sclerosis.

Note 5: Open source software progress requires feedback from scientists.

The health of open-source scientific software depends on interaction between the core development team and scientists outside of that team. Such interactions help all of those involved; developers gain insight into software usability and, potentially, identify bugs; scientists gain the opportunity to interact with individuals who are experts in a technical field on which their work may rely. The ANTs team, in particular, welcomes feedback, contributions, criticism, and discussion via the issues links for individual projects that can be found at https://github.com/ANTsX. We encourage readers to reach out to us via these public venues for questions with respect to the work described in this protocol.

References

1. de Haan B, Karnath H-O (2018) A hitchhiker's guide to lesion-behaviour mapping. Neuropsychologia 115:5–16. https://doi.org/10.1016/j.neuropsychologia.2017.10.021
2. Varjačić A, Mantini D, Levenstein J, Slavkova ED, Demeyere N, Gillebert CR (2018) The role of left insula in executive set-switching: lesion evidence from an acute stroke cohort. Cortex 107:92–101. https://doi.org/10.1016/j.cortex.2017.11.009
3. Liew S-L, Anglin JM, Banks NW, Sondag M, Ito KL, Kim H, Chan J, Ito J, Jung C, Khoshab N et al (2018) A large, open source dataset of stroke anatomical brain images and manual lesion segmentations. Sci Data 5. https://doi.org/10.1038/sdata.2018.11
4. Maier O, Menze BH, von der Gablentz J, Häni L, Heinrich MP, Liebrand M, Winzeck S, Basit A, Bentley P, Chen L et al (2017) ISLES 2015 – a public evaluation

benchmark for ischemic stroke lesion segmentation from multispectral MRI. Med Image Anal 35:250–269. https://doi.org/10.1016/j.media.2016.07.009

5. Avants BB, Tustison NJ, Stauffer M, Song G, Wu B, Gee JC (2014) The insight ToolKit image registration framework. Front Neuroinform 8. https://doi.org/10.3389/fninf.2014.00044
6. Avants BB, Tustison NJ, Song G, Cook PA, Klein A, Gee JC (2011) A reproducible evaluation of ANTs similarity metric performance in brain image registration. NeuroImage 54: 2033–2044. https://doi.org/10.1016/j.neuroimage.2010.09.025
7. Muschelli J, Gherman A, Fortin J-P, Avants B, Whitcher B, Clayden JD, Caffo BS, Crainiceanu CM (2018) Neuroconductor: an R platform for medical imaging analysis. Biostatistics 20:218–239. https://doi.org/10.1093/biostatistics/kxx068
8. Tustison NJ, Shrinidhi KL, Wintermark M, Durst CR, Kandel BM, Gee JC, Grossman MC, Avants BB (2014) Optimal symmetric multimodal templates and concatenated random forests for supervised brain tumor segmentation (simplified) with ANTsR. Neuroinformatics 13:209–225. https://doi.org/10.1007/s12021-014-9245-2
9. Pustina D, Coslett HB, Turkeltaub PE, Tustison N, Schwartz MF, Avants B (2016) Automated segmentation of chronic stroke lesions using LINDA: lesion identification with neighborhood data analysis. Hum Brain Mapp 37:1405–1421. https://doi.org/10.1002/hbm.23110
10. R Core Team (2019) R: a language and environment for statistical computing. R Foundation for Statistical Computing 3.6
11. Tustison NJ, Avants BB, Lin Z, Feng X, Cullen N, Mata JF, Flors L, Gee JC, Altes TA, Mugler Iii JP, Qing K (2019) Convolutional neural networks with template-based data augmentation for functional lung image quantification. Acad Radiol 26:412–423. https://doi.org/10.1016/j.acra.2018.08.003
12. Souza R, Lucena O, Garrafa J, Gobbi D, Saluzzi M, Appenzeller S, Rittner L, Frayne R, Lotufo R (2018) An open, multi-vendor, multi-field-strength brain MR dataset and analysis of publicly available skull stripping methods agreement. NeuroImage 170: 482–494. https://doi.org/10.1016/j.neuroimage.2017.08.021
13. Puccio B, Pooley JP, Pellman JS, Taverna EC, Craddock RC (2016) The preprocessed connectomes project repository of manually corrected skull-stripped T1-weighted anatomical MRI data. GigaScience 5. https://doi.org/10.1186/s13742-016-0150-5
14. Kalavathi P, Prasath VBS (2016) Methods on skull stripping of MRI head scan images – a review. J Digit Imaging 29:365–379. https://doi.org/10.1007/s10278-015-9847-8
15. Novosad P, Collins DL (2018) An efficient and accurate method for robust inter-dataset brain extraction and comparisons with 9 other methods. Hum Brain Mapp 39:4241–4257. https://doi.org/10.1002/hbm.24243
16. Ronneberger O, Fischer P, Brox T (2015) U-net: convolutional networks for biomedical image segmentation. In: Medical image computing and computer-assisted intervention (MICCAI). Springer, Cham, pp 234–241
17. Hesamian MH, Jia W, He X, Kennedy P (2019) Deep learning techniques for medical image segmentation: achievements and challenges. J Digit Imaging 32:582–596. https://doi.org/10.1007/s10278-019-00227-x
18. McClure P, Rho N, Lee JA, Kaczmarzyk JR, Zheng CY, Ghosh SS, Nielson DM, Thomas AG, Bandettini P, Pereira F (2019) Knowing what you know in brain segmentation using Bayesian deep neural networks. Front Neuroinform 13. https://doi.org/10.3389/fninf.2019.00067
19. Klein A, Andersson J, Ardekani BA, Ashburner J, Avants B, Chiang M-C, Christensen GE, Collins DL, Gee J, Hellier P, Song JH, Jenkinson M, Lepage C, Rueckert D, Thompson P, Vercauteren T, Woods RP, Mann JJ, Parsey RV (2009) Evaluation of 14 nonlinear deformation algorithms applied to human brain MRI registration. NeuroImage 46:786–802. https://doi.org/10.1016/j.neuroimage.2008.12.037
20. Miller KL, Alfaro-Almagro F, Bangerter NK, Thomas DL, Yacoub E, Xu J, Bartsch AJ, Jbabdi S, Sotiropoulos SN, Andersson JLR et al (2016) Multimodal population brain imaging in the UK biobank prospective epidemiological study. Nat Neurosci 19: 1523–1536. https://doi.org/10.1038/nn.4393
21. Horn A, Li N, Dembek TA, Kappel A, Boulay C, Ewert S, Tietze A, Husch A, Perera T, Neumann W-J, Reisert M, Si H, Oostenveld R, Rorden C, Yeh F-C, Fang Q, Herrington TM, Vorwerk J, Kühn AA (2019) Lead-DBS v2: towards a comprehensive pipeline for deep brain stimulation imaging. NeuroImage 184:293–316. https://doi.org/10.1016/j.neuroimage.2018.08.068
22. Crinion J, Ashburner J, Leff A, Brett M, Price C, Friston K (2007) Spatial normalization of lesioned brains: performance evaluation

and impact on fMRI analyses. NeuroImage 37: 866–875. https://doi.org/10.1016/j.neuroimage.2007.04.065

23. Brett M, Leff AP, Rorden C, Ashburner J (2001) Spatial normalization of brain images with focal lesions using cost function masking. NeuroImage 14:486–500. https://doi.org/10.1006/nimg.2001.0845
24. Andersen SM, Rapcsak SZ, Beeson PM (2010) Cost function masking during normalization of brains with focal lesions: still a necessity? NeuroImage 53:78–84. https://doi.org/10.1016/j.neuroimage.2010.06.003
25. Nachev P, Coulthard E, Jäger HR, Kennard C, Husain M (2008) Enantiomorphic normalization of focally lesioned brains. NeuroImage 39: 1215–1226. https://doi.org/10.1016/j.neuroimage.2007.10.002
26. Xiong H, Wang C, Barnett M, Wang C (2020) Multiple sclerosis lesion filling using a non-lesion attention based convolutional network. In: International Conference on Neural Information Processing. Springer, Cham, pp 448–460
27. Fleishman GM, Valcarcel A, Pham DL, Roy S, Calabresi PA, Yushkevich P, Shinohara RT, Oguz I (2017) Joint intensity fusion image synthesis applied to multiple sclerosis lesion segmentation. In: International MICCAI brainlesion workshop. Springer, Cham, pp 43–54

Chapter 5

Voxel-Based Lesion Symptom Mapping

Juliana V. Baldo, Maria V. Ivanova, Timothy J. Herron, Stephen M. Wilson, and Nina F. Dronkers

Abstract

Lesion-behavior mapping has been used for over 200 years as a method for understanding the relationship between distinct brain regions and the functions they subserve. A relatively recent advancement in this area has been the development of voxel-based statistical techniques that allow for whole-brain analysis of lesion data in relation to specific behaviors and symptoms in neurologic patients. In this chapter, we describe current best practices for one of these techniques, voxel-based lesion-symptom mapping (VLSM), and review key studies in the literature that have made use of this methodology to advance our understanding of brain-behavior relationships. Best practices include rigorous statistical correction for family-wise error, covarying for potentially confounding variables, and limiting interpretation to brain regions with sufficient statistical power. In some cases, VLSM has confirmed earlier findings of lesion-behavioral correlates from small group and case studies, but in other cases, VLSM findings have challenged classical models of brain-behavior relationships. Together with newer multivariate techniques, VLSM is a powerful new tool that can be applied to large datasets to further our understanding of the relationship between brain and behavior.

Key words Voxel-based lesion-symptom mapping (VLSM), MRI, Stroke, Aphasia, Neuroimaging, Brain mapping

1 History of Lesion-Behavior Mapping

Lesion-behavior mapping is a method that identifies the brain correlates of behaviors and symptoms, based on findings from individuals who have suffered a neurologic event like stroke or a progressive disorder like dementia. Early approaches in the nineteenth century by pioneers such as [20, 23, 24, 112] relied on autopsy data to identify the underlying cause of deficits observed in their patients. This interest in examining how specific brain regions were associated with different behaviors set the stage for the field of brain-behavior mapping. Since the advent of modern neuroimaging in the 1970s

The original version of this chapter was revised. The correction to this chapter is available at https://doi.org/10.1007/978-1-0716-2225-4_16

Dorian Pustina and Daniel Mirman (eds.), *Lesion-to-Symptom Mapping: Principles and Tools*, Neuromethods, vol. 180, https://doi.org/10.1007/978-1-0716-2225-4_5, © Springer Science+Business Media, LLC, part of Springer Nature 2022, Corrected Publication 2022 and 1980s, clinicians and scientists have been able to visualize the living brain with imaging technologies such as CT and MRI. These advances have allowed for more systematic exploration of different theoretical models based on the observed correlation between lesion site and behavioral changes in a single patient or group of patients [14, 39, 64, 75, 79, 81, 83, 120 ; *see* Chapter 1].

2 Development of Voxel-Based Lesion-Symptom Mapping (VLSM)

As interest in the area of lesion-behavior mapping grew, standardized methods were developed that allowed researchers to systematically identify critical brain regions involved in a particular behavior. One of these methods involved reconstructing lesions on a brain atlas template so that a group of patients' lesions could be overlaid on top of each other in a common space [34, 35]. In this way, brain regions associated with a particular behavioral deficit or symptom could be readily identified [e.g., 36, 41, 69, 70]. While the lesion overlay method led to a number of important findings, one of its biggest limitations was the reliance on nominal patient labels that categorized patients by overly broad and sometimes arbitrary divisions (e.g., fluent vs. non-fluent aphasia). What was needed was a more quantitative, statistical approach that could capture the graded/continuous nature of behavioral data in relation to lesion site. New approaches to lesion mapping were thus developed in the early 1990s that related aggregated, normalized lesion data to fully parametric behavioral scores [1, 3, 45; *see* Chapter 7 on voxel-based morphometry]. Building on these techniques, Bates and colleagues [15] developed voxel-based lesion-symptom mapping (VLSM; also called voxel-based lesion-behavior mapping (VLBM) and univariate lesion-symptom mapping (ULSM); https://aphasialab.org/vlsm/). The goal was to create a new lesion analysis method that was statistically rigorous and that made full use of the complexity of both behavioral and neuroimaging data acquired from neurologic patients. Bates et al. were inspired by univariate analysis approaches to fMRI data, with the aim of being able to more readily compare results from lesion studies to functional imaging findings using a common brain coordinate space. As shown in Fig. 1, the number of studies using VLSM has steadily increased since the introduction of the technique in 2003, and the diversity of applications (e.g., multiple sclerosis, resection, TBI) has expanded as well.

3 VLSM Method and Approach

VLSM uses univariate models to relate normalized lesion data from a group of patients to their corresponding scores, in order to identify brain regions underlying a particular behavior, task, or

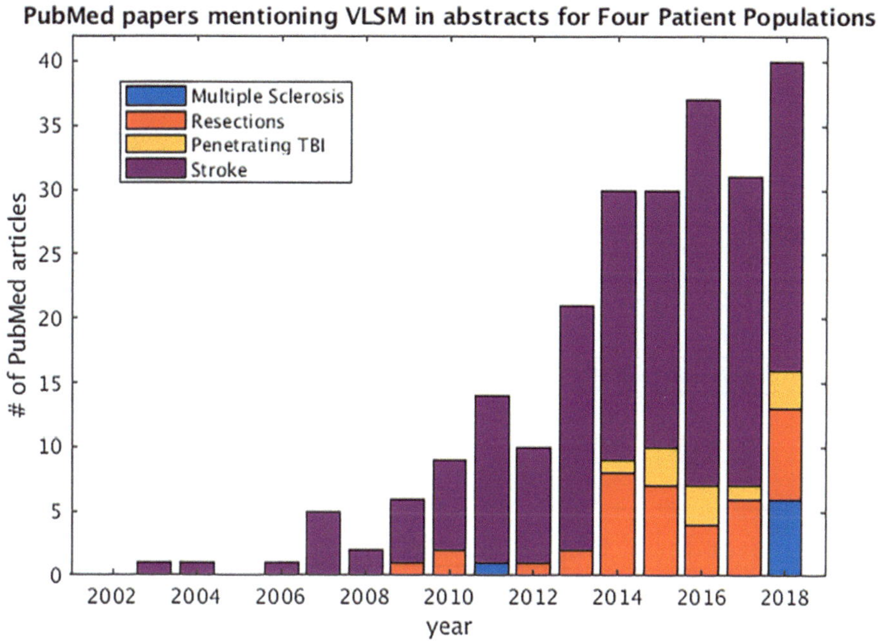

Fig. 1 Number of VLSM studies in PubMed per year across four patient populations since the technique was introduced in 2003

clinical symptom. A schematic of the steps involved in VLSM is shown in Fig. 2. First, patients' lesions are traced on digital brain images (e.g., MRI T1, T2, and/or FLAIR) and normalized into a standard brain space, so that all patients' brains are aligned in a comparable coordinate system (e.g., Montreal Neurologic Institute or MNI template; [21] (*see* Chapters 2, 3, and 4). These lesion data (also referred to as lesion masks) are then read into the VLSM program along with each patient's corresponding behavioral score. A test statistic is calculated at each brain voxel, comparing scores between patients with and without damage to that voxel. These tests are typically one-tailed, with the assumption that lesioned voxels are associated with worse performance or greater symptom severity. The parameters in VLSM can also be set to process continuous lesion data using regression, where the degree of lesion in a voxel ranges from 0 to 1 [108, 116]; *see* Chapter 7]. The statistical test is repeated for every voxel. A correction is applied to account for the large number of statistical tests, in order to minimize false positives (described in detail below; also *see* Chapter 6). The VLSM program then generates maps based on the values of the statistics at each voxel and highlights those brain regions most strongly associated with the given behavior/symptom (*see* bottom row of Fig. 2). These VLSM maps can be visualized in freely available software (e.g., ITK-SNAP http://www.itksnap.org/pmwiki/pmwiki.php and MRIcron https://people.cas.sc.edu/rorden/mricron/index.html; nitrc.org/projects/mricron). A yoked brain atlas with anatomical labels can be used to identify

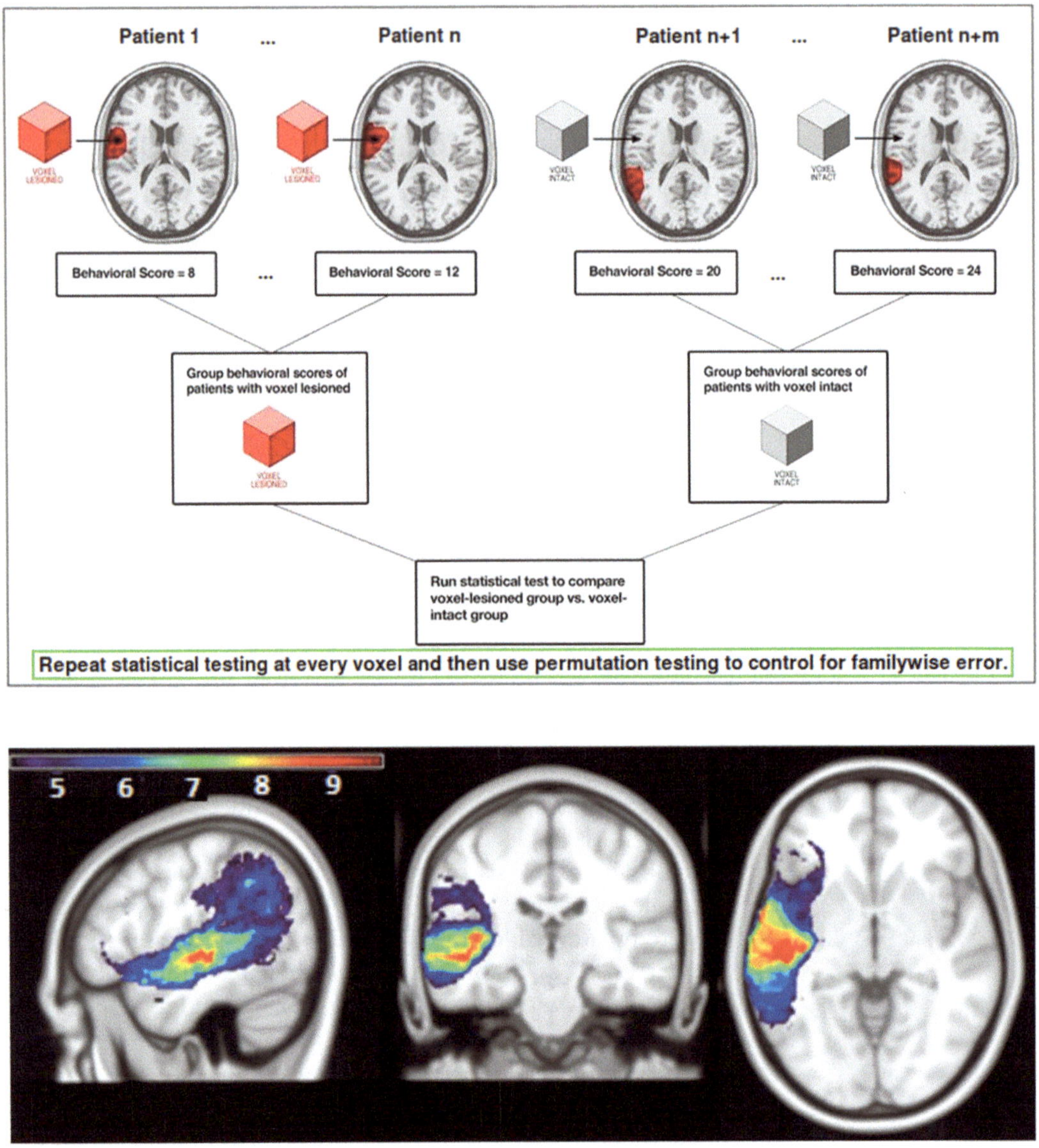

Fig. 2 A schematic of the VLSM workflow. On the top row, a series of different patients' normalized lesions and their corresponding behavioral scores are shown. For the specific voxel shown (cube), some patients' lesions encompass this voxel, and others' lesions do not. Those with lesions encompassing the voxel have numerically lower scores than those whose lesions spare this voxel. A statistical test (typically a *t*-test or regression) is run at this voxel to determine whether these relative score differences are statistically significant. This procedure is repeated across all voxels, and statistical correction for multiple tests is applied. The resulting VLSM map (bottom row) shows which voxels are significantly related to behavioral performance, based on a predefined statistical threshold and correction for multiple comparisons. Cooler colors on the map correspond to relatively lower (but still significant) *t*-values, and hotter colors correspond to higher *t*-values, as indicated by the legend

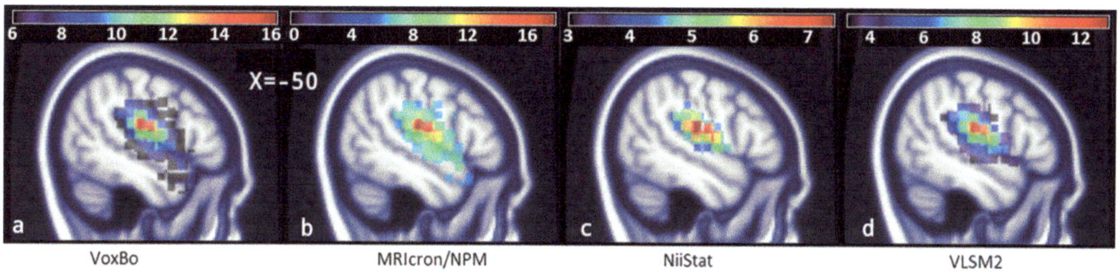

Fig. 3 VLSM maps generated from synthetic data and lesion masks from 131 patients in the Moss Rehabilitation Research Institute (MRRI) database, using four distinct statistical tests and permutation targets to identify the critical brain region associated with a synthetic behavior variable targeting a ventral frontal ROI: (**a**) VoxBo using a Welch *t*-test with maximum statistic permutation thresholding (www.nitrc.org/projects/voxbo); (**b**) MRIcron/NPM using the Brunner-Munzel statistic and a maximum statistic permutation test [93]; (**c**) NiiStat using the General Linear Model plus maximum-based permutation testing (www.nitrc.org/projects/niistat); (**d**) VLSM2 using a linear regression and cluster size-based (fixed $p < 0.001$) permutation test thresholding (https:/aphasialab.org/vlsm/). All maps covaried lesion volume and were derived using permutation testing to establish a $p = 0.05$ family-wise alpha. The color bars in the legend indicate increasing *t*-values (**a**, **c**, and **d**) or *z*-values (**b**), from cooler to hotter colors

the specific gray and white matter regions corresponding to the voxels significantly associated with the behavior/symptom. Such atlases include the automated anatomical atlas (AAL [109], Harvard-Oxford structural atlases (https://neurovault.org/collections/262/), Johns Hopkins University white matter atlases (https://neurovault.org/collections/264/), and Natbrainlab white matter atlases (https://www.natbrainlab.co.uk/atlas-maps).

Early versions of VLSM relied mostly on parametric statistics (e.g., *t*-tests for continuous behavioral data and chi-square for binomial data), but since assumptions of these tests can be easily violated in numerous voxels, other nonparametric tests were proposed. These include the Brunner-Munzel test for continuous behavioral data and the Liebermeister test for binomial data [91]. Ideally, the data should be evaluated to determine the most appropriate statistic, although this option is not yet commonly found among available VLSM software packages. Whether *t*-tests or other types of statistical tests are used in VLSM, the output is relatively stable. Figure 3 shows VLSM results using four different statistical tests from four different software packages on the same synthetic dataset. As can be seen, all four VLSM maps (a–d) show a relatively similar distribution with the same critical focus in left lateral frontal cortex regardless of the specific test used, showing the robustness and consistency of the VLSM technique.

There are also a variety of critical threshold criteria that can be used to identify significant foci in VLSM [118]. These include (1) maximum voxel statistic to identify localized critical hotspots (e.g., x-y-z location of peak *t*-value), (2) submaximal voxel statistic to identify moderate-sized clusters (typically ≥50 voxels to be considered reliable), and (3) largest spatially contiguous clusters of modest voxel-level significance to identify multiple regions of potential loci of behavior [114].

4 Advantages of VLSM

The VLSM method has many advantages over earlier lesion overlay techniques. Most prominently, VLSM takes full advantage of all the dimensions of the available behavioral and lesion data, making it a powerful, sensitive tool that minimizes bias. VLSM was specifically designed to handle continuous independent variables, such as scores on language and cognitive tests, eliminating the need for grouping patients by clinical syndromes or cutoff scores that indicate a presence or absence of a specific impairment. Such groupings can be problematic, both because they constrain the variability often present in a given syndrome (e.g., patients with Broca's aphasia can demonstrate dramatically different language profiles) and because such groupings can introduce subjectivity when designating a cutoff score.

VLSM also does not require dividing patients based on lesion location (e.g., frontal vs. temporal, anterior vs. posterior). Since patients' lesions rarely fall into neat borders and are rarely limited to a circumscribed brain region, grouping lesions by broad location when using lesion overlay methods not only leads to an exclusion of a large number of patients but also limits the precision of any spatial inferences that can be derived from the data. In contrast, VLSM can identify neural dissociations in closely neighboring regions. For example, Fig. 4 shows VLSM findings showing adjacent but distinct regions for speech production versus recognition [76].

VLSM also has the flexibility to handle different types of lesion and imaging data. Lesion data are typically traced on digital MRI T1, T2, and/or FLAIR images but can also be reconstructed from clinical scans such as CT and lower-resolution MRI scans, thus increasing potential sample sizes. VLSM studies in stroke patients have most often focused on the chronic phase of recovery when lesion boundaries are relatively clear (i.e., residual edema and bleeding have cleared; *see* Chapter 1 and 2), but some recent studies have applied VLSM to the analysis of data from acute stroke patients using MRI diffusion-weighted imaging, e.g., [43, 55, 71, 99]. Analyses of acute patient data have the benefit of mapping brain-behavior relationships prior to any significant neural reorganization [63, 73], providing an important complement to chronic stroke studies.

While the majority of VLSM studies have been conducted on stroke lesion patients, VLSM has also been used to analyze data from other types of neurologic patients, such as individuals with multiple sclerosis or traumatic brain injury and patients who

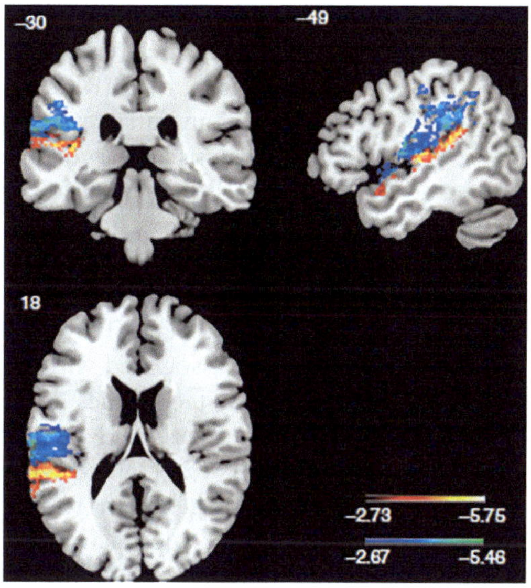

Fig. 4 VLSM findings from Mirman et al. [76] showing distinct regions for speech production (blue-green) versus speech recognition (red-yellow) in closely adjacent brain regions

undergo surgical resection for tumors or epilepsy [33, 94, 96, 100, 116]. These latter studies have assessed both the behavioral effects of lesions prior to surgery and the postsurgical impact of the resulting resections. VLSM findings from resection patients have generally been consistent with VLSM findings from stroke patients, reinforcing the reliability of the VLSM method [116]. Since stroke lesion patterns are constrained by the vascular tree and thus do not include all possible brain regions, analyzing data from these other types of lesions is critical to provide a more comprehensive analysis of brain-behavior relationships.

Another advantage of VLSM is its focus on identifying brain regions that are more centrally related to the behavior/symptom being measured. This stands in contrast to functional neuroimaging methods in healthy individuals such as fMRI, which identify a more extensive network of regions correlated with performance on a given task, but which might not be causally related (for an extended discussion, *see* [60, 110]). Therefore, lesion analysis techniques such as VLSM and functional neuroimaging findings are complementary. Since both VLSM and fMRI studies report results in terms of significant voxels on a common brain template (e.g., MNI space), results from the two complementary methodologies can be directly compared. Confirmation of findings goes in both directions, with results from fMRI used to bolster VLSM findings and vice versa. For example, a common area of involvement in the right inferior parietal sulcus was identified on tasks of spatial selective attention, using VLSM in stroke patients and fMRI in healthy participants ([80]; also *see* [110] for an example of VLSM-fMRI comparisons in the area of executive functioning). Given the more extensive nature of fMRI activation maps, fMRI findings might be considered as an initial "spatial guide" for areas to be investigated in a VLSM analysis and accordingly ensure sufficient lesion coverage.

While VLSM has clear advantages over earlier lesion overlay techniques, there is limited research on the advantages of VLSM over more recent multivariate lesion symptom mapping (MLSM) techniques [115] (*see* Chapter 11). Recent research from our group has suggested that VLSM is better at identifying a single critical lesion focus associated with a particular behavior/symptom, while most MLSM techniques are better at identifying multiple critical foci in a network and more diffuse brain-behavior relationships [59, 74, 88, 121]. VLSM also typically requires a smaller sample size [59, 103], but more work in this area is needed to directly compare these approaches.

5 VLSM Recommendations and Caveats

As with any method, there are several issues that need to be considered during the planning, execution, and interpretation of VLSM analyses (also *see* Chapter 6). Here, we lay out best practice recommendations currently in the field, but these guidelines continue to evolve over time as new improvements and increased computing power become available. First, VLSM requires a large patient sample to provide rigorous, repeatable results. The minimum sample size for reliable VLSM results depends on effect sizes and variability in the behavioral data, but modeling work done by our group suggests that a sample size of at least 50 patients is typically needed for sufficient power and robust results [59, 72]. This relatively large sample size can be achieved by aggregating patient data over time or by conducting multicenter studies. The size of individual lesions also has an impact on power, as 50 small lesions may have little overlap and thus low statistical power, while 50 large lesions may be sufficient.

As described above, a rigorous statistical correction is necessary to control family-wise error, given the large number of statistical tests in VLSM. The current gold standard is permutation test thresholding of voxel-wise parameters. (Procedures such as false discovery rate (FDR) and Bonferroni correction are inappropriate correction procedures for VLSM, the former due to the presence of negative correlations among voxels in lesion data and the latter due to the severe reduction in statistical power [9, 16, 68, 77, 91].) Permutation testing involves randomly reassigning the behavioral data across all patients' lesions to identify the incidence of extreme observations under the null hypothesis (i.e., no relationship between behavioral score and lesioned voxel). Typically,

1000–10,000 permutations are run, and a critical threshold is identified (e.g., alpha = 0.05), below which the values are likely spurious. With permutation testing, the assumption of a normal distribution does not have to be satisfied, so rather than having to choose a statistic based on a probability distribution to fit the general linear model, the most behaviorally meaningful statistic can be chosen. This flexibility is especially important when using clinical behavioral measures that have ceiling/floor effects that cannot be transformed into unimodal distributions (e.g., with zero-inflated distributions). VLSM maps generated with permutation testing control false positives regardless of data distribution; however, the statistic should take into account the data distribution in order to boost detection power. Figure 5 shows a series of VLSM maps of single-word auditory comprehension at varying levels of statistical correction, highlighting the significant impact of permutation-based correction in focusing the critical voxel cluster from a more diffuse distribution to a core region in the left mid-posterior middle temporal gyrus [12].

Another critical issue in VLSM is adjusting for potentially confounding variables that could affect interpretation of the data. Nuisance variables such as demographic data (e.g., age, gender, etc.) should be included as covariates, power-permitting (*see* Fig. 5). The most important covariate that should routinely be included in all VLSM analyses is lesion volume. Figure 6 shows the impact of covarying for lesion volume in a VLSM analysis: the left side of the figure shows a circumscribed cluster of significant voxels when lesion volume is included as a covariate, and the right side of the figure shows much more diffuse results in the same data without covarying for lesion volume. Importantly, the location of the peak *t*-values in both maps is the same (shown in red), indicating the stability of this peak *t* metric.

Covarying for lesion volume minimizes the effect of non-spatially-specific gray matter damage, thus significantly reducing spatial bias [37, 102]. This is because some voxels are more likely to be present in larger lesions, and larger lesions are more likely to affect overall behavioral performance. Furthermore, including lesion volume as a covariate allows one to partially account for any vasculature distribution effects. Figure 7 shows the high degree of correlation in many brain regions between a patient's lesion volume and whether a particular voxel is lesioned. These spatially inhomogeneous correlations indicate the importance of using lesion size as a covariate in VLSM analyses to minimize false positives and potential cluster shift (*see* [102]).

Behavioral variables can also be included as covariates in VLSM to provide a more focused analysis. For example, [11] conducted a VLSM analysis of picture naming in 96 left hemisphere stroke patients. The VLSM map associated with overall picture naming performance consisted of a large network of frontotemporal

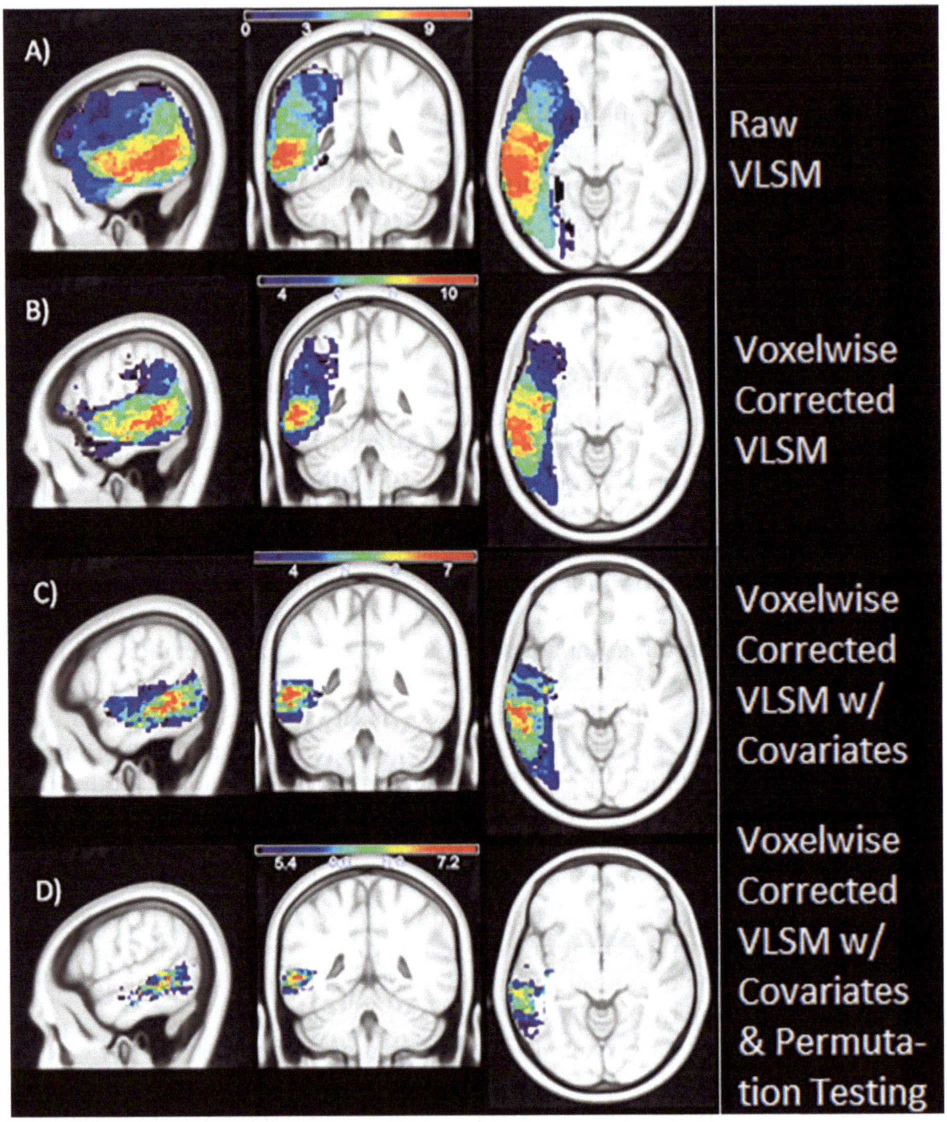

Fig. 5 VLSM maps of single-word auditory comprehension in 109 stroke patients with varying levels of correction and covariates: (**a**) raw t-map with no voxel-wise correction, no permutation testing, and no covariates; (**b**) voxel-wise-corrected *t*-map ($p < 0.001$) with no permutation testing and no covariates; (**c**) voxel-wise-corrected *t*-map ($p < 0.001$) with lesion volume, education, and age as covariates but no permutation testing; and (**d**) permutation testing-derived *t*-map with lesion volume, education, and age as covariates. The colored bars represent the range of significant *t*-values for each analysis, from lower (though still significant) *t*-values shown in purple to higher *t*-values shown in red. (Reprinted from [12])

cortical regions and underlying white matter (*see* Fig. 8, top row). It was of greater interest, however, to isolate regions specifically associated with lexical-semantic retrieval processes. Thus, a VLSM map of the same naming data was generated, using speech production and visual recognition scores as covariates to account for variability

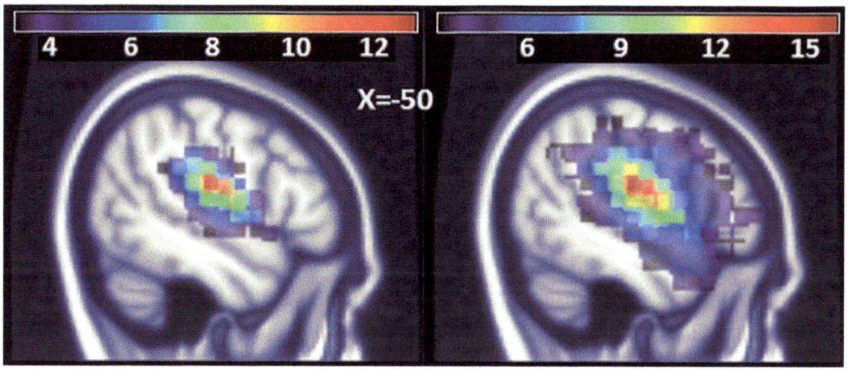

Fig. 6 VLSM results with and without covarying for lesion volume (left and right, respectively), using linear regression and cluster-sized based (fixed $p < 0.001$) permutation test thresholding. Synthetic behavior derived from the MRRI dataset and processed at 4 mm isotropic voxel resolution

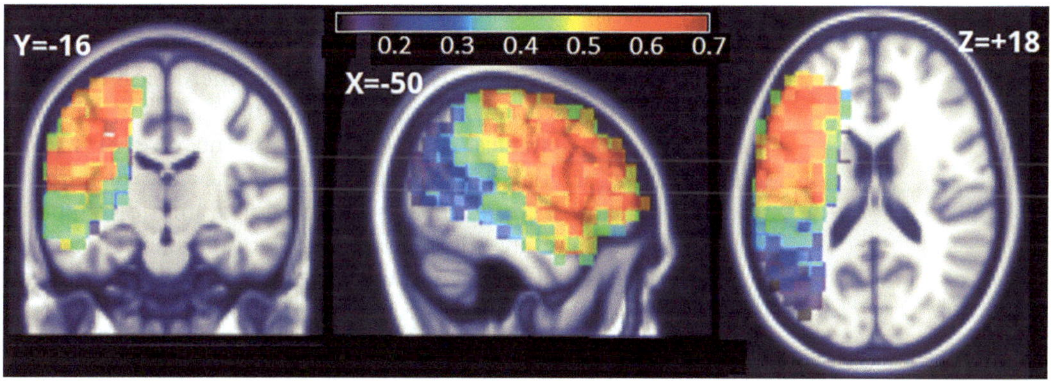

Fig. 7 Voxel-wise Pearson correlations between lesion volume and lesion status at each voxel, using 131 lesion masks from the MRRI database

due to extraneous processes outside of lexical-semantic retrieval. As shown in the bottom row of Fig. 8, the resultant VSLM map highlighted a more circumscribed region of significant voxels in the left mid-posterior middle temporal gyrus, the same area that has been suggested to serve as a hub for lexical-semantic processing [107].

Lesion status in particular voxels or brain regions can also be used as covariates in VLSM to isolate the most critical areas for a particular behavior [117]. For example, the original VLSM paper by Bates et al. [15] identified the left insula and the left middle temporal gyrus as critical regions for speech production and comprehension, respectively. Since these processes are traditionally associated with Broca's and Wernicke's area, respectively, Bates et al. included voxels in these regions as covariates in the VLSM analysis. They found that the left insula was still critical for speech fluency after voxels in Broca's area were covaried, but that Broca's area was not strongly related to fluency after voxels in the left insula

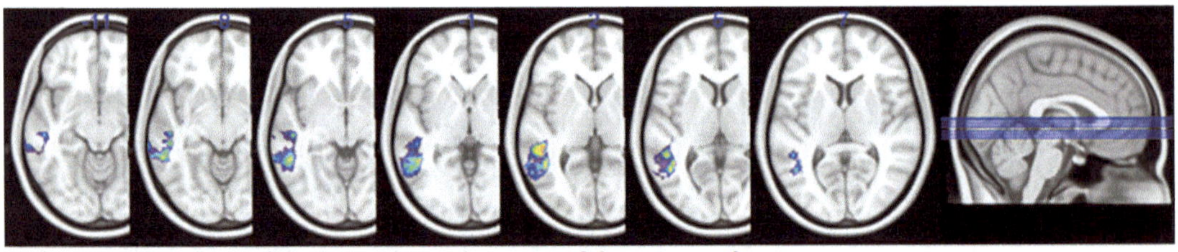

Fig. 8 VLSM map showing significant voxels associated with overall picture naming scores after covarying for speech production and visual recognition scores. All voxels shown in color exceeded the critical threshold for significance, and the colors reflect increasing t-values from purple to red

were covaried. Similarly, the left middle temporal gyrus was still found to be strongly related to auditory comprehension after voxels in Wernicke's area were covaried, but Wernicke's area was not strongly related to auditory comprehension once voxels in the middle temporal gyrus were covaried.

Another consideration with VLSM (and other lesion mapping techniques) is that lesion distributions are not random, which is problematic for statistical analyses [74]. In the case of stroke, lesions follow arterial distributions, and in the case of surgical resections, the most commonly resected region is anterior temporal cortex. Thus, there are substantial spatial contingencies between voxels: if one voxel is lesioned, then the adjacent voxel is likely to be lesioned as well, but not necessarily vice versa. Due to this issue, there has been concern that VLSM and other lesion mapping techniques may be systematically biased toward specific areas where stroke lesions tend to converge due to vascular configuration [57, 74, 101]. However, recent simulation work by our group and others using synthetic behavioral data has empirically demonstrated the robust ability of VLSM to identify critical foci well beyond the common "stroke zones" [59, 88, 102]. Furthermore, VLSM studies with real behavioral data have shown distinct regions critical for different cognitive and language processes that do not overlap with common lesion convergence zones [11, 15, 38, 42, 60, 76]. For example, in the picture naming study described above [11], the stroke convergence zone was left fronto-insular cortex (top row, Fig. 9), but the VLSM focus associated with lexical-semantic retrieval was in a completely distinct region in left mid-posterior middle temporal cortex (bottom row).

A related issue is whether VLSM can account for the potential impact of remote lesion effects, such as white matter fiber disconnection [26, 27, 53]. That is, lesions can lead to disconnection between distant areas that are intact but not functioning properly as a result of a disconnection. This concern is somewhat allayed by the fact that many VLSM studies have identified white matter foci that are significantly related to distinct behaviors and symptoms,

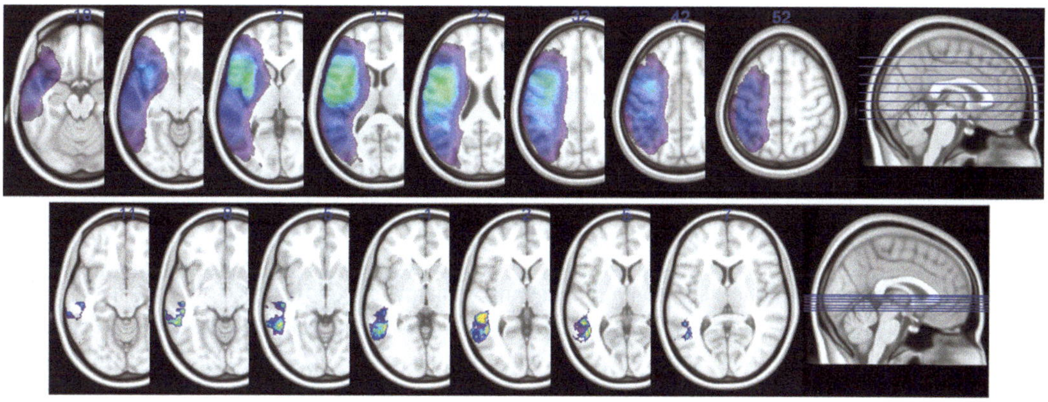

Fig. 9 Top row shows lesion overlay of 96 left stroke patients in a picture naming study [11] with a typical pattern of maximal lesion convergence in left fronto-insular cortex. The bottom row shows the VLSM map of impaired picture naming after covarying for speech production and visual recognition, in which significant voxels were confined to the left mid-posterior middle temporal gyrus and underlying white matter. The colors show increasing *t*-values from purple to red

beginning with the initial VLSM study [15] (also *see* [62, 66, 94]). Also, counter to some theoretical concerns about the potentially problematic scenario of lesions intersecting at oblique angles in disperse directions [27], stroke lesions typically overlap to a large extent in wide swaths of white matter so that VLSM can detect critical white matter sites. Still, detection of white matter disconnection effects can be strengthened with the application of additional methods (e.g., [10, 38, 44]; also *see* Chapter 9). For example, combined VLSM and probabilistic tractography was used to identify a network of cortical regions and white matter pathways underlying different aspects of syntactic processing in the left hemisphere [38].

Another important consideration in VLSM analyses is constraining interpretation to brain regions with sufficient lesion coverage and statistical power. Analyses should be limited to voxels that are lesioned in at least 5–10% of the patient sample (and no more than 90–95% of the sample) to avoid lopsided/unstable statistical tests and low power [102]. Lesion coverage and power maps should be provided in publications to show the extent of brain regions under consideration in a particular analysis, either a priori (e.g., based on a given effect size) or post-analysis (based on actual effect sizes in the data). Examples of both types of power maps are shown in Fig. 10. Conclusions drawn from VLSM maps should be limited to those brain regions with sufficient statistical power, avoiding interpretation of null findings outside these boundaries. Also, if there are a priori hypotheses about involvement of specific brain areas, sufficient lesion coverage/power in those regions should be confirmed. Large sample sizes with diverse lesion coverage can help ensure sufficient statistical power.

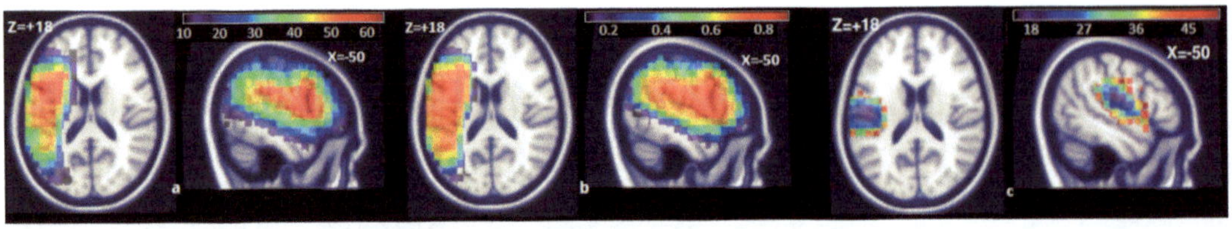

Fig. 10 Maps generated from 131 MRRI lesion masks using synthetic behavioral data and targeting a left lateral frontal region: (**a**) lesion overlay showing the number of patients who have a lesion in each voxel; (**b**) pre-analysis power map showing the fraction of times that a standardized effect size of 1 would result in a $p < 0.0001$ result at a given voxel; (**c**) post-analysis power map showing the number of participants needed at a given voxel to have an 80% chance of achieving a $p < 0.01$ regression result, given patients drawn from the same population

Another important metric that can be displayed using VLSM is the amount of statistical variance explained by the data, that is, how much variance is accounted for at each voxel by each behavioral variable. This approach can also provide information as to the location and amount of variance accounted for by each of the covariates. A general recommendation is that each voxel should account for at least 5% of the variance to be considered in data interpretation. As an example, Fig. 11 shows the amount of variance explained by the main variable of interest in a picture naming study, overall naming score (blue), versus the variance explained by a fluency covariate (red).

6 Contributions of VLSM to Brain-Behavior Mapping

VLSM has allowed scientists to address a wide range of research questions, adding to our theoretical understanding of brain-behavior relationships across a variety of different behavioral domains [4–13, 15, 17, 19, 22, 28, 29, 31, 32, 43, 50, 60, 61, 65, 82, 89, 92, 95, 113, 116, 119]. In particular, VLSM has made a large impact on the area of speech and language mapping (*see* Table 1 for a sample of recent VLSM studies of language). As shown in Table 1, VLSM has been used to assess a wide range of questions, from identifying neural correlates of lower-level processes such as phonology and articulation to higher-level processes such as reading and working memory. In some cases, VLSM studies have confirmed previously identified brain regions critical for particular aspects of speech and language, and in other cases, VLSM studies have challenged those findings.

The original VLSM paper [15] not only introduced this new technique but also updated our understanding of the neural correlates of speech and language. In that study, Bates and colleagues found that speech production was most strongly associated with the left insula and underlying white matter, and auditory

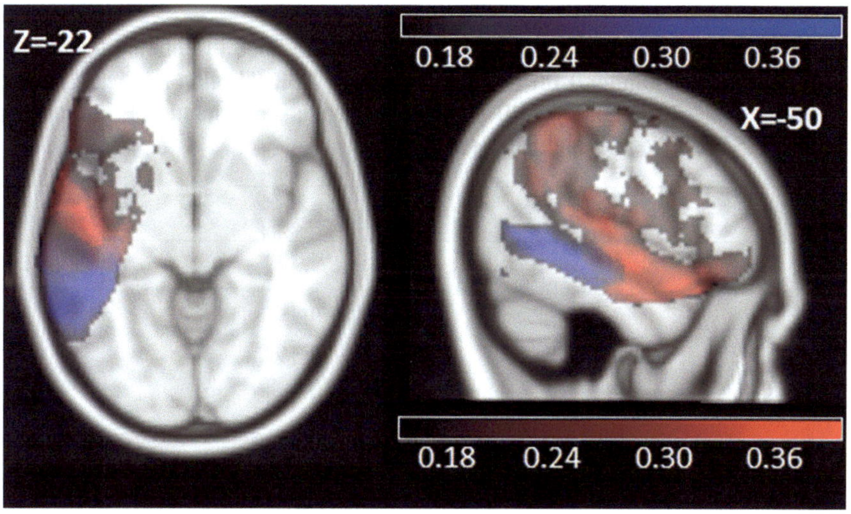

Fig. 11 VLSM maps showing the fraction of variance explained at each voxel by the overall picture naming variable (blue) and a fluency covariate (red). Variance was determined with a tenfold cross validation predictive residual sums squared (PRESS) vs. residual sums squared [56]

comprehension was most strongly associated with the left middle temporal gyrus. Counter to previous findings, traditional Wernicke's area (i.e., posterior superior temporal gyrus and inferior parietal cortex) was not among the key regions identified for comprehension at the sentence level. In this way, VLSM elaborated on classic models of speech and language that tend to focus primarily on Broca's and Wernicke's areas for speech production and comprehension, respectively. In a follow-up study, VLSM was used to identify a broad network of left hemisphere regions underlying distinct aspects of auditory comprehension based on specific grammatical constructions [42]. Comprehension of basic possession (e.g., *The dog has spots*) and simple declarative sentences (e.g., *The girl is sitting*) was associated with the left middle temporal gyrus and underlying white matter. More complex grammatical constructions such as double embedding (e.g., *The clown that is little has the flower that is red*) and subject relatives (e.g., *The girl who is pushing the boy is happy*) were additionally associated with left lateral prefrontal cortex, possibly due to its role in verbal working memory.

Another important study identified two dissociations in the language network with VLSM in a large cohort of chronic stroke patients with a range of aphasia severity [76]. The first dissociation was the identification of distinct but parallel networks for speech production and recognition that run superior and inferior to the left Sylvian fissure, respectively. The second dissociation was separate networks in left peri-Sylvian and extra-Sylvian regions for phonological versus semantic processing, respectively (also *see* [78] for a review of this type of approach). Similar findings were obtained in a recent VLSM study by [46], highlighting a

Table 1
VLSM studies of speech and language in stroke patients since 2009 in PubMed using the following search string: (("Lesion Parameter") OR ("Lesion Symptom") OR ("Lesion Behavior") OR ("Lesion Behaviour") OR ("VLSM" OR "VLBM")) AND Voxel AND (Stroke OR Ischemic OR Hemorrhagic). For a review of earlier VLSM studies of language, *see* [10]

Language domain	References	PubMed ID
Reasoning and language	Baldo et al. [7]	20206985
Articulation	Baldo et al. [8]	20691968
Repetition and verbal working memory	Baldo et al. [9]	24976669
Picture naming	Baldo et al. [11]	22482693
Reading and writing	Baldo et al. [13]	29572061
Semantic and phonemic fluency	Biesbroek et al. [17]	25939335
Surface errors in acquired dyslexia	Binder et al. [18]	26966139
Phonologic and semantic verbal fluency	Chouiter et al. [30]	27217213
Syntactic processing	den Ouden et al. [38]	30666767
Recovery from acute aphasia	Forkel and Catani [43]	29605593
Correlates of Broca's aphasia	Fridriksson et al. [47]	25016386
Multiple linguistic domains	Fridriksson et al. [48]	27956600
Inner speech	Geva et al. [49]	21975590
Lexical and semantic access	Harvey and Schnur [51]	25880795
Multiple linguistic domains	Henseler et al. [52]	24525451
Apraxia of speech in acute stroke	Itabashi et al. [58]	26645260
Verbal working memory	Ivanova et al. [60]	29526647
Syntactic comprehension	Kim et al. [67]	20543453
Complex syntactic processing	Magnusdottir et al. [73]	22522937
Multiple linguistic domains	Mirman et al. [76]	25879574
Speech comprehension	Pillay et al. [84]	28179469
Word and number reading	Piras and Marangolo [85]	19428427
Speech repetition	Rogalsky et al. [90]	25777496
Semantic word retrieval	Schwartz et al. [97]	19942676
Phonological retrieval in object naming	Schwartz et al. [98]	23171662
Error patterns in connected speech	Stark et al. [104]	31005024
Reversible sentence comprehension	Thothathiri et al. [105]	21861679
Praxis and language	Weiss et al. [111]	25352157

dissociation between speech comprehension and production tasks along the two streams. These findings complement earlier findings from progressive dementia patients and also refined earlier theoretical models of speech and language proposing separate dorsal and ventral pathways [54].

7 Summary and Future Directions

Historically, lesion studies have formed the basis of our understanding of the neural correlates of behavior. Even in the current era of neuroimaging techniques such as fMRI, lesion mapping continues to provide critical insights into the effects of neurologic disease, as well as our conceptualization of the functional organization of the healthy brain. Recent techniques such as VLSM have enabled us to more rigorously identify the relationship between behavior and different brain structures, including both gray and white matter regions. It is critical that we continue to develop and refine these methods that provide increasingly reliable findings. Furthermore, it is critical to integrate these methods with other imaging techniques [38, 40, 93, 106, 107, 110] (*see* Chapters 10, 11, 14, 15). Multimodal neuroimaging within the realm of a single study can help to corroborate findings and mitigate limitations of each method. For example, VLSM can be combined with techniques such as resting-state fMRI and tractography to address the question of functional disconnection and remote lesion effects, respectively (*see* Fig. 12) [38, 44, 109]. Indeed, there is a growing interest to use more functional atlases that do not rely solely on brain morphology but rather reflect functional specialization of distinct brain regions. Future VLSM studies should also strive to integrate lesion data derived from multiple etiologies with distinct and complementary lesion distributions, in order to better characterize the role of brain regions across the entire brain. Finally, future endeavors should include collaboration across multiple research centers to generate larger patient samples that will provide greater statistical power [86].

In addition to furthering our basic understanding of brain-behavior relationships, VLSM also holds promise in the realm of clinical applications. It has been used in longitudinal recovery studies of stroke to identify lesion sites that are predictive of better/worse long-term behavioral outcomes [43, 62]. Such information can help guide treatment planning and provide more accurate education to patients and their caregivers. VLSM has also been used to assess the effects of surgical resection in epilepsy and tumor patients, which can lead to more accurate prognosis and better clinical outcomes [100, 116]. Finally, VLSM can be combined with techniques that have potential for treatment intervention, such as transcranial stimulation, deep brain stimulation, and

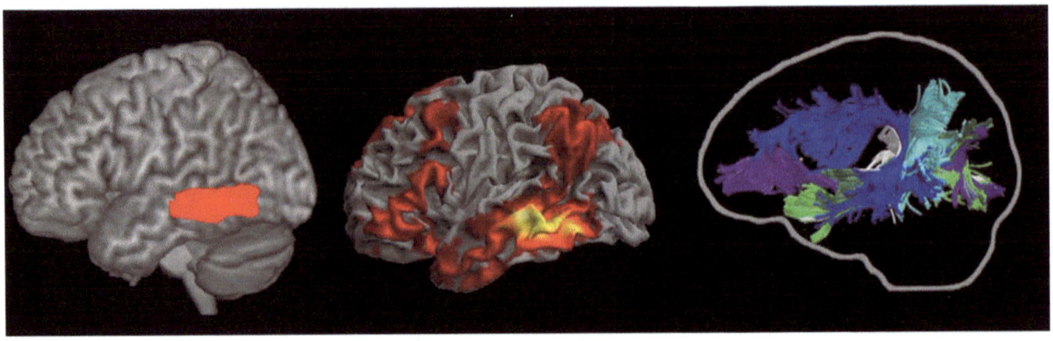

Fig. 12 Turken and Dronkers [110] used VLSM to first identify a critical region associated with auditory comprehension (left) and then mapped the associated functional and structural connectivity networks associated with that region using resting-state fMRI (middle) and diffusion tensor imaging (right), respectively

robot-assisted motor rehabilitation (Chapters 14 and 15) [25, 28, 82]. For example, VLSM was used in combination with transcranial direct current stimulation (tDCS) in a group of non-fluent aphasic patients receiving speech therapy, in order to identify which specific frontal regions were associated with a better behavioral outcome [25]. Such information can help tailor the use of specific treatments to those individuals who can most readily benefit from them.

Future advances in the development of newer statistical techniques, software applications, automated pipelines, and increasing computing power will continue to fuel progress in the field of brain-behavior mapping [2, 87]. However, as we explore new computational horizons, we should not forgo the basic study design principles underlying lesion analysis: appropriate patient selection, precise lesion delineation, assurance of sufficient lesion coverage, and the use of sensitive and reliable behavioral measures all help to form the foundations of any lesion study, irrespective of the chosen lesion-symptom mapping technique. Such carefully designed and executed lesion-symptom mapping studies will continue to expand our knowledge about the neural foundations of behavior in the coming decades.

Acknowledgments

This material is based on work supported in part by the US Department of Veterans Affairs, Office of Research and Development CSR&D Grant (CX000254-04). MI contribution was also funded by the Center for Language and Brain NRU Higher School of Economics, RF Government grant, ag. No. 14.641.31.0004. The contents reported/presented within do not represent the views of the Department of Veterans Affairs or the United States Government. We would like to thank our lab colleagues for their comments on earlier versions of the manuscript.

References

1. Adolphs R, Damasio H, Tranel D, Cooper G, Damasio A (2000) A role for somatosensory cortices in the visual recognition of emotion as revealed by three-dimensional lesion mapping. J Neurosci 20:2683–2690
2. Arnoux A, Toba MN, Duering M, Diouf M, Daouk J, Constans JM, Puy L, Barbay M, Godefroy O (2018) Is VLSM a valid tool for determining the functional anatomy of the brain? Usefulness of additional Bayesian network analysis. Neuropsychologia 121:69–78
3. Ashburner J, Friston K (2000) Voxel-based morphometry—the methods. NeuroImage 11:805–821
4. Baldo JV, Dronkers NF, Wilkins D, Ludy C, Raskin P, Kim J (2005) Is problem solving dependent on language? Brain Lang 92: 240–250
5. Baldo J, Schwartz S, Wilkins D, Dronkers N (2006) Role of frontal versus temporal cortex in verbal fluency as revealed by voxel-based lesion symptom mapping. J Int Neuropsychol Soc 12:896–900
6. Baldo J, Dronkers N (2007) Neural correlates of arithmetic and language comprehension: a common substrate? Neuropsychologia 45: 229–235
7. Baldo JV, Bunge SA, Wilson SM, Dronkers NF (2010) Is relational reasoning dependent on language? A voxel-based lesion symptom mapping study. Brain Lang 113(2):59–64
8. Baldo JV, Wilkins DP, Ogar J, Willock S, Dronkers NF (2011) Role of the precentral gyrus of the insula in complex articulation. Cortex 47(7):800–807
9. Baldo JV, Katseff S, Dronkers NF (2012) Brain regions underlying repetition and auditory-verbal short-term memory deficits in aphasia: evidence from voxel-based lesion symptom mapping. Aphasiology, 26(3–4): 338–354. https://doi.org/10.1080/02687038.2011.602391 [PubMed: 24976669]
10. Baldo J, Wilson S, Dronkers N (2012) Uncovering the neural substrates of language: a voxel-based lesion symptom mapping approach. To appear in M. Faust (ed), Advances in the neural substrates of language: Toward a synthesis of basic science and clinical research. Wiley-Blackwell, Oxford
11. Baldo JV, Arévalo A, Patterson JP, Dronkers NF (2013) Grey and white matter correlates of picture naming: evidence from a voxel-based lesion analysis of the Boston Naming Test. Cortex 49(3):658–667
12. Baldo JV, Dronkers NF (2018) Lesion studies. In: de Groot A, Hagoort P (eds) Research methods in psycholinguistics and the neurobiology of language. Wiley Blackwell, Hoboken
13. Baldo JV, Kacinik N, Ludy C, Paulraj S, Moncrief A, Piai V, Curran B, Turken A, Herron T, Dronkers NF (2018) Voxel-based lesion analysis of brain regions underlying reading and writing. Neuropsychologia 115: 51–59
14. Basso A, Lecours AR, Moraschini S, Vanier M (1985) Anatomoclinical correlations of the aphasias as defined through computerized tomography: exceptions. Brain Lang 26: 201–229
15. Bates E, Wilson S, Saygin A, Dick F, Sereno MI, Knight R, Dronkers N (2003) Voxel-based lesion-symptom mapping. Nat Neurosci 6:448–450
16. Benjamini Y, Yekutieli D (2001) The control of the false discovery rate in multiple testing under dependency. Ann Stat 29(4): 1165–1188. https://doi.org/10.1214/aos/1013699998. MR 1869245
17. Biesbroek JM, van Zandvoort MJ, Kappelle LJ, Velthuis BK, Biessels GJ, Postma A (2016) Shared and distinct anatomical correlates of semantic and phonemic fluency revealed by lesion-symptom mapping in patients with ischemic stroke. Brain Struct Funct 221(4):2123–2134
18. Binder JR, Pillay SB, Humphries CJ, Gross WL, Graves WW, Book DS (2016) Surface errors without semantic impairment in acquired dyslexia: a voxel-based lesion–symptom mapping study. Brain 139(5): 1517–1526
19. Borovsky A, Saygin AP, Bates E, Dronkers NF (2007) Lesion correlates of conversational speech production deficits. Neuropsychologia 45(11):2525–2533
20. Bouillaud MJ (1825) Recherches cliniques propres a demontrer que la perte de la parole correspond a la lesion des lobules anterieurs du cerveau, et a confirmer l'opinion de M. Gall, sur le siege de l'organe du langage articule. Archives Generales de Medecine 3: 25–45
21. Brett M, Leff AP, Rorden C, Ashburner J (2001) Spatial normalization of brain images with focal lesions using cost function masking. NeuroImage 14:486–500
22. Bright P, Moss HE, Longe O, Stamatakis EA, Tyler LK (2007) Conceptual structure

modulates anteromedial temporal involvement in processing verbally presented object properties. Cereb Cortex 17:1066–1073

23. Broca P (1861) Remarques sur le sie'ge de la faculte´ du langage articule´, suivies d'une observation d'aphe'mie (perte de la parole). Bulletins de la Socie'te´ d'Anatomie (Paris), 2e serie 6:330–357

24. Broca P (1865) Sur le siege de la faculte du langage articule. Bulletin de la Societe d'Anthropologie 6:337–393

25. Campana S, Caltagirone C, Marangolo P (2015) Combining voxel-based lesion-symptom mapping (VLSM) with A-tDCS language treatment: predicting outcome of recovery in nonfluent chronic aphasia. Brain Stimul 8(4): 769–776

26. Carrera E, Tononi G (2014) Diaschisis: past, present, future. Brain 137(9):2408–2422. https://doi.org/10.1093/brain/awu101

27. Catani M, Mesulam MM (2008) The arcuate fasciculus and the disconnection theme in language and aphasia: history and current state. Cortex 44(8):953–961. https://doi.org/10.1016/j.cortex.2008.04.002.The

28. Cerasa A, Pignolo L, Gramigna V, Serra S, Olivadese G, Roccca F, Perrotta P, Dolce G, Quattrone A, Tonin P (2018) Exoskeleton-robot assisted therapy in stroke patients: a lesion mapping study. Front Neuroinform 12:44

29. Chatterjee A (2008) The neural organization of spatial thought and language. Semin Speech Lang 29:226–238

30. Chouiter L, Holmberg J, Manuel AL, Colombo F, Clarke S, Annoni JM, Spierer L (2016) Partly segregated cortico-subcortical pathways support phonologic and semantic verbal fluency: a lesion study. Neuroscience 329:275–283

31. Committeri G, Pitzalis S, Galati G, Patria F, Pelle G, Sabatini U, Castriota-Scanderbeg A, Piccardi L, Guariglia C, Pizzamiglio L (2007) Neural bases of personal and extrapersonal neglect in humans. Brain 130:431–441

32. Corbetta M, Ramsey L, Callejas A, Baldassarre A, Hacker CD, Siegel JS, Serguei A, Rengachary J, Zinn K, Lang C, Connor L, Fucetola R, Strube M, Carter A, Shulman G (2015) Common behavioral clusters and subcortical anatomy in stroke. Neuron 85(5):927–941

33. Dal Monte O, Krueger F, Solomon JM, Schintu S, Knutson KM, Strenziok M, Pardini M, Leopold A, Raymont V, Grafman J (2013) A voxel-based lesion study on facial emotion recognition after penetrating brain injury. Soc Cogn Affect Neurosci 8(6): 632–639

34. Damasio H, Damasio A (1979) "Paradoxic" ear extinction in dichotic listening: possible anatomic significance. Neurology 29(5): 644–644

35. Damasio H, Damasio AR (1980) The anatomical basis of conduction aphasia. Brain 103: 337–350

36. Damasio H, Grabowski TJ, Tranel D, Hichwa RD, Damasio A (1996) A neural basis for lexical retrieval. Nature 380:499–505

37. DeMarco AT, Turkeltaub PE (2018) A multivariate lesion symptom mapping toolbox and examination of lesion-volume biases and correction methods in lesion-symptom mapping. Hum Brain Mapp 39(11):4169–4182

38. Den Ouden D-B, Malyutina S, Basilakos A, Bonilha L, Gleichgerrcht E, Yourganov G, Hillis AE, Hickok G, Rorden C, Fridriksson J (2019) Cortical and structural-connectivity damage correlated with impaired syntactic processing in aphasia. Hum Brain Mapp 40: 2153–2173. https://doi.org/10.1002/hbm.24514

39. DeWitt LD, Grek AJ, Buonanno FS, Levine DN, Kistler JP (1985) MRI and the study of aphasia. Neurology 35:861–865

40. Dick F, Saygin AP, Galati G, Pitzalis P, Bentrovato S, D'Amico S, Wilson S, Bates E, Pizzamiglio L (2007) What is involved and what is necessary for complex linguistic and non-linguistic auditory processing: evidence from fMRI and lesion data. J Cogn Neurosci 19:799–816

41. Dronkers NF (1996) A new brain region for coordinating speech articulation. Nature 384: 159–161

42. Dronkers NF, Wilkins DP, Van Valin RD Jr, Redfern BB, Jaeger JJ (2004) Lesion analysis of the brain areas involved in language comprehension. Cognition 92(1–2):145–177

43. Forkel SJ, Catani M (2018) Lesion mapping in acute stroke aphasia and its implications for recovery. Neuropsychologia 115:88–100

44. Foulon C, Cerliani L, Kinkingnehun S, Levy R, Rosso C, Urbanski M, Volle E, Thiebaut de Schotten M (2018) Advanced lesion symptom mapping analyses and implementation as BCBtoolkit. GigaScience 7(3):giy004

45. Frank RJ, Damasio H, Grabowski TJ (1997) Brainvox: an interactive, multimodal visualization and analysis system for neuroanatomical imaging. NeuroImage 5(1):13–30

46. Fridriksson J, den Ouden DB, Hillis AE, Hickok G, Rorden C, Basilakos A,

Yourganov G, Bonilha L (2018) Anatomy of aphasia revisited. Brain 141(3):848–862
47. Fridriksson J, Fillmore P, Guo D, Rorden C (2015) Chronic Broca's aphasia is caused by damage to Broca's and Wernicke's areas. Cereb Cortex 25(12):4689–4696
48. Fridriksson J, Yourganov G, Bonilha L, Basilakos A, Den Ouden DB, Rorden C (2016) Revealing the dual streams of speech processing. Proc Natl Acad Sci 113(52): 15108–15113
49. Geva S, Jones PS, Crinion JT, Price CJ, Baron JC, Warburton EA (2011) The neural correlates of inner speech defined by voxel-based lesion–symptom mapping. Brain 134(10): 3071–3082
50. Glascher J, Tranel D, Paul LK et al (2009) Lesion mapping of cognitive abilities linked to intelligence. Neuron 61:681–691
51. Harvey DY, Schnur TT (2015) Distinct loci of lexical and semantic access deficits in aphasia: evidence from voxel-based lesion-symptom mapping and diffusion tensor imaging. Cortex 67:37–58
52. Henseler I, Regenbrecht F, Obrig H (2014) Lesion correlates of patholinguistic profiles in chronic aphasia: comparisons of syndrome-, modality-and symptom-level assessment. Brain 137(3):918–930
53. Herbet G, Lafargue G, Duffau H (2015) Rethinking voxel-wise lesion-deficit analysis: a new challenge for computational neuropsychology. Cortex 64:413–416. https://doi.org/10.1016/j.cortex.2014.10.021
54. Hickok G, Poeppel D (2007) The cortical organization of speech processing. Nat Rev Neurosci 8(5):393
55. Hickok G, Rogalsky C, Chen R, Herskovits EH, Townsley S, Hillis AE (2014) Partially overlapping sensorimotor networks underlie speech praxis and verbal short-term memory: evidence from apraxia of speech following acute stroke. Front Hum Neurosci 8:649
56. Holiday D, Ballard J, McKeown B (1995) PRESS-related statistics: regression tools for cross-validation and case diagnostics. Med Sci Sports Exerc 27:612–620
57. Inoue K, Madhyastha T, Rudrauf D, Mehta S, Grabowski T (2014) What affects detectability of lesion-deficit relationships in lesion studies? NeuroImage Clin 6:388–397. https://doi.org/10.1016/j.nicl.2014.10.002
58. Itabashi R, Nishio Y, Kataoka Y, Yazawa Y, Furui E, Matsuda M, Mori E (2016) Damage to the left precentral gyrus is associated with apraxia of speech in acute stroke. Stroke 47(1):31–36
59. Ivanova M, Herron T, Dronkers N, Baldo J (2021) An empirical comparison of univariate versus multivariate methods for the analysis of brain-behavior mapping. Hum Brain Mapp 42(4):1070–1101. https://doi.org/10.1002/hbm.25278
60. Ivanova MV, Dragoy O, Kuptsova SV, Akinina SY, Petrushevskii AG, Fedina ON, Turken A, Shklovsky V, Dronkers NF (2018) Neural mechanisms of two different verbal working memory tasks: a VLSM study. Neuropsychologia 115:25–41
61. Karnath H, Berger M, Kuker W, Rorden C (2004) The anatomy of spatial neglect based on voxelwise statistical analysis: a study of 140 patients. Cereb Cortex 14:1164–1172
62. Karnath HO, Rennig J, Johannsen L, Rorden C (2011) The anatomy underlying acute versus chronic spatial neglect: a longitudinal study. Brain 134(3):903–912
63. Karnath HO, Rennig J (2017) Investigating structure and function in the healthy human brain: validity of acute versus chronic lesion-symptom mapping. Brain Struct Funct 222(5):2059–2070
64. Kertesz A, Harlock W, Coates R (1979) Computer tomographic localization, lesion size, and prognosis in aphasia and nonverbal impairment. Brain Lang 8:34–50
65. Kessner SS, Schlemm E, Cheng B, Bingel U, Fiehler J, Gerloff C, Thomalla G (2019) Somatosensory deficits after ischemic stroke: time course and association with infarct location. Stroke 50(5):1116–1123
66. Kim DH, Kyeong S, Do KH, Lim SK, Cho HK, Jung S, Kim HW (2018) Brain mapping for long-term recovery of gait after supratentorial stroke: a retrospective cross-sectional study. Medicine 97(16):e0453
67. Kim MJ, Jeon HA, Lee KM (2010) Impairments of syntactic comprehension in Korean and the location of ischemic stroke lesions: a voxel-based lesion-symptom mapping study. Behav Neurol 22(1–2):3–10
68. Kimberg DY, Coslett HB, Schwartz MF (2007) Power in voxel-based lesion-symptom mapping. J Cogn Neurosci 19:1067–1080
69. Knight RT, Grabowecky MF, Scabini D (1995) Role of human prefrontal cortex in attention control. Adv Neurol 66:21–36
70. Knight RT, Scabini D, Woods DL, Clayworth C (1988) The effects of lesions of superior temporal gyrus and inferior parietal lobe on temporal and vertex components of the

human AEP. Electroencephalogr Clin Neurophysiol 70:499–509

71. Kümmerer D, Hartwigsen G, Kellmeyer P, Glauche V, Mader I, Klöppel S, Suchan J, Karnath H, Weiller C, Saur D (2013) Damage to ventral and dorsal language pathways in acute aphasia. Brain 136(2):619–629
72. Lorca-Puls DL, Gajardo-vidal A, White J, Seghier ML, Le AP, Green DW, Crinion J, Ludersdorfer P, Hope T, Bowman H, Price CJ (2018) The impact of sample size on the reproducibility of voxel-based lesion-deficit mappings. Neuropsychologia 115 (March):101–111. https://doi.org/10.1016/j.neuropsychologia.2018.03.014
73. Magnusdottir S, Fillmore P, Den Ouden DB, Hjaltason H, Rorden C, Kjartansson O, Bonilha L, Fridriksson J (2013) Damage to left anterior temporal cortex predicts impairment of complex syntactic processing: a lesion-symptom mapping study. Hum Brain Mapp 34(10):2715–2723
74. Mah YH, Husain M, Rees G, Nachev P (2014) Human brain lesion-deficit inference remapped. Brain 137(9):2522–2531. https://doi.org/10.1093/brain/awu164
75. Mazzocchi F, Vignolo LA (1979) Localisation of lesions in aphasia: clinical-CT scan correlations in stroke patients. Cortex 15(4): 627–653
76. Mirman D, Chen Q, Zhang Y, Wang Z, Faseyitan OK, Coslett HB, Schwartz MF (2015) Neural organization of spoken language revealed by lesion–symptom mapping. Nat Commun 6:6762
77. Mirman D, Landrigan J-F, Kokolis S, Verillo S, Ferrara C, Pustina D (2018) Corrections for multiple comparisons in voxel-based lesion-symptom mapping. Neuropsychologia 115:112–123. https://doi.org/10.1016/j.neuropsychologia.2017.08.025
78. Mirman D, Thye M (2018) Uncovering the neuroanatomy of core language systems using lesion-symptom mapping. Curr Dir Psychol Sci 27(6):455–461. https://doi.org/10.1177/0963721418787486
79. Mohr JP, Pessin MS, Finkelstein S, Funkenstein HH, Duncan GW, Davis KR (1978) Broca aphasia: pathologic and clinical. Neurology 28:311–324
80. Molenberghs P, Gillebert CR, Peeters R, Vandenberghe R (2008) Convergence between lesion-symptom mapping and functional magnetic resonance imaging of spatially selective attention in the intact brain. J Neurosci 28(13):3359–3373
81. Naeser MA, Hayward RW (1978) Lesion localization in aphasia with cranial computed tomography and the Boston Diagnostic Aphasia Exam. Neurology 28:545–551
82. Nyffeler T, Vanbellingen T, Kaufmann BC, Pflugshaupt T, Bauer D, Frey J, Chechlacz M, Bohlhalter S, Muri R, Nef T, Cazzoli D (2019) Theta burst stimulation in neglect after stroke: functional outcome and response variability origins. Brain 142(4): 992–1008
83. Pettit JM, Duffy JR (1991) Left-hemisphere lesions: a comparison of subjects with and without aphasia. In: Clinical aphasiology. Pro-Ed, Austin, pp 55–68
84. Pillay SB, Binder JR, Humphries C, Gross WL, Book DS (2017) Lesion localization of speech comprehension deficits in chronic aphasia. Neurology 88(10):970–975
85. Piras F, Marangolo P (2009) Word and number reading in the brain: evidence from a voxel-based lesion-symptom mapping study. Neuropsychologia 47(8–9):1944–1953
86. Price CJ, Seghier ML, Leff AP (2010) Predicting language outcome and recovery after stroke: the PLORAS system. Nat Rev Neurol 6(4):202
87. Pustina D, Coslett H, Turkeltaub P, Tustison N, Schwartz M, Avants B (2016) Automated segmentation of chronic stroke lesions using LINDA: lesion identification with neighborhood data analysis. Hum Brain Mapp 37(4):1405–1421
88. Pustina D, Avants B, Faseyitan OK, Medaglia JD, Coslett HB (2018) Improved accuracy of lesion to symptom mapping with multivariate sparse canonical correlations. Neuropsychologia 115:154–166. https://doi.org/10.1016/j.neuropsychologia.2017.08.027
89. Ramsey LE, Siegel JS, Lang CE, Strube M, Shulman GL, Corbetta M (2017) Behavioural clusters and predictors of performance during recovery from stroke. Nat Hum Behav 1(3): 0038
90. Rogalsky C, Poppa T, Chen KH, Anderson SW, Damasio H, Love T, Hickok G (2015) Speech repetition as a window on the neurobiology of auditory–motor integration for speech: a voxel-based lesion symptom mapping study. Neuropsychologia 71:18–27
91. Rorden C, Karnath HO, Bonhila L (2007) Improving lesion-symptom mapping. J Cogn Neurosci 19:1081–1088
92. Rudrauf D, Mehta S, Bruss J, Tranel D, Damasio H, Grabowski TJ (2008) Thresholding lesion overlap difference maps: application

to category-related naming and recognition deficits. NeuroImage 41:970–984
93. Rudrauf D, Mehta S, Grabowski T (2008) Disconnection's renaissance takes shape: formal incorporation in group-level lesion studies. Cortex 44:1084–1096
94. Ruggieri S, Fanelli F, Castelli L, Petsas N, De Giglio L, Prosperini L (2018) Lesion symptom map of cognitive–postural interference in multiple sclerosis. Mult Scler J 24(5): 653–662
95. Saygin AP, Wilson SM, Dronkers N, Bates E (2004) Action comprehension in aphasia: linguistic and non-linguistic deficits and their lesion correlates. Neuropsychologia 42: 1788–1804
96. Schoch B, Dimitrova A, Gizewski ER, Timmann D (2006) Functional localization in the human cerebellum based on voxelwise statistical analysis: a study of 90 patients. NeuroImage 30(1):36–51
97. Schwartz MF, Kimberg DY, Walker GM, Faseyitan O, Brecher A, Dell GS, Coslett HB (2009) Anterior temporal involvement in semantic word retrieval: voxel-based lesion-symptom mapping evidence from aphasia. Brain 132(12):3411–3427
98. Schwartz MF, Faseyitan O, Kim J, Coslett HB (2012) The dorsal stream contribution to phonological retrieval in object naming. Brain 135(12):3799–3814
99. Shahid H, Sebastian R, Schnur TT, Hanayik T, Wright A, Tippett DC, Fridriksson J, Rorden C, Hillis AE (2017) Important considerations in lesion-symptom mapping: illustrations from studies of word comprehension. Hum Brain Mapp 38(6): 2990–3000
100. Sierpowska J, Gabarrós A, Fernández-Coello A, Camins À, Castañer S, Juncadella M, Francois C, Rodríguez-Fornells A (2019) White-matter pathways and semantic processing: intrasurgical and lesion-symptom mapping evidence. NeuroImage Clin 22:101704
101. Sperber C, Karnath HO (2016) Topography of acute stroke in a sample of 439 right brain damaged patients. NeuroImage Clin 10: 124–128. https://doi.org/10.1016/j.nicl.2015.11.012
102. Sperber C, Karnath HO (2017) Impact of correction factors in human brain lesion-behavior inference. Hum Brain Mapp 38(3): 1692–1701. https://doi.org/10.1002/hbm.23490
103. Sperber C, Wiesen D, Karnath HO (2019) An empirical evaluation of multivariate lesion behaviour mapping using support vector regression. Hum Brain Mapp 40(5): 1381–1390
104. Stark BC, Basilakos A, Hickok G, Rorden C, Bonilha L, Fridriksson J (2019) Neural organization of speech production: a lesion-based study of error patterns in connected speech. Cortex 117:228–246
105. Thothathiri M, Kimberg DY, Schwartz MF (2012) The neural basis of reversible sentence comprehension: evidence from voxel-based lesion symptom mapping in aphasia. J Cogn Neurosci 24(1):212–222
106. Turken A, Whitfield-Gabrieli S, Bammer R, Baldo J, Dronkers N, Gabrieli J (2008) Cognitive processing speed and the structure of white matter pathways: convergent evidence from normal variation and lesion studies. NeuroImage 42:1032–1044
107. Turken A, Dronkers NF (2011) The neural architecture of the language comprehension network: converging evidence from lesion and connectivity analyses. Front Syst Neurosci 5:1
108. Tyler LK, Marslen-Wilson WD, Stamatakis EA (2005) Dissociating neuro-cognitive component processes: voxel-based correlational methodology. Neuropsychologia 43: 771–778
109. Tzourio-Mazoyer N, Landeau B, Papathanassiou D, Crivello F, Etard O, Delcroix N, Mazoyer B, Joliot M (2002) Automated anatomical labeling of activations in SPM using a macroscopic anatomical parcellation of the MNI MRI single-subject brain. NeuroImage 15(1):273–289
110. Varjacic A, Mantini D, Demeyere N, Gillebert CR (2018) Neural signatures of trail making test performance: evidence from lesion-mapping and neuroimaging studies. Neuropsychologia 115:78–87
111. Weiss PH, Ubben SD, Kaesberg S, Kalbe E, Kessler J, Liebig T, Fink GR (2016) Where language meets meaningful action: a combined behavior and lesion analysis of aphasia and apraxia. Brain Struct Funct 221(1): 563–576
112. Wernicke C (1874) Der aphasische Symptomenkomplex. Cohn, Weigert, Breslau
113. Wilson SM, Saygin AP (2004) Grammaticality judgment in aphasia: deficits are not specific to syntactic structures, aphasic syndromes or lesion sites. J Cogn Neurosci 16:238–252
114. Wilson SM, Henry ML, Besbris M, Ogar JM, Dronkers NF, Jarrold W, Miller B, Gorno-Tempini ML (2010) Connected speech

production in three variants of primary progressive aphasia. Brain 133(7):2069–2088

115. Wilson SM, Hula W (2019) Multivariate approaches to understanding aphasia and its neural substrates. Curr Neurol Neurosci Rep 19:53. https://doi.org/10.1007/s11910-019-0971-6

116. Wilson SM, Lam D, Babiak MC, Perry DW, Shih T, Hess CP, Berger M, Chang EF (2015) Transient aphasias after left hemisphere resective surgery. J Neurosurg 123(3):581–593

117. Wilson SM (2017) Lesion-symptom mapping in the study of spoken language understanding. Lang Cogn Neurosci 32(7):891–899

118. Woo CW, Krishnan A, Wager TD (2014) Cluster-extent based thresholding in fMRI analyses: pitfalls and recommendations. NeuroImage 91:412–419

119. Wu DH, Waller S, Chatterjee A (2007) The functional neuroanatomy of thematic role and locative relational knowledge. J Cogn Neurosci 19:1542–1555

120. Yarnell F, Monroe P, Sobel L (1976) Aphasia outcome in stroke. Stroke 7:516–522

121. Zhang Y, Kimberg DY, Coslett HB, Schwartz MF, Wang Z (2014) Multivariate lesion-symptom mapping using support vector regression. Hum Brain Mapp 35(12):5861–5876

Chapter 6

Statistical Considerations in Voxel-Based Lesion-Behavior Mapping

Christoph Sperber and Hans-Otto Karnath

Abstract

Voxel-based lesion-behavior mapping (VLBM) is a popular method to investigate the functional anatomy of the brain. It is based on the statistical parametric mapping framework, and this framework and its application on lesion data come with several statistical challenges. In this chapter, we outline what statistical aspects emerge in voxel-based lesion-behavior mapping and by what means they can be dealt with. We address the choice of statistical tests, skewed data, multiple comparisons, required sample sizes, statistical power, control of covariates, the effect of lesion size, and the resolution of anatomical data.

Key words Stroke, Inference, VLSM, Voxel-based lesion-symptom mapping, Statistical parametric mapping, Mass univariate

1 Introduction

For more than one and a half centuries, scientists have been investigating lesion-deficit inference to map the functions of the human brain. Today, paralleled with more recent methods such as functional magnetic resonance imaging, lesion analysis mapping is still a major method in neuroscience [18]. In this chapter, we aim to discuss the technique that dominated the lesion-behavior mapping method during the last decade: *voxel-based lesion-behavior mapping (VLBM)*. It is a statistical analysis method that aims to map the neural correlates of healthy human behavior by finding areas in the brain where damage leads to a deficit in this behavior. For this purpose, it performs a large amount of univariate statistical tests, which is why it has been termed a "mass univariate approach."

In the following, we focus on the statistical aspects of the method. First, we illustrate the general idea behind VLBM: statistical parametric mapping. Both the framework of statistical mapping and the analysis of lesion imaging pose statistical challenges. Finally, we will provide practical advice on the use of analysis software and the limitations of VLBM.

Dorian Pustina and Daniel Mirman (eds.), *Lesion-to-Symptom Mapping: Principles and Tools*, Neuromethods, vol. 180, https://doi.org/10.1007/978-1-0716-2225-4_6, © Springer Science+Business Media, LLC, part of Springer Nature 2022

2 Statistical Parametric Mapping in Lesion-Behavior Mapping

The methodological principle underlying VLBM is statistical parametric mapping. It has already been mentioned in the context of lesion-behavior mapping in the 1990s [6] and was first implemented only a few years later [1]. Statistical parametric mapping tests each point in the image space, i.e., each pixel or, in 3D data, each volumetric pixel or "voxel," with a statistical test. This test investigates if the imaging information in this voxel is related to a non-imaging variable (*see* Chapter 5 in this volume). The resulting test statistic and its corresponding p-level are then mapped back into image space. In the context of lesion-behavior mapping, we interpret voxels containing significant statistics as the neural correlates of the investigated behavior.

VLBM requires at least the following data for each patient: (i) a behavioral variable measuring the deficit of interest and (ii) CT or MR brain imaging that depicts the focal brain damage. In particular, every single statistical test in a VLBM analysis is performed on the voxel level and is based on the measured behavioral performance and the lesion status of the tested voxel. Both can be either binomial or continuous. The voxel status indicates the damage status of the voxel (binomial, lesioned/not lesioned; continuous, e.g., probability of voxel being lesioned). The behavioral variable is supposed to measure the behavioral pathology that we want to map in the brain (binomial, e.g., aphasia/no aphasia; continuous, performance score resulting from an aphasia test). Figure 1 illustrates the VLBM approach in an example with binomial lesion status (lesioned/not lesioned) and a continuous behavioral variable using *t*-tests.

Before we go into details the statistics behind VLBM, we have to mention some important basics in study design. A VLBM analysis does not include healthy controls; all recruited subjects are patients with focal brain damage. However, it is crucial to include not only brain-damaged patients showing a deficit of interest but also brain-damaged patients without this pathological behavior [18]. For continuous behavioral variables, this means that a VLBM analysis needs to be based on patients with a wide range of different scores related to the deficit of interest. Moreover, all statistics that we discuss in this chapter require lesion data that were transferred into common space. This is usually done by spatial normalization of lesion images (cf. [3, 10]; *see* Chapter 4 in this volume). In spatially normalized brain images, a voxel coordinate will correspond to the same brain region in all subjects.

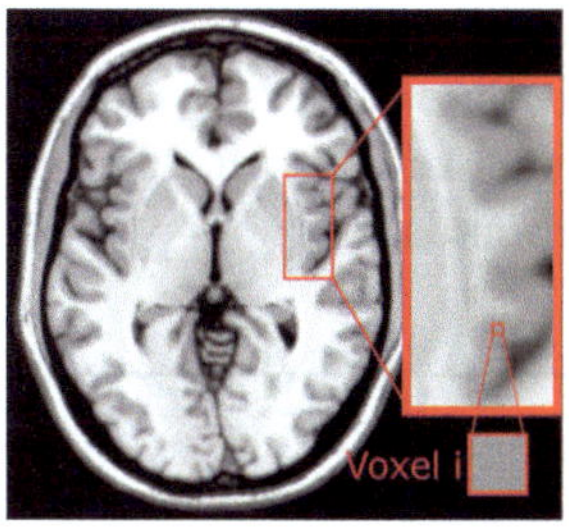

Patient	1	2	3	4	5	...	n
Behavioural Score	12	2	11	4	6	...	Score n
Status Voxel i	1	0	1	1	0	...	Status n

Fig. 1 Principle behind voxel-based lesion-behavior mapping. Each voxel in the brain is tested individually with a statistical test. Here, the data underlying a single voxel-wise univariate test are depicted. For each patient, two aspects are known: (i) the behavioral score to be mapped and (ii) the damage status of the tested voxel, which is binary here, i.e., damaged (1) or not damaged (0). The damage information is contained in a normalized lesion map. Using the binary damage information, the patient sample can be divided into two groups that are tested for differences in the behavioral variable, e.g., by a *t*-test. This procedure is done for each voxel, and the resulting statistics are mapped on the brain

3 Different Statistical Tests and Skewed Data

The main reasons for the popularity of the statistical parametric mapping framework are its simplicity and its flexibility. The latter means that we can choose different statistical tests depending on what we aim to analyze. This is also the case in VLBM, where many different tests are commonly used. To decide which tests to use, we have to consider two aspects: (i) are our lesion data and behavioral data binomial (i.e., binary) or continuous, and (ii) do we want to perform parametric or nonparametric statistics? If the voxel status is binomial (lesioned/not lesioned), then a statistical comparison between groups is performed. In such case, we can use t-tests or the Brunner-Munzel test for continuous behavioral scores (*see* Fig. 1), or the χ^2- or the Liebermeister test for binomial behavioral scores. If, on the other hand, voxel status is continuous, a regression approach has to be chosen, such as linear regression or logistic regression. Many of these tests can be formulated as a general linear model, and thus some software tools provide general linear models for different situations.

The next factor is the choice between parametric and nonparametric tests. Statistical tests such as *t*-tests or χ^2-tests make assumptions about the data, e.g., about data distribution. If such assumptions are violated, the test might not perform at the intended α-level, and test results might be biased. Especially, the distribution of behavioral data is a major problem in VLBM. While patients without a deficit in a behavior of interest often show normally distributed performance in a small range, pathological scores have high variance and are non-normally distributed within a wide range. In other words, the distribution of behavioral data in a patient sample recruited for a VLSM analysis is typically heavily skewed. This violates, e.g., the assumptions of t-tests. To overcome this issue, nonparametric tests should be used [19]. These include the Brunner-Munzel test for datasets with binomial lesion data/continuous behavioral data and the Liebermeister test for datasets with binomial lesion data/binomial behavioral data.

4 Corrections for Multiple Comparisons

A statistical test has a probability to provide a false positive decision, i.e., to reject the null hypothesis when it is actually true. In the context of lesion-behavior mapping, this is the association of damage in a voxel with the target behavior, although there is no relation between both. In statistical inference, it is a convention to minimize the probability of such false positives (and not the occurrence of false negatives) up to a certain α-level that the field deems to be acceptable. Often, it is set to $\alpha < 0.05$ (= 5%). Thus, if we perform one statistical test on data without true signal, we have a 5% probability to obtain a false positive. If we instead perform 100 tests, the probability to obtain at least 1 false positive will be 99.41% ($p = 1 - (1 - \alpha)^n$ or $p = 1 - (0.95)^{100}$), and we should expect to find on average 5 false positives. This inflation of false positive rates in multiple tests is called the 'multiple comparison problem', and it is present in any situation where multiple tests are performed. It is especially problematic in statistical parametric mapping, where we statistically test many thousands of voxels.

Imagine we test 100,000 voxels with $\alpha < 0.05$ and we find 5000 significant results. In this situation, the number of significant voxels is just what we would expect if there were no true lesion-behavior relation at all. Thus, we would likely interpret the analysis as a non-result. The situation becomes more difficult if we find 13,000 out of 100,000 voxels to be significant. There will likely be true positives but also some false positives among the significant voxels. Luckily, solutions to this problem exist.

There are two major strategies to deal with the multiple comparison problem. On the one hand, family-wise error (FWE) correction makes sure that the probability of finding *at least one* false positive across all tests will be set to an a priori chosen level. On the other hand, false discovery rate (FDR) correction accepts a certain amount of false positives across all significant voxels, and we a priori choose a proportion of false positives among all positive findings that we deem to be acceptable. Note that this section does not provide an extensive list of all possible correction algorithms, and some more recent approaches exist (e.g., [15]).

A popular FWE correction is the Bonferroni method. For this correction, we a priori choose an acceptable probability of having a single false positive across all tests (e.g., α (a priori) < 0.05) and then adjust the α-level of each test by dividing it with the number of tests n: $\alpha(\text{adjusted}) = \alpha\ (\text{a priori})/n$. For the first example in this chapter with 100 tests and $\alpha < 0.05$, we would thus perform each test with $\alpha < 0.05/100$ or $\alpha < 0.0005$. This correction effectively corrects for false positives, but it is generally very conservative, i.e., it has the downside of eventually producing many false negatives. This problem is enhanced in statistical parametric mapping due to violations of independence between voxels (*see* Section 2.2 in [2]). The problem here is that Bonferroni correction assumes that all statistical tests are performed independently; however, the lesion information across voxels is not independent but highly collinear (*see* supplementary materials in [17]). For this reason, Bonferroni correction can be more conservative than it is intended to be in the setting of lesion-behavior mapping.

A better-suited FWE approach is maximum statistic permutation testing (*see* section "Single Threshold Test" in [16]). Its main advantage is that—contrary to Bonferroni correction—it is valid even if statistical tests are not independent. In short, this approach asks how large test statistics can become in data without true signal (i.e., under the null hypothesis, where all positives are false positives) and then compares these largest statistics with the statistics that we obtained in the actual analysis. If the true test statistics are larger than most test statistics in the samples without true signal, we can consider them truly significant. In detail, first, we again choose an α-level, e.g., $\alpha < 0.05$. Then, a standard lesion-behavior mapping analysis (e.g., by t-tests) is performed with the original dataset. Next, the same lesion-behavior mapping analysis is performed on a large number of permutated samples, i.e., samples where the behavioral scores have been randomly exchanged across lesions/patients. These permutation samples provide us with whole-brain topographical lesion mapping results that can result by chance in a sample where the null hypothesis is by definition true. In the next step, we look at what maximum test statistic was obtained in each analysis (e.g., for an analysis by t-tests, the largest t-statistic found in each complete analysis). We can now plot the distribution of maximum statistics across all analyses. Using this maximum test statistic distribution, we can subsequently obtain a cutoff maximum test statistic value, with maximum test statistic values surpassing this cutoff in less than 5% (for $\alpha < 0.05$) of the permutation analyses. In other words, we choose the value at the 95th percentile of the distribution for an $\alpha < 0.05$. That means that any maximum test statistic value that exceeds this cutoff had a probability of less than 5% of occurring under the null hypothesis (remember, each permutation assesses the null hypothesis). With this cutoff, we can now go back to our analysis using original data. Any voxel with a t-statistic that is larger than this cutoff can be considered significant. Importantly, this cutoff was derived from whole analyses, i.e., one max statistic from one analyzed imaging volume each (and not from individual voxels, which would have resulted in many statistics obtained per analysis). The probability to obtain such a high value for a statistic just by chance (e.g., by chance in permuted data, i.e., when the null hypothesis is true) in an *entire* analysis is thus $p < 0.05$. This makes it a family-wise error correction, or, in other words, any published results that are thresholded with the max statistic procedure only have a probability of less than 5% to contain any false positive at all. Permutation thresholding is considered to be a gold standard of correction for multiple comparisons [9].

Contrary to FWE corrections, FDR does not aim to eliminate all false positives in the analysis. Instead, a researcher using FDR accepts a certain rate q of false positive results in all positive results. We do this by a priori choosing a false discovery rate q. If we choose an FDR of $q = 0.05$, this means that up to 5% of all positive findings are expected to be false positives. Thus, if we find 5000 significant voxels after FDR correction, we expect about 250 voxels to be false positives. FDR thus offers a trade-off between the ability to find true signal and obtaining some false positive findings. Caution is advised in lesion-behavior mapping using smaller samples of up to 30 patients, as FDR might be too anti-conservative in such a situation [15], thus producing more false positives than it should. Furthermore, FDR also assumes independence of statistical tests (*see* above the paragraph on Bonferroni correction). This is especially important as the correction of one voxel depends on the significance level of other voxels. For example, if some voxels exhibit high significance, the correction in others is more lenient.

The multiple comparison problem in VLBM results from the many voxels that are individually tested by a statistical test. However, it should be noted that in a VLBM, we might perform fewer tests than there are voxels in the brain. Often, neighboring voxels are lesioned in the very same patients. Therefore, virtually the same test with the same test result would be performed in these voxels. Such groups of voxels that contain the same lesion information were termed "unique patches," and efficient analysis tools such as NPM or LESYMAP [17] only perform a single test per unique patch, which ultimately reduces the number of tests that are performed in the VLBM.

5 Sample Sizes

Acquisition of clinical data for a VLBM analysis demands a lot of time and work. Accordingly, a researcher might like to know how many patients have to be tested to obtain a valid VLBM analysis. Unfortunately, a definitive answer to this question cannot be provided yet.

Sample sizes that are commonly found in published studies range from roughly 20 to several hundred patients. This tells us what is common in the field, but not necessarily what sample size is appropriate. Until recently, studies investigating the proper sample size for lesion-behavior mapping studies were lacking. Two recent studies, however, shed some light on required sample sizes [13, 17]. These studies suggested that VLBM on sample sizes in the range of 20–60 patients performs suboptimal and that such samples can both over- or underestimate effects in the data. However, VLBM performance becomes relatively stable with sample sizes of about 80–90 patients.

Still, we cannot conclude from these studies a definitive number of required patients as general guidance. Individual characteristics of a patient sample and study design influence the VLBM analysis and thus the number of patients. Imagine we test an unselected sample of stroke patients for two different behaviors. One behavior is deficient in many patients, and it likely follows damage to areas in the center of the middle cerebral artery territory. With the high clinical variance, mapping of this behavior should be feasible with relatively small samples. In contrast, the second behavior is deficient only in few patients, and the neural correlates are situated in a border zone of the middle cerebral artery territory, where lesions are rare. In that case, we will likely require more patients to obtain a replicable and valid mapping. Scientists should thus make sure that the behavioral data include a decent amount of pathological variance, i.e., patients where the neural correlate of the behavior is damaged (e.g., *see* Fig. 2). A way to investigate such issues post hoc is power analysis [12], which we discuss in the next paragraph.

6 Voxel-Wise Power and Rarely or Never Damaged Voxels

A major challenge in VLBM is best explained if we first look at a simplified (fictional) example of a functional brain imaging study based on statistical parametric mapping. Imagine that we tested 30 subjects with functional magnetic resonance imaging to identify differences in brain activity viewing between known tools and unknown tools. Each subject will then view some known tools and some unknown tools. In the statistical parametric mapping analysis in a single voxel, we can then compare the activity of 30 subjects in this voxel in one condition versus the activity of the same 30 subjects in the other condition. We will end up performing t-tests comparing 2 groups of 30 versus 30 data points in every voxel.

This is different in VLBM. In VLBM (with binomial lesion data), the groups that are compared are assigned for each voxel individually depending on the patients' lesion status in this voxel. A region's susceptibility to stroke and the frequency of lesions in this region varies across the brain. Many voxels are affected by only a few lesions, and in a sample of, e.g., 80 patients, we might have

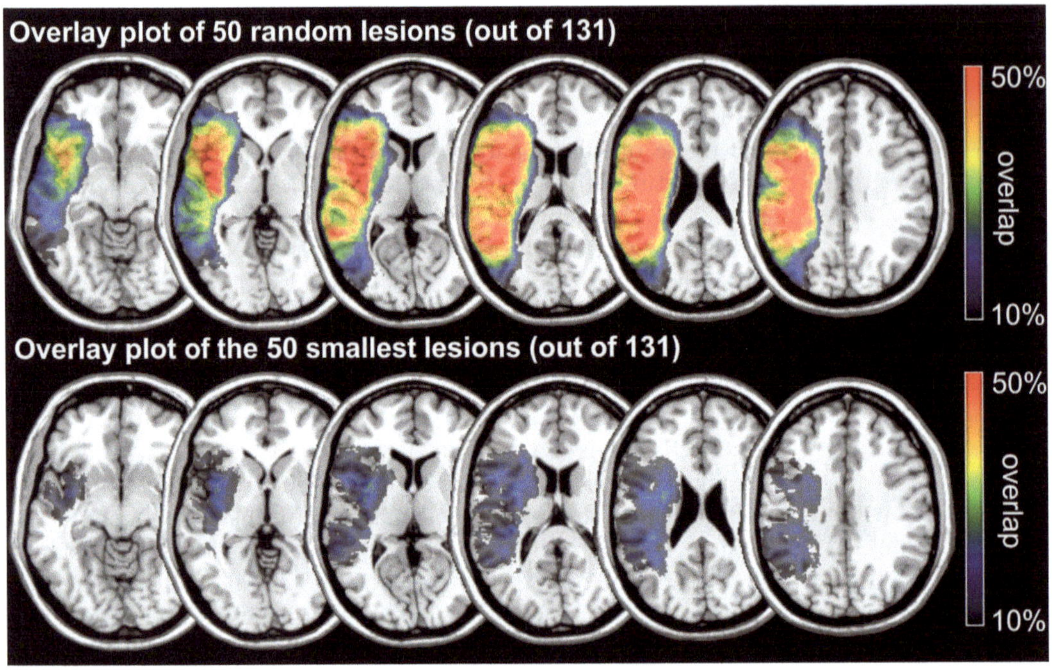

Fig. 2 Why choosing a good sample size in lesion-behavior mapping is not trivial. Both panels show lesion overlay topographies of equally sized subsamples in the 131 left hemisphere lesions in the Moss Rehabilitation Research Institute dataset (as included in the LESYMAP software package). Colors code the number of patients with a lesion in a voxel in relation to the total number of patients (i.e., 50% would be equal to an overlap of 25 out of 50 patients in a voxel). Only values above 10% are shown, which would correspond to a VLBM analysis that only analyzes voxels damaged in at least five patients. In the upper panel, the overlay plot is based on the first 50 lesions (i.e., subjects 1–50; average lesion size 118,736 mm^3). In the lower panel, the overlay plot is based on the smallest 50 lesions out of all 131 (average lesion size 30,489 mm^3). Voxel-wise statistical power is highest if the amount of patients with a lesion and patients without a lesion is equal, i.e., if the overlap is 50% of the sample [12]. The random, unselected sample fluctuates around this value in many voxels in the middle cerebral artery (MCA) territory. Neuropsychological deficits that emerge after damage to a structure in this territory could be found in many of the investigated patients, and a VLBM analysis would likely be feasible with such a sample size. On the other hand, the sample of small lesions shows much smaller overlap frequencies, and many voxels do not even surpass the minimum criterion of five lesions. Only a few patients would show a deficit that emerges after damage to a circumscribed MCA territory region. This lesion sample would not be large enough for a VLBM analysis. Importantly, both samples would be too small to investigate a deficit that emerges after damage to the posterior cerebral artery (PCA) territory, as patients in neither sample show lesions to PCA territory areas. As this example illustrates, it is not possible to provide a universally valid guideline on required sample sizes in VLBM

voxels with only 10 lesions. The statistical test for such voxels would compare unequal groups of 70 versus 10 patients. In more extreme cases, we might have voxels with only a single lesion or no lesion at all in our sample. A statistical group comparison obviously does not make sense here. Such voxels thus need to be excluded from the analysis. The question is what minimum number of lesions we accept in a single voxel. Clear guidelines on this threshold do not exist, but in most publications, it has been set to 5–10% of the total patient sample [21].

Post hoc power analyses [12] can help find a threshold for a particular study. Such power analyses map for each voxel how much power the statistical test has with an absolute scale ranging from 0 to 1. Higher values mean that the statistical test has a higher probability of yielding a true positive in voxels with true signal. Commonly, a power of at least 0.8 is considered acceptable. It has been shown that statistics in voxels with uneven groups—which are almost always voxels where only few patients do have a lesion—have a reduced power to find effects. If voxels with low power are included in the analysis, the current threshold might thus be adjusted post hoc.

Nevertheless, setting a threshold for a minimum of lesion affection increases the validity of VLBM analyses for those brain areas analyzed [21]. However, there is also a downside of this procedure: the voxels that we need to exclude are simply not tested. Therefore, VLBM analyses are not usually performed "whole-brain." A study of, e.g., unilateral left hemisphere stroke patients that investigates the neural correlates of a certain left hemispheric function likely only tests parts of the left hemisphere. Some parts of the hemisphere will not be included for the reason outlined above. For these parts, we cannot decide whether or not they contain neural correlates of the behavior or disturbance of interest. These areas remain a "terra incognita" in our mapping process (cf. Figure 3). For this reason, it is helpful for readers of any report of a VLBM study to show a simple overlap map of all lesions, designed to identify which voxels were tested in the final analysis.

7 Controlling for Further Variables

In some cases, we might be interested in including other variables than only the behavioral target variable in our analysis. For example, think of a VLBM study that aims to map the neural correlates of object perception. A patient might be unable to correctly name an object simply due to a disturbance of the language system. Aphasia might be correlated with the performance in the object perception task, and it might directly and causally affect task performance. More generally, we can think of different reasons for two behavioral variables to correlate. First, they might at least partially rely on the same basic cognitive functions. Here, we might think of object perception and face perception, where both functions likely share some cognitive processes, while face perception might be a more specialized cognitive function including additional processes. Second, correlating behavior might represent additional processes required for the task. For example, abilities such as primary vision, attention, naming, and primary motor abilities of the motor speech system might be involved in, but not be specific for, object perception. Third, both behavioral variables might have no direct relation

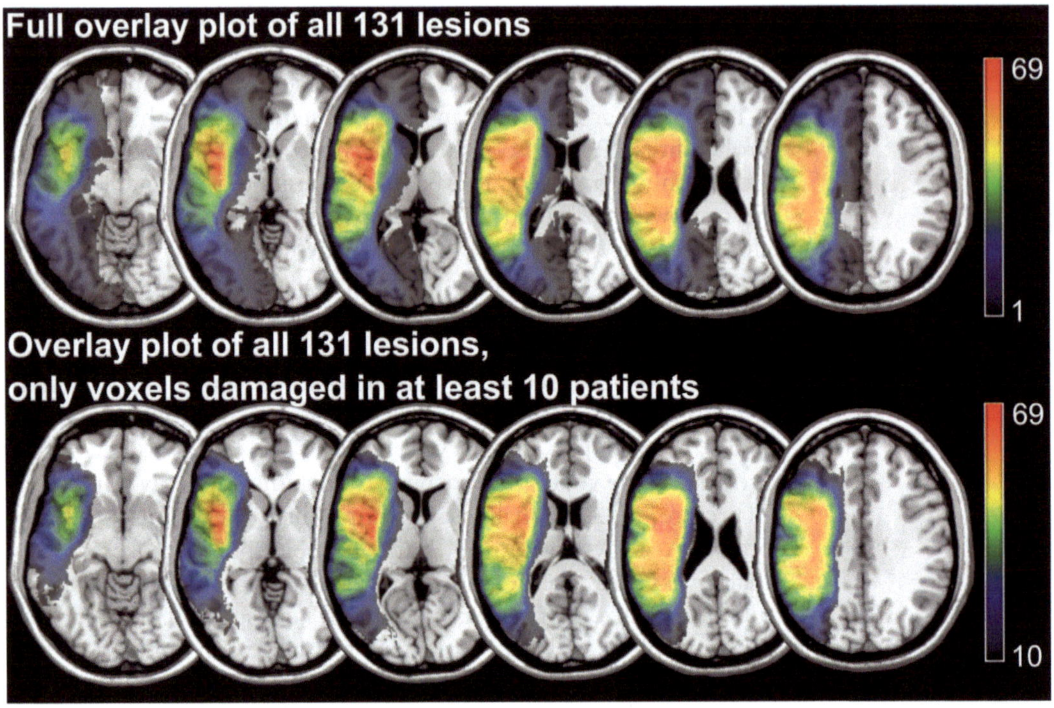

Fig. 3 The "terra incognita" in lesion-behavior mapping. Both panels show lesion overlay topographies of the 131 left hemisphere lesions in the Moss Rehabilitation Research Institute dataset (as included in the LESYMAP software package). The *upper panel shows* the full range of overlay frequencies from 1 to a maximum of 69. This illustration suggests that lesions in nearly the whole left hemisphere were included in the sample. However, if lesion-behavior mapping is restricted to only those voxels affected in at least ten patients, many voxels would be excluded from the analysis, i.e., they would not be tested whether or not they are associated with the target variable. The *lower panel* shows which voxels would remain in such analysis. Thus, we cannot tell whether or not areas such as the prefrontal cortex, frontal pole, head of caudate nucleus, thalamus, and large parts of the posterior cerebral artery territory are associated with the investigated behavior

at all, but still, they might correlate simply because lesions often damage the neural correlates of both functions at once. Think of aphasia and primary motor function of right-sided extremities; both hemiparesis and aphasia are common consequences of stroke to the center of the middle cerebral artery territory and thus often co-occur.

In some cases, our variable might be controlled to map only some aspects of the measured behavior in several ways. A first option is to adapt the study design or recruitment strategies when we collect the data for the VLBM analysis to prevent the confounding influence (e.g., excluding all patients with aphasia in the example above if you are studying object perception). This might, however, be difficult or might bias the study sample. For example, if a behavioral variable is controlled for a second variable that is localized more posteriorly, the results of the VLBM for the first variable might be (wrongfully) shifted toward anterior areas. An alternative option to control for the effect of additional, correlated variables would be to include such variables into the statistical analysis. The flexibility of statistical parametric mapping comes into play here again: theoretically, we can perform more complex statistical tests that allow us to include multiple variables in each voxel, as, for example, ANCOVA or multiple regression. Some tools like NiiStat utilize the Freedman-Lane method to perform a GLM analysis that includes nuisance regressors (as described by [24]). An option allowing to include multiple variables in each voxel is, however, not included in all publicly available software tools and thus requires custom modifications to the code. Instead, a commonly implemented approach in many tools is nuisance regression. In a nuisance regression, before the actual analysis, we regress out variance explained by the second variable (e.g., aphasia) from the behavioral target variable (e.g., object perception). Then, we perform the actual VLBM with the residuals. In software tools that do not offer this functionality, a nuisance regression can also be applied beforehand, i.e., before loading the behavioral data into the software. However, we advise caution, as covariate control is not always appropriate and might likewise bias results. This might especially be the case when deficits only co-occur without causal relation between both [20].

8 Controlling for the Effect of Lesion Size

The severity of behavioral deficits after stroke often highly correlates with lesion size. This is likely no direct effect of lesion size itself, but it results from the fact that larger lesions are more likely to damage the neural correlates of the investigated behavior in general. Additionally, larger lesions leave less neural tissue for functional reorganization (which is relevant in chronic patient studies). They are also more likely to induce damage to domain-general neural resources, which are not specific to the investigated behavior but are still engaged during that behavior. Thus, the problem emerges that many patients with a certain deficit have larger lesions, and therefore their lesions do not provide us with spatially specific information. Additionally, lesion size varies across the brain, and some voxels are more likely to be affected by larger lesions than others. If VLBM associates such voxels with the behavioral variable, it is not clear whether the damage in this voxel itself was the reason or if it was only mediated by lesion size. Previous studies have controlled this effect by design/sampling strategies or by statistical means. A common statistical approach is to regress the effect of lesion size out of the behavioral variable [8]. It has been shown that correction for lesion size is on average beneficial in VLBM analyses [21]. Beyond, other statistical strategies to control for the effect of lesion size, such as direct total lesion volume control (dTLVC;

[26]), exist, which are elaborated and utilized in multivariate lesion-behavior mapping studies [4, 26]. Application of such approaches is also possible in univariate VLBM [23].

Nevertheless, the procedure is ambiguous [25]. Because lesion size varies across the brain, regions that are typically affected by larger lesions might be unfairly penalized. In other words, if the neural correlate of a particular behavior is often affected by larger lesions and less frequently by smaller lesions, a correction for lesion size might conceal positive findings in this area. Another downside of lesion size control is that—given the high correlation between lesion size and the behavioral variable—much of the variance is regressed out. The VLBM might thus have low power and might yield null results, especially with smaller samples. Further discussion on the topic of lesion volume confounds or corrections can be found in Chapters 11 and 13.

9 Resolution of Anatomical Data: Investigating Regions of Interest

So far, we only have discussed voxel-wise statistics. Although this is the dominant approach in the field, some tools also offer region-wise approaches. In such region-wise approaches, the anatomical information is merged for a priori chosen regions of interest (ROIs). The ROIs can be taken from brain atlases, and they might be parcellated based on anatomical or functional data. For example, an ROI could be the angular gyrus, and instead of voxel-wise information, the lesion-behavior mapping analysis then utilizes the lesion load of the entire region. For a patient with 40% of all voxels in the angular gyrus damaged, the lesion load would be 0.40. The statistical analysis could mirror the analysis of voxel-wise data, using a regression in the framework of statistical parametric mapping. The main idea behind such an approach is to merge voxel-wise anatomical data that are imprecise due to, e.g., noise in normalization or small interindividual differences in brain anatomy into less noisy anatomical variables. The possible advantage of such an approach is that statistical analyses might gain statistical power, and the number of statistical tests performed is dramatically reduced. If a researcher has clear a priori hypotheses about the role of certain well-defined brain regions (e.g., due to preceding observations in the field), such a procedure is certainly beneficial.

However, we are skeptical about ROI approaches in lesion-behavior mapping as it comes with some pitfalls. First, the same lesion load of, e.g., 0.40 in the same ROI may derive from very different anatomical situations. For example, it may be the case that in one patient with a lesion load of 0.40, the anterior 40% of the angular gyrus is lesioned, while in another patient with a lesion load of 0.40 of the same sample, the posterior 40% of this structure is injured. The lesions of these two patients would not overlap at all, but they would be treated identically by such an ROI analysis approach. The underlying assumption of such an approach is that anterior and posterior parts of the angular gyrus have identical functions in terms of representation of the function under interest. If this should, however, not be the case, we in turn introduce new noise into the data. A differentiation of areas inside the ROI is not possible after merging voxel-wise data into an ROI. A possible way to overcome this problem would be to use an atlas that only contains regions that are uniform cognitive modules instead of atlases based on anatomical boundaries. Some recent atlases proposed to provide such fine-graded cognitive architecture [5, 7]. However, so far it is not known if lesion-behavior mapping indeed profits from performing lesion analysis on a ROI level, and at least post-stroke deficit prediction does not improve by doing so [11].

10 Caution: Available Statistical Software Tools Perform Very Differently

So far, we have discussed several options that are available in statistical lesion-behavior mapping, which hopefully allows the reader to choose the best for his or her research. However, there is one major caveat. With a large number of different approaches and without many generally accepted gold standards, statistical software tools do not all offer the same options. Some tools might 'clandestinely' just run default options; other tools might entirely miss some options mentioned in this chapter or might even provide some alternative approaches that we did not discuss.

For example, both NPM and LESYMAP offer the nonparametric Brunner-Munzel test, while this test is not included in NiiStat. NiiStat, on the other hand, utilizes a general linear model as a default statistical test, which opens up more flexible analysis options to the experienced user. As another example, NPM can control for nuisance variables and lesion size by nuisance regression. NiiStat's default control for lesion size is also based on nuisance regression; control for nuisance variables, however, is performed using the Freedman-Lane method [24].

11 Limitations of the Univariate Statistical Approach

In each voxel, statistical parametric mapping performs a statistical test independently of all other tests. However, voxels in the brain are not independent. First, many cognitive functions are represented in configurations of multiple voxels in larger brain regions or even in brain networks. Second, damage between voxels systematically co-occurs following typical patterns of stroke. In other words, due to the typical vascularization of the brain, voxels are often damaged or not damaged together. Both these violations of statistical independence limit the validity of lesion-behavior mapping [14]. A consensus on the perfect solution to these issues has not been found yet. Multivariate lesion-behavior mapping ([9, 14]; Chapter 11 in this volume) might provide options for some of these issues; however, such methods only emerged recently, and the field is still searching for widely accepted procedures [4, 17, 22, 23].

Acknowledgments

This work was supported by the Deutsche Forschungsgemeinschaft (KA 1258/23-1). We thank Stefan Smaczny for his comments on the manuscript.

References

1. Bates E, Wilson SM, Saygin AP et al (2003) Voxel-based lesion-symptom mapping. Nat Neurosci 6:448–450. https://doi.org/10.1038/nn1050
2. Brett M, Penny W, Kiebel S. Introduction to random field theory. In: RSJ Frackowiak, KJ Friston, CD Frith, RJ Dolan, CJ Price, S Zeki, JT Ashburner, WD. Penny (eds.) Human Brain Function (Second Edition), Academic Press, 2004: 867–879. https://doi.org/10.1016/B978-012264841-0/50046-9
3. de Haan B, Karnath H-O (2018) A hitchhiker's guide to lesion-behaviour mapping. Neuropsychologia 115:5–16. https://doi.org/10.1016/j.neuropsychologia.2017.10.021
4. DeMarco AT, Turkeltaub PE (2018) A multivariate lesion symptom mapping toolbox and examination of lesion-volume biases and correction methods in lesion-symptom mapping. Hum Brain Mapp 21:2461–2467. https://doi.org/10.1002/hbm.24289
5. Eickhoff SB, Constable RT, Yeo BTT (2018) Topographic organization of the cerebral cortex and brain cartography. NeuroImage 170:332–347. https://doi.org/10.1016/j.neuroimage.2017.02.018
6. Friston KJ, Holmes AP, Worsley KJ, Poline J-P, Frith CD, Frackowiak RSJ (1994) Statistical parametric maps in functional imaging: a general linear approach. Hum Brain Mapp 2:189–210. https://doi.org/10.1002/hbm.460020402
7. Glasser MF, Coalson TS, Robinson EC et al (2016) A multi-modal parcellation of human cerebral cortex. Nature 536:171–178. https://doi.org/10.1038/nature18933
8. Karnath H-O, Fruhmann Berger M, Küker W, Rorden C (2004) The anatomy of spatial neglect based on voxelwise statistical analysis: a study of 140 patients. Cereb Cortex 14:1164–1172. https://doi.org/10.1093/cercor/bhh076
9. Karnath H-O, Sperber C, Rorden C (2018) Mapping human brain lesions and their functional consequences. NeuroImage 165:180–189. https://doi.org/10.1016/j.neuroimage.2017.10.028
10. Karnath H-O, Sperber C, Wiesen D, de Haan B (2020) Lesion-behavior mapping in cognitive neuroscience: a practical guide to univariate and multivariate approaches. In: Pollmann S (ed) Spatial learning and attention guidance, Neuromethods, vol 151. Humana Press, New York, pp 209–238. https://doi.org/10.1007/7657_2019_18
11. Kasties V, Karnath H-O, Sperber C (2021). Strategies for feature extraction from structural brain imaging in lesion-deficit modelling. Hum Brain Mapp 42: 5409-5422. https://doi.org/10.1002/hbm.25629
12. Kimberg DY, Coslett HB, Schwartz MF (2007) Power in voxel-based lesion-symptom mapping. J Cogn Neurosci 19:1067–1080. https://doi.org/10.1162/jocn.2007.19.7.1067
13. Lorca-Puls DL, Gajardo-Vidal A, White J et al (2018) The impact of sample size on the reproducibility of voxel-based lesion-deficit mappings. Neuropsychologia 115:101–111. https://doi.org/10.1016/j.neuropsychologia.2018.03.014
14. Mah Y-H, Husain M, Rees G, Nachev P (2014) Human brain lesion-deficit inference

remapped. Brain 137:2522–2531. https://doi.org/10.1093/brain/awu164

15. Mirman D, Landrigan J-F, Kokolis S, Verillo S, Ferrara C, Pustina D (2018) Corrections for multiple comparisons in voxel-based lesion-symptom mapping. Neuropsychologia 115:112–123. https://doi.org/10.1016/j.neuropsychologia.2017.08.025
16. Nichols TE, Holmes AP (2002) Nonparametric permutation tests for functional neuroimaging: a primer with examples. Hum Brain Mapp 15:1–25. https://doi.org/10.1002/hbm.1058
17. Pustina D, Avants B, Faseyitan OK, Medaglia JD, Coslett HB (2018) Improved accuracy of lesion to symptom mapping with multivariate sparse canonical correlations. Neuropsychologia 115:154–166. https://doi.org/10.1016/j.neuropsychologia.2017.08.027
18. Rorden C, Karnath H-O (2004) Using human brain lesions to infer function: a relic from a past era in the fMRI age? Nat Rev Neurosci 5:812–819. https://doi.org/10.1038/nrn1521
19. Rorden C, Karnath H-O, Bonilha L (2007) Improving lesion-symptom mapping. J Cogn Neurosci 19:1081–1088. https://doi.org/10.1162/jocn.2007.19.7.1081
20. Sperber C, Nolingberg C, Karnath H-O (2020) Post-stroke cognitive deficits rarely come alone: handling co-morbidity in lesion-behaviour mapping. Hum Brain Mapp 41:1387–1399. https://doi.org/10.1002/hbm.24885
21. Sperber C, Karnath H-O (2017) Impact of correction factors in human brain lesion-behavior inference. Hum Brain Mapp 38:1692–1701. https://doi.org/10.1002/hbm.23490
22. Sperber C, Wiesen D, Karnath H-O (2019) An empirical evaluation of multivariate lesion behaviour mapping using support vector regression. Hum Brain Mapp 40:1381–1390. https://doi.org/10.1002/hbm.24476
23. Wiesen D, Sperber C, Yourganov G, Rorden C, Karnath H-O (2019) Using machine learning-based lesion behavior mapping to identify anatomical networks of cognitive dysfunction: spatial neglect and attention. NeuroImage 201:116000. https://doi.org/10.1016/j.neuroimage.2019.07.013
24. Winkler AM, Ridgway GR, Webster MA, Smith SM, Nichols TE (2014) Permutation inference for the general linear model. NeuroImage 92:381–397. https://doi.org/10.1016/j.neuroimage.2014.01.060
25. Xu T, Jäger HR, Husain M, Rees G, Nachev P (2018) High-dimensional therapeutic inference in the focally damaged human brain. Brain 141:48–54. https://doi.org/10.1093/brain/awx288
26. Zhang Y, Kimberg DY, Coslett HB, Schwartz MF, Wang Z (2014) Multivariate lesion-symptom mapping using support vector regression. Hum Brain Mapp 35:5861–5876. https://doi.org/10.1002/hbm.22590

Chapter 7

Voxel-Based Brain-Behavior Mapping in Neurodegenerative Diseases

Sladjana Lukic, Valentina Borghesani, Maria Luisa Gorno-Tempini, and Giovanni Battistella

Abstract

Over the last decades, sophisticated neuroimaging techniques have become a key methodology in defining brain areas affected by progressive illnesses, helping the identification of syndrome-specific anatomic patterns and/or the resulting changes in cognitive function and behavior. This chapter provides a review of the contemporary research on neurodegenerative diseases, spanning a variety of different clinical and brain atrophy patterns, along with the methodologies that helped to uncover them. We start by outlining the brain-behavior mapping method commonly used, followed by a direct comparison of automated techniques of brain morphometry, voxel- versus surface-based. We then present the results of seminal studies applying these techniques to investigate clinical group differences or brain-language relationships. These studies have improved our ability to identify clinical symptoms and perform differential diagnoses and greatly deepened our understanding of the neurobiology of language. We conclude by highlighting how, thanks to these methodological advances, the field can now tackle unresolved anatomical-functional questions embracing clinical neurodegenerative data.

Key words Neurodegenerative diseases, Voxel-based morphometry, Surface-based morphometry, Primary progressive aphasia, Language

1 Brain-Behavior Mapping in Neurodegenerative Disease: A Tool for Cognitive and Clinical Neuroscience

Our understanding of the relation between brain structures and functions is rooted in neuropsychological studies and promoted by advances in brain imaging techniques. Since the emblematic case of Phineas Gage in 1848, whose injury to the frontal lobe caused dramatic personality change, mapping of injury- and/or disease-related changes to discrete brain regions has been crucial in pushing further the field of cognitive neuroscience. The vast majority of structure-function mapping studies have focused on brain lesions caused by strokes. However, for more than a century, neurodegenerative diseases, such as frontotemporal lobar degeneration (FTLD)

Dorian Pustina and Daniel Mirman (eds.), *Lesion-to-Symptom Mapping: Principles and Tools*, Neuromethods, vol. 180, https://doi.org/10.1007/978-1-0716-2225-4_7,

and Alzheimer's disease (AD), have provided a useful clinical model to address issues related to the neurobiology of behavior and cognition. Pivotal lessons on the organization of cortical and subcortical structures stem from neuropathological, neuropsychological, and neuroimaging investigations of the major FTLD syndromes: the behavioral variant (known as bvFTD), the motor variants (corticobasal degeneration, progressive supranuclear palsy, and dementia with motor neuron disease), and the language variants (primary progressive aphasia (PPA)).

For decades, the most commonly used lesion-symptom imaging analysis method has been voxel-based morphometry (VBM), which allowed testing of key hypotheses on the cognitive and behavioral sequelae of the neural tissue dysfunctions (i.e., cortical atrophy) associated with specific FTLD syndromes (*see* [1] for a review). VBM is an automated quantification method of magnetic resonance imaging (MRI) scans. It performs a given statistical *t*-test across all voxels in the MRI images, thus allowing to identify volume differences between groups (between-group *t*-test) or to assess the neuroanatomical correlates of cognitive or behavioral deficits (within-group regression analysis). For instance, VBM allowed the isolation of atrophy epicenters in PPA, revealing a remarkable asymmetry toward the left hemisphere where the disease impacts major language networks within the temporal, frontal, and parietal lobes [2–4]. Critically, in these syndromes, the anatomical specificity is paralleled by neuropsychological specificity. For example, cortical atrophy in the anterior temporal cortex is correlated with loss of semantic knowledge in the semantic variant of PPA (svPPA), a relationship that has been neglected by traditional stroke aphasia models as this area is not susceptible to focal vascular lesions.

Overall, this line of research has had a critical, twofold impact. On one hand, it has greatly contributed to our understanding of brain dysfunction in dementia. The clinical use of VBM in neurodegenerative disorders has helped (1) to diagnose AD and FTLD syndromes by identifying cortically related abnormalities and (2) to identify and track the progression of neurodegenerative diseases [5–7]. One the other hand, VBM studies have provided novel insights into the neural mechanisms of cognitive functions, allowing refinement of theoretical models by complementing functional magnetic resonance imaging (fMRI) and stroke-based lesion-symptom mapping findings.

2 Brain-Behavior Mapping: The Analytical Framework and the Available Software

Traditionally, studies on brain morphometry involved the calculation of the average volume of manually delineated regions of interest (ROIs) in structural MRI images. This approach limited the application of morphometric studies only to brain regions that could be clearly defined and manually traced. Over the past 20 years, enormous advances in computational neuroanatomy overcame this limitation by enabling the unprecedented capability to comprehensively study neuroanatomical differences using automatic approaches. We generally refer to these techniques as unbiased, since they consider each single voxel in the image as an independent entity. The possibility to acquire MRI data with increasing spatial resolution and tissue contrasts, in parallel with the development of sophisticated image processing techniques, led to the proliferation of approaches aimed at characterizing brain structure with high accuracy. These approaches can be broadly classified into those that measure *shape* and those that use information about the *local composition* of brain tissue. This chapter describes methods for voxel-based and surface-based quantifications of *local composition* of brain tissues that are released within the Computational Anatomy Toolbox (CAT12: http://www.neuro.uni-jena.de/cat/) and integrated in the latest version of the Statistical Parametric Mapping (SPM12) software.

VBM is considered the standard technique to perform voxel-by-voxel comparison of local concentration of gray matter (GM). It relies on the probability estimates of each voxel to be assigned to GM given a high-resolution T1-weighted image. The GM probability maps are then normalized into a standard template space and smoothed to ultimately perform voxel-wise parametric tests to assess how the estimated GM probability values relate to, for instance, a disease status or the score in a neuropsychological test. Since its first appearance in 1995 [8], VBM has been widely used to study neurodevelopment [9] and to assess the effect of a specific training or specialized activities [10], as well as in several neurological and psychiatric disorders (for a meta-analysis, *see* [11]). Surface-based morphometry (SBM) is a complementary technique that aims at characterizing properties of the cortical surface such as cortical folding and cortical thickness. Within the CAT12 framework, the two approaches share most of the preprocessing steps. *See* Fig. 1 for a summary on voxel-based and/or surface-based morphometry methods.

2.1 Preprocessing of MRI Data

Prior to actual volume-based or surface-based quantifications, three preprocessing steps are performed: (1) *unified segmentation*, which includes two separate processes, i.e., *segmentation* of T1-weighted MR images into the three main tissue classes—gray matter (GM), white matter (WM), and cerebrospinal fluid (CSF)—and *normalization* (i.e., *registration*) of T1-weighted MR images to a common space, and (2) *smoothing* of segmented MRI-derived maps.

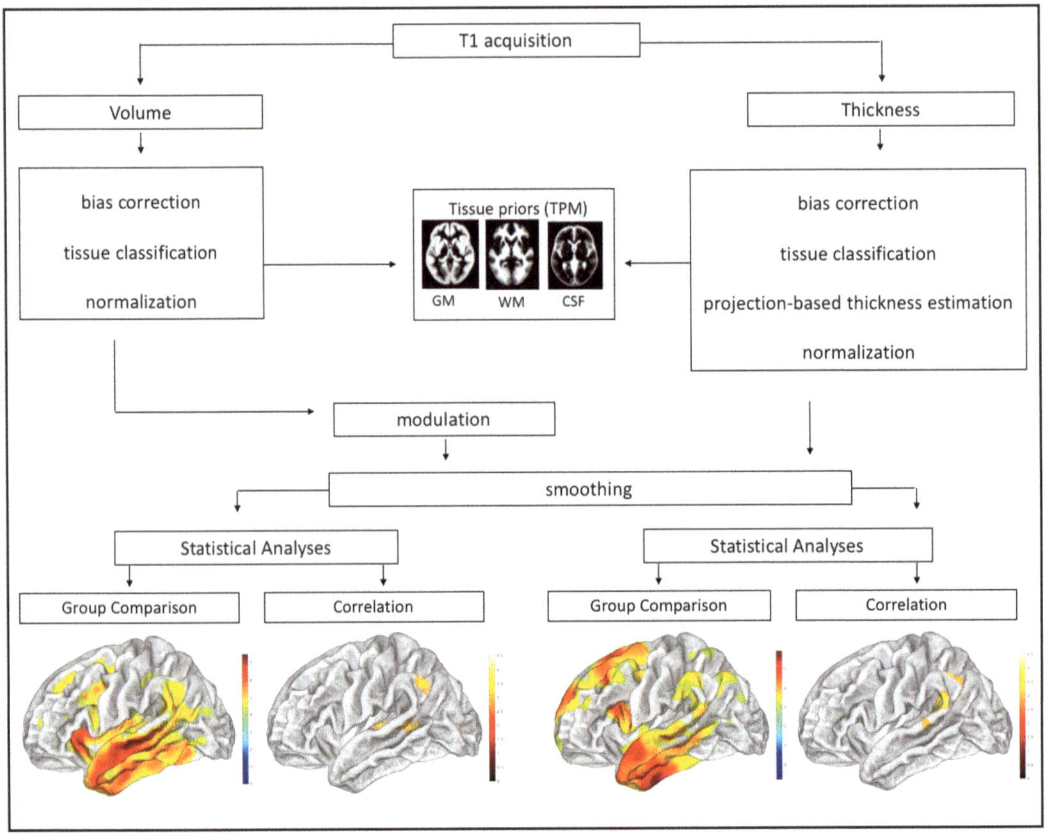

Fig. 1 Comparisons of voxel-based (VBM) and surface-based morphometry (SBM) methods within the Computational Anatomy Toolbox (CAT12) in SPM12

1. Unified Segmentation

This step aims at extracting tissue probability maps of GM, WM, and CSF from the high-resolution T1-weighted image. This step is achieved by the so-called unified segmentation approach [12], which permits image registration, tissue classification, and correction of intensity inhomogeneity artifacts (technically called bias correction) to be combined within the same generative model. Brain damage quantification methods used for stroke lesions differ from those described here. For example, steps such as tissue segmentation have been automated in the case of neurodegeneration but are more frequently manual for strokes due to the higher qualitative variability in stroke lesions. There are a number of ways in the literature to perform the segmentation; SPM uses a combination of voxel intensity and prior tissue probability maps to guide it. The GM and WM tissue probability maps are then normalized to the MNI standard template to enable (1) signal averaging across participants for final statistical tests and (2) the report of the results in standard coordinate system. This image registration is accomplished through the algorithm called DARTEL (fast diffeomorphic image registration algorithm; *see* [13]), which improves registration accuracy by creating group-specific templates from the GM and WM segmentation maps in the subjects' native space (but *see* [14] on the standard and optimized VBM methods). The normalization step expands and contracts some regions in order to register them to the coordinate system of the MNI template. It also accounts for the amount of contractions and expansions via the "Modulation" step, so that the total amount of GM in the "Modulated GM" remains the same as it would have been in the original image. In the latest versions of SPM12, the segmentation algorithm is essentially the same as the one described in the unified segmentation paper [12], with an improved registration model; the ability to use multispectral data, e.g., simultaneous segmentation of other imaging modalities, like T2-weighted or proton density; and an extended set of tissue probability maps, including bone, soft tissue, and air/background, which allows a different model of the voxels outside the brain. The modulation is only applied to the volume-based data but not to the surface-based data. Additionally, for the surface-based data, the estimation of the cortical thickness (CT) relies on the tissue segmentation maps using the so-called projection-based thickness (PBT, [15]). The algorithm includes topology correction of the surface, using spherical harmonics [16]; the registration to a standard coordinate system is performed using an adaptation of the DARTEL algorithm to the surface.

2. Smoothing

Segmented T1-weighted MRI images are then smoothed by the convolution of the images with an isotropic Gaussian kernel. This step is able to correct for residual small misregistrations and increases the signal-to-noise ratio [17]. By the central limit theorem, smoothing makes the data more normally distributed, thus increasing the validity of the parametric statistical tests performed subsequently [18, 19]. There is no standard smoothing size, but a value of 8 and 15 mm full width at half maximum (FWHM) is generally applied to the volume- and surface-based MR images, respectively.

2.2 Statistical Analysis of MRI Data

Statistical analysis of MR data uses univariate voxel-by-voxel parametric tests based on the general linear model (GLM). This typically proceeds in two stages: first, the GLM design matrix is specified by including one row per subject and one column for each explanatory variable and then followed by estimation of GLM parameters using classical or Bayesian approaches. Correction for multiple comparisons is achieved using the random field theory. As the Modulation step only controls for regional expansions and contractions during the normalization step, a measure of total GM or total intracranial volume (TIV) may be used as a covariate of interest in volume-based data to correct for individual brain sizes or global atrophy effects in the statistical model. Including the overall amount of GM as a covariate in the model enables the detection of GM that is regionally specific, beyond any global differences. This is important when studying neurodegeneration, since patients' brains are characterized by a different extent of GM local atrophy. The output is a statistical parametric map showing voxels that are significantly related to the particular explanatory variable under study.

2.3 Methodological Advantages and Challenges

The VBM is a relatively simple technique for assessing local GM volume status, yet several methodological choices can influence the results (e.g., the amount of smoothing, the type of segmentation, the software version; [18]). Thus, it is crucial to detail each of the parameters used to perform the analyses in publications and be consistent when wishing to compare results to other previously published findings (e.g., for reproducibility purposes). Relying on the T1-weighted intensity value at each voxel, VBM has proven to have high sensitivity in capturing GM differences across groups but to be lacking in specificity. In fact, VBM provides global information about GM volume in the cortex without being able to separate the contribution of different GM properties such as cortical surface area, cortical folding, and cortical thickness. Thus, the detection of subtle cortical changes and the assessment of each of the abovementioned cortical metrics requires the use of dedicated algorithms—more computationally demanding—that reconstruct the cortical surface [20]. Compared to voxel-wise measures, these advanced approaches follow more closely the anatomical boundaries and thus permit a good estimation of several cortical properties [21]. The surface-based pipeline consists of several steps that ultimately are able to produce maps of several cortical measures, like cortical thickness, surface area, sulcation, and a gyrification index [22]. This type of approach has been used to study normal aging and suggests that cortical thickness has a higher sensitivity in measuring age-related decline in GM compared to VBM [20] (see also Box 1).

Finally, we would like to stress the importance of combining multiple modalities when conducting neuroimaging research, especially if aiming at characterizing diseases. VBM and SBM do not only differ in their algorithmic procedures; they also provide complementary information of brain structure and function. Contrasting the morphometric characterization of the cortex provided by the two approaches, one can rely on VBM for a more a general account of local tissue properties, while SBM ensures a more specific characterization of cortical measures.

Box 1. Surface-Based Analyses Using FreeSurfer

FreeSurfer (http://surfer.nmr.mgh.harvard.edu) has always been considered the main software to perform surface-based morphometry. The processing steps include bias field correction, white matter classification, hemisphere separation with brainstem and cerebellum removal, and reconstruction of white and gray matter surfaces. The cortical thickness is measured as a distance between the pial surface—defined as the boundary between gray matter and cerebrospinal fluid—and the white matter surface. Yet CAT12, despite some algorithmic differences in the estimation of the cortical surface, produces comparable cortical measures while leading to a significant reduction of computation time. A recent study by Seiger and colleagues [23] evaluated gray matter changes in patients with Alzheimer's disease (34 AD) compared with healthy controls (44 HC) using cortical thickness measurements. To assess putative superiority of one method over the other in detecting cortical thickness abnormalities in distinct brain regions between the two groups, two linear mixed models were separately calculated for FreeSurfer and CAT12 with Group (HC, AD) and 34 ROIs being fixed factors and cortical thickness being the dependent variable. Furthermore, the authors conducted a test-retest analysis using 19 HC subjects that were measured at two time points. The study showed that both methods were able to detect reliably atrophic temporal brain areas in patients with AD and that both methods had excellent test-retest variability scores. However, CAT12 showed significantly higher thickness estimations in almost all regions of the brain compared to FreeSurfer, which is considered the state-of-the-art approach, supporting the use of CAT12 as a reliable tool for CT estimation when computational resources are limited. Moreover, the graphical user interface of CAT12 integrated in SPM12 eases the conduction of this type of analyses, while FreeSurfer requires familiarity with command line. Yet, due to the discrepancy in the absolute CT values estimation between the two software, researchers should use caution when comparing CT values across studies.

3 Brain-Behavior Mapping in Neurodegenerative Diseases: Examples of Applications

As mentioned above, voxel-based and/or surface-based morphometry can be used to detect changes in gray matter in neurodegenerative disorders [2, 23] and/or to relate these changes to covariates of interest such as cognitive scores [24–27]. Here, we summarize and provide examples of brain mapping findings in neurodegenerative disorders such as primary progressive aphasia (PPA) and Alzheimer's disease (AD).

A seminal comprehensive cognitive, neuroimaging, and genetic study in the field of behavioral neurology leveraged VBM in patients with PPA ($N = 31$) and age- and gender-matched healthy controls ($N = 64$) to isolate the neurocognitive profiles of the three variants of PPA [2]. The study used the standard VBM procedure (as outlined above) and examined differences in gray matter volumes between groups of subjects (i.e., pairwise comparison between each PPA subgroup and healthy controls) using SPM99 software package. The total intracranial volume age and gender were entered into the design matrix as nuisance variables. Areas of significant atrophy in the three variants of PPA relative to controls are indicated in three different colors (red for the non-fluent variant, green for the semantic variant, and blue for the logopenic variant) in Fig. 2.

In another between-group comparison study, Seiger and colleagues [23] compared different cortical thickness measurements while evaluating gray matter changes in patients with AD ($N = 34$) compared with healthy controls ($N = 44$). The authors were

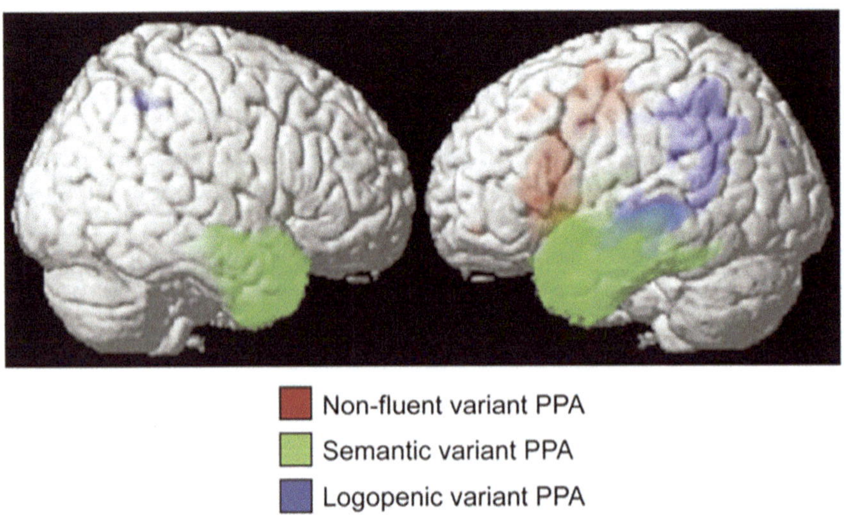

Fig. 2 Atrophy maps in all primary progressive aphasia (PPA) patients versus controls. (Adapted from Gorno-Tempini et al. [2])

particularly interested in determining whether there were any group differences in ROI-wise cortical thickness, when measurements were derived from the two different methods: the FreeSurfer software and CAT12 (*see* above for the workflow of the CAT12). Furthermore, the authors examined a test-retest reliability by measuring the thickness of the 19 healthy controls at two time points. The study showed that both methods were able to reliably detect atrophic temporal brain areas in patients with AD and that both methods had excellent test-retest variability scores (close to or even below 1). These findings suggest that CAT12 can be considered a fast, easy-to-use, and reliable alternative to FreeSurfer.

Voxel-based and/or surface-based morphometry has been also used to investigate within-group and across-group brain-behavior relationships. For example, the researchers identified regions where structural changes (e.g., brain atrophy) correlated with language impairments in PPA, like motor speech and linguistic features using connected speech [24, 28], naming deficits [27, 29, 30], and sentence comprehension and verbal working memory deficits [31–35].

In one voxel-based brain-symptom mapping study, Wilson and colleagues [24] provided the most comprehensive quantitative description to date of connected speech production and investigated the neural correlates of different aspects of abnormal connected speech in neurodegenerative aphasias. Specifically, the authors investigated the utility of automated structural MR image analysis to discriminate PPA variants ($N = 86$) from each other and from healthy controls ($N = 115$) but, more importantly, identified regions where atrophy correlated significantly with motor speech and linguistic features using connected speech. The authors adopted the standard VBM procedure in SPM5, also accounting for the potential confounds such as age, gender, TIV, and scanner (1.5 or 4T). Critically, while many previous studies have analyzed WM and GM separately [36, 37], in this procedure, they were added together, and the final results were then interpreted according to whether the effect fell in gray or white matter according to the template.

Similarly, in a recent study [26], we used surface-based morphometry (SBM) as implemented in CAT12 and correlated cortical thickness and repetition scores across PPA variants. In addition, for the purposes of this chapter, we also conducted VBM analyses (as also implemented in CAT12) on the same data. We wanted to simply test the capability of both VBM and SBM to identify structural differences in PPA patients ($N = 59$) as compared to healthy controls ($N = 25$) and associate these structural changes with repetition deficits. As shown in Fig. 1, for between-group comparison, both voxel- and surface-based methods produce comparable maps of significant brain atrophy in PPA patients compared to the healthy controls (p(FWE) > 0.05). Moreover, both analyses showed the correlation between the left temporoparietal areas and repetition deficits across PPA patients. Differences can also be appreciated. While the surface-based analyses revealed the two clusters of significance, i.e., inferior parietal (supramarginal/angular gyri) and superior temporal areas (p(FWE) > 0.05, $k > 90$), in the voxel-based methods, these clusters emerged with a reduced threshold ($p > 0.001$ uncorrected). Overall, while no extra computation time was required to run SBM analyses, visual comparison of the results suggests that this method can be used reliably to detect atrophic tissue and/or associate it with impairments.

These fully automated methods for analyzing MRI data provide highly reproducible and reliable measurement of brain atrophy in multiple neurodegenerative disorders. Another advantage of these quantitative MRI methods is that they offer objective measurement of brain atrophy over time, essential to the determination of the severity and progression of multiple neurodegenerative disorders. However, one limitation of these and other VBM and/or SBM studies should be acknowledged. Given the strong association between the anatomical and cognitive features of these syndromes, one can never rule out the possibility that the correlation detected by VBM and SBM methods is spurious or epiphenomenal. The atrophy/damage to a given region could be the direct cause of the symptoms or only happen to be the consistent neurological finding associated with it. These issues should be addressed in future work.

Overall, both research into the group comparisons and brain-behavioral correlations have contributed greatly to our understanding of the neural correlates of many cognitive and behavioral symptoms. For instance, the role of specific brain areas, such as the anterior temporal lobe (ATL), was first unveiled, thanks to the studies of neurodegenerative syndromes. Now included as a key region in virtually all language models, in particular as a major hub for word comprehension [3], the ATL has long been neglected by functional neuroimaging and classical aphasiology. SvPPA patients have been instrumental in revealing its pivotal role in semantic knowledge [38–44], a finding now corroborated by fMRI, TMS, and MEG.

Similarly, the role of so-called Wernicke's area in language comprehension has been primarily based on stroke aphasia literature, leaving its location and function uncertain. The controversy can be addressed by clinical neurodegenerative data. A recent study by Mesulam et al. [45] analyzed structural MRI in 72 patients with PPA and showed that neuronal loss in temporoparietal areas, traditionally included within Wernicke's area, is correlated with impaired sentence comprehension but intact single-word comprehension. This finding was supported by a subsequent meta-analytic study and review by Ardila, Bernal, and Rosselli [46, 47], who suggested that Wernicke's area includes not only temporal areas (BAs 21 and 22) but also parietal ones (BAs 41 and 42), implicating Wernicke's area in both phonological and lexical recognition of words. These studies provide novel evidence of unique structure-behavior relationships and further our insights into brain areas that cannot be examined in classic stroke aphasiology.

References

1. Whitwell JL, Josephs KA (2012) Recent advances in the imaging of frontotemporal dementia. Curr Neurol Neurosci Rep 12(6): 715–723
2. Gorno-Tempini ML, Dronkers NF, Rankin KP, Ogar JM, Phengrasamy L, Rosen HJ et al (2004) Cognition and anatomy in three variants of primary progressive aphasia. Ann Neurol 55(3):335–346
3. Mesulam MM (2013) Primary progressive aphasia and the language network: the 2013 H. Houston Merritt Lecture. Neurology 81(5):456–462
4. Battistella G, Henry M, Gesierich B, Wilson SM, Borghesani V, Shwe W et al (2019) Differential intrinsic functional connectivity changes in semantic variant primary progressive aphasia. NeuroImage Clin 22:101797
5. Cajanus A, Hall A, Koikkalainen J, Solje E, Tolonen A, Urhemaa T et al (2018) Automatic MRI quantifying methods in behavioral-variant frontotemporal dementia diagnosis. Dement Geriatr Cogn Disord Extra 8(1): 51–59
6. Rabinovici GD, Seeley WW, Kim EJ, Gorno-Tempini ML, Rascovsky K, Pagliaro TA et al (2008) Distinct MRI atrophy patterns in autopsy-proven Alzheimer's disease and frontotemporal lobar degeneration. Am J Alzheimers Dis Other Dement 22(6):474–488
7. McCarthy J, Collins DL, Ducharme S (2018) Morphometric MRI as a diagnostic biomarker of frontotemporal dementia: a systematic review to determine clinical applicability. NeuroImage Clin 20:685–696
8. Wright IC, McGuire PK, Poline JB, Travere JM, Murray RM, Frith CD et al (1995) A voxel-based method for the statistical analysis of gray and white matter density applied to schizophrenia. NeuroImage 2(4):244–252
9. Kaaden S, Quesada CM, Urbach H, Koenig R, Weber B, Schramm J et al (2011) Neurodevelopmental disruption in early-onset temporal lobe epilepsy: evidence from a voxel-based morphometry study. Epilepsy Behav 20(4): 694–699
10. Maguire EA, Gadian DG, Johnsrude IS, Good CD, Ashburner J, Frackowiak RS, Frith CD (2000) Navigation-related structural change in the hippocampi of taxi drivers. Proc Natl Acad Sci 97(8):4398–4403
11. Goodkind M, Eickhoff SB, Oathes DJ, Jiang Y, Chang A, Jones-Hagata LB et al (2015) Identification of a common neurobiological substrate for mental illness. JAMA Psychiat 72(4):305 315
12. Ashburner J, Friston KJ (2005) Unified segmentation. NeuroImage 26(3):839–851
13. Ashburner J (2007) A fast diffeomorphic image registration algorithm. NeuroImage 38(1): 95–113
14. Senjem ML, Gunter JL, Shiung MM, Petersen RC, Jack CR Jr (2005) Comparison of different methodological implementations of voxel-based morphometry in neurodegenerative disease. NeuroImage 26(2):600–608
15. Ashburner J, Friston K (1997) Multimodal image coregistration and partitioning—a unified framework. NeuroImage 6(3):209–217
16. Dahnke R, Yotter RA, Gaser C (2013) Cortical thickness and central surface estimation. NeuroImage 65:336–348
17. Yotter RA, Dahnke R, Thompson PM, Gaser C (2011) Topological correction of brain surface meshes using spherical harmonics. Hum Brain Mapp 32(7):1109–1124
18. Ashburner J, Friston KJ (2000) Voxel-based morphometry—the methods. NeuroImage 11(6):805–821
19. Nichols T, Hayasaka S (2003) Controlling the familywise error rate in functional neuroimaging: a comparative review. Stat Methods Med Res 12(5):419–446
20. Hutton C, Draganski B, Ashburner J, Weiskopf N (2009) A comparison between voxel-based cortical thickness and voxel-based morphometry in normal aging. NeuroImage 48(2): 371–380
21. Zatorre RJ, Fields RD, Johansen-Berg H (2012) Plasticity in gray and white: neuroimaging changes in brain structure during learning. Nat Neurosci 15(4):528
22. Dale AM, Fischl B, Sereno MI (1999) Cortical surface-based analysis: I. Segmentation and surface reconstruction. Neuroimage 9(2): 179–194

23. Seiger R, Ganger S, Kranz GS, Hahn A, Lanzenberger R (2018) Cortical thickness estimations of FreeSurfer and the CAT12 toolbox in patients with Alzheimer's disease and healthy controls. J Neuroimaging 28(5):515–523
24. Wilson SM, Henry ML, Besbris M, Ogar JM, Dronkers NF, Jarrold W et al (2010) Connected speech production in three variants of primary progressive aphasia. Brain 133(7): 2069–2088
25. Wilson SM, Dronkers NF, Ogar JM, Jang J, Growdon ME, Agosta F et al (2010) Neural correlates of syntactic processing in the nonfluent variant of primary progressive aphasia. J Neurosci 30(50):16845–16854
26. Lukic S, Mandelli M, Welch A, Jordan K, Shwe W, Miller Z, Neuhaus J, Hubbard IH, Henry M, Miller BL, Dronkers NF, Gorno-Tempini ML (2019) Neurocognitive basis of repetition deficits in primary progressive aphasia. Brain Lang 194:35–45
27. Lukic S, Borghesani V, Weis E, Welch A, Bogley R, Neuhaus J, Deleon J, Miller ZA, Kramer JH, Miller BL, Dronkers NF, Gorno-Tempini ML (2021) Dissociating nouns and verbs in temporal and perisylvian networks: evidence from neurodegenerative diseases. Cortex 142:47–61
28. Catani M, Mesulam MM, Jakobsen E, Malik F, Martersteck A, Wieneke C et al (2013) A novel frontal pathway underlies verbal fluency in primary progressive aphasia. Brain 136(8): 2619–2628
29. Grossman M, McMillan C, Moore P, Ding L, Glosser G, Work M, Gee J (2004) What's in a name: voxel-based morphometric analyses of MRI and naming difficulty in Alzheimer's disease, frontotemporal dementia and corticobasal degeneration. Brain 127(3):628–649
30. Brambati SM, Myers D, Wilson A, Rankin KP, Allison SC, Rosen HJ et al (2006) The anatomy of category-specific object naming in neurodegenerative diseases. J Cogn Neurosci 18(10):1644–1653
31. Grossman M, Mickanin J, Onishi K, Hughes E, D'Esposito M, Ding XS et al (1996) Progressive nonfluent aphasia: language, cognitive, and PET measures contrasted with probable Alzheimer's disease. J Cogn Neurosci 8(2): 135–154
32. Turner RS, Kenyon LC, Trojanowski JQ, Gonatas N, Grossman M (1996) Clinical, neuroimaging, and pathologic features of progressive nonfluent aphasia. Ann Neurol 39(2): 166–173
33. Amici S, Brambati SM, Wilkins DP, Ogar J, Dronkers NL, Miller BL, Gorno-Tempini ML (2007) Anatomical correlates of sentence comprehension and verbal working memory in neurodegenerative disease. J Neurosci 27(23): 6282–6290
34. Peelle JE, Troiani V, Gee J, Moore P, McMillan C, Vesely L, Grossman M (2008) Sentence comprehension and voxel-based morphometry in progressive nonfluent aphasia, semantic dementia, and nonaphasic frontotemporal dementia. J Neurolinguistics 21(5): 418–432
35. Wilson SM, Galantucci S, Tartaglia MC, Gorno-Tempini ML (2012) The neural basis of syntactic deficits in primary progressive aphasia. Brain Lang 122(3):190–198
36. Good CD, Johnsrude IS, Ashburner J, Henson RN, Friston KJ, Frackowiak RS (2001) A voxel-based morphometric study of ageing in 465 normal adult human brains. NeuroImage 14(1):21–36
37. Good CD, Scahill RI, Fox NC, Ashburner J, Friston KJ, Chan D et al (2002) Automatic differentiation of anatomical patterns in the human brain: validation with studies of degenerative dementias. NeuroImage 17(1):29–46
38. Snowden JS, Goulding PJ, Neary D (1989) Semantic dementia: a form of circumscribed cerebral atrophy. Behav Neurol 2:124043
39. Hodges JR, Patterson K, Oxbury S, Funnell E (1992) Semantic dementia: progressive fluent aphasia with temporal lobe atrophy. Brain 115(6):1783–1806
40. Rogers TT, Ralph L, Matthew A, Garrard P, Bozeat S, McClelland JL et al (2004) Structure and deterioration of semantic memory: a neuropsychological and computational investigation. Psychol Rev 111(1):205
41. Jefferies E, Lambon Ralph MA (2006) Semantic impairment in stroke aphasia versus semantic dementia: a case-series comparison. Brain 129(8):2132–2147
42. Patterson K, Nestor PJ, Rogers TT (2007) Where do you know what you know? The representation of semantic knowledge in the human brain. Nat Rev Neurosci 8(12):976
43. Hurley RS, Paller KA, Rogalski EJ, Mesulam MM (2012) Neural mechanisms of object naming and word comprehension in primary progressive aphasia. J Neurosci 32(14): 4848–4855
44. Mesulam MM, Wieneke C, Hurley R, Rademaker A, Thompson CK, Weintraub S, Rogalski EJ (2013) Words and objects at the tip of the left temporal lobe in primary progressive aphasia. Brain 136(2):601–618
45. Mesulam MM, Thompson CK, Weintraub S, Rogalski EJ (2015) The Wernicke conundrum

and the anatomy of language comprehension in primary progressive aphasia. Brain 138(8): 2423–2437
46. Ardila A, Bernal B, Rosselli M (2016) The role of Wernicke's area in language comprehension. Psychol Neurosci 9(3):340
47. Ardila A, Bernal B, Rosselli M (2016) How localized are language brain areas? A review of Brodmann areas involvement in oral language. Arch Clin Neuropsychol 31(1):112–122

Chapter 8

Lesion-Network Mapping: From a Topologic to Hodologic Approach

Amy Kuceyeski and Aaron Boes

Abstract

Historically, topological lesion-based studies have provided a direct way to map brain regions to their corresponding functions; however, these approaches fail to consider the impact of the lesion on broader brain networks. Focus has shifted in the latter half of the nineteenth century from topological to hodological, or network-based, approaches that link networks of brain regions to function. More recent advances in neuroimaging, including diffusion and resting-state functional MRI, have begun to reveal network-behavior maps in both healthy and clinical populations. An understanding of these relationships is imperative if we are to improve diagnostics, prognostics, and treatments of neurological and neuropsychiatric disorders. In this chapter, we begin by introducing general concepts in connectomics, including graph theory and connectome imaging techniques. We then outline strategies for quantifying lesions' impact on the connectome and, finally, discuss clinical applications and possible avenues of future work.

Key words Neuroimaging, Connectome, Brain-behavior mapping, Network-based analysis, Lesion-network mapping

1 Introduction

The human brain contains a vast 100 billion neurons connected through perhaps 100 trillion synapses. Electrochemical signals traversing this enormously complex interconnected network give rise to an individual's thought, emotion, and behavior. The consideration of the brain as a series of interacting networks, i.e., the hodological approach, emerged as a natural avenue of inquiry when two ideas became prevalent in neurology and psychology in the second half of the nineteenth century. One was anatomical—the focus on mapping white matter connections, or projection pathways, from the cortical surface. Such studies accentuated the heterogeneous nature of brain regions and assumed specific functions for each region based on differing patterns of structural connectivity. The other idea was functional—the shift in thinking of cognitive functions as requiring the successful integration of

Dorian Pustina and Daniel Mirman (eds.), *Lesion-to-Symptom Mapping: Principles and Tools*, Neuromethods, vol. 180, https://doi.org/10.1007/978-1-0716-2225-4_8, © Springer Science+Business Media, LLC, part of Springer Nature 2022

signals from various heterogeneous areas in the brain. Historically, topological lesion-based studies have provided a direct route for one-to-one mapping of regions necessary for certain functions. The observation of a lesion in a particular gray matter region of the brain in a person with a specific impairment enabled conclusions about that region's functional specialization, e.g., language production mapping to Broca's area [1] or language comprehension mapping to Wernicke's area [2]. However, this topological mapping approach fails to consider the networks in which the site of the lesion is embedded—whether a focal lesion disrupts white matter connections between regions in the network or damages gray matter regions that integrate signals from multiple regions in the network. One of the first examples of mapping a lesion's hodologic effects was the observation that lesions in the white matter connecting Broca's and Wernicke's areas impaired speech repetition [1]. Interestingly, a recent study reexamined the brains of Broca's original patients and also identified damage in the superior longitudinal fasciculus, an important white matter connection, that likely also contributed to the loss of speech production.

Connectivity mapping advances in the late nineteenth century had a reemergence in the mid-twentieth century with Norman Geschwind's seminal papers, where he coined the term "disconnexion syndromes" to describe such disorders as aphasia, apraxia, agnosia, and hemispatial neglect [3]. In the disconnection paradigm, gray matter regions on either end of the connection remain structurally intact, but their functionality is altered due to loss of the connection between them. In a similar vein, the destruction of gray matter regions that act as integration or coordination sites would also impact negatively its associated cognitive function. The modern conceptualization of the brain as an interconnected network of regions working in concert to produce complex cognitive processes has shifted approaches in brain-behavior mapping from a topological focus on localization of functions to discrete regions to the investigation of how lesions disrupt functional networks to produce symptoms [3–5]. Modern in vivo imaging techniques, including MRI, EEG, and PET, have provided methods with which to observe and quantify lesions' effects on the macroscopic brain network, i.e., the connectome [6, 7], and to map these network changes to cognitive, behavioral, or physical impairments [6, 8] as well as recovery [9, 10]. The impact of the lesion on the brain's connectivity network is critical to understand, as having a quantitative prediction of lesions' anatomical and physiological implications can help inform prognoses and possibly develop novel treatments. This chapter begins by introducing general concepts in connectomics, including graph theory and connectome imaging techniques, outlines strategies for quantifying lesions' impact on the connectome, and wraps up by discussing clinical applications and possible avenues of future work.

1.1 Connectomics and Graph Theory

The brain's connectivity network can be represented as a graph, consisting of a set of nodes connected through a system of edges that quantify nodal relationships, interactions, or physical connections (*see* Fig. 1). The nodes are generally defined by non-overlapping regions of cortical and subcortical gray matter, ranging from large-scale macroscopic regions to single neurons, depending on the experimental paradigm and methods used. In the structural connectome (SC; *see* Chapter 9), nodes are parcellated brain regions, and edges represent anatomical white matter fiber bundles—the brain's wiring—that allow communication between brain regions [11]. In the functional connectome (FC; *see* Chapter 10), edges represent some measure of similarity of temporal activation patterns within these brain regions. The SC and FC networks can be represented by an adjacency matrix, or graph, wherein the entry in the *i*th row and *j*th column gives the strength of connection between the *i*th and *j*th nodes. A graph's properties, i.e., how connected a particular node is to the rest of the network or how efficiently information flows between nodes, can be quantified using graph theory. Graph theory is a branch of pure mathematics originally developed by Euler when tackling the Konigsberg bridge problem [12], in which he defined a set of edges (in his case bridges) and nodes (in his case locations) that had certain properties depending on their topology or organization (in his case how efficiently one could travel from location to location). Graph theory is especially useful when assessing complex systems of interacting parts—exactly the brain's network under the hodologic approach. Graph theoretical metrics can be as simple as describing the number of connections per node (node degree) to quantifying the average length of the shortest path needed to travel between node pairs (characteristic path length). More complex measures like clustering coefficient, e.g., the number of connected triangles divided by the number of connected triples, can be used to quantify FC and SC networks. An in-depth discussion of specific graph theoretical measures is beyond the scope of this chapter; for a thorough review, *see* [13]. Complex graph theoretical measures have revealed brain networks to be "small world," i.e., their structure contains many shorter-range connections and a few long-range connections [14]. Small-worldness arises in many networks, from social to gene, and is more resource efficient and robust to attacks than other types of architectures [15, 16]. One particularly important nodal measure that has been shown by countless studies to be cognitively or behaviorally important is the so-called "hubs" of the small-world network—essentially those regions that are highly connected to the rest of the network and therefore may be integral in coordinating activity both within and between networks [17]. In the following sections, we provide a brief overview of the imaging modalities used to construct the brain's connectivity networks.

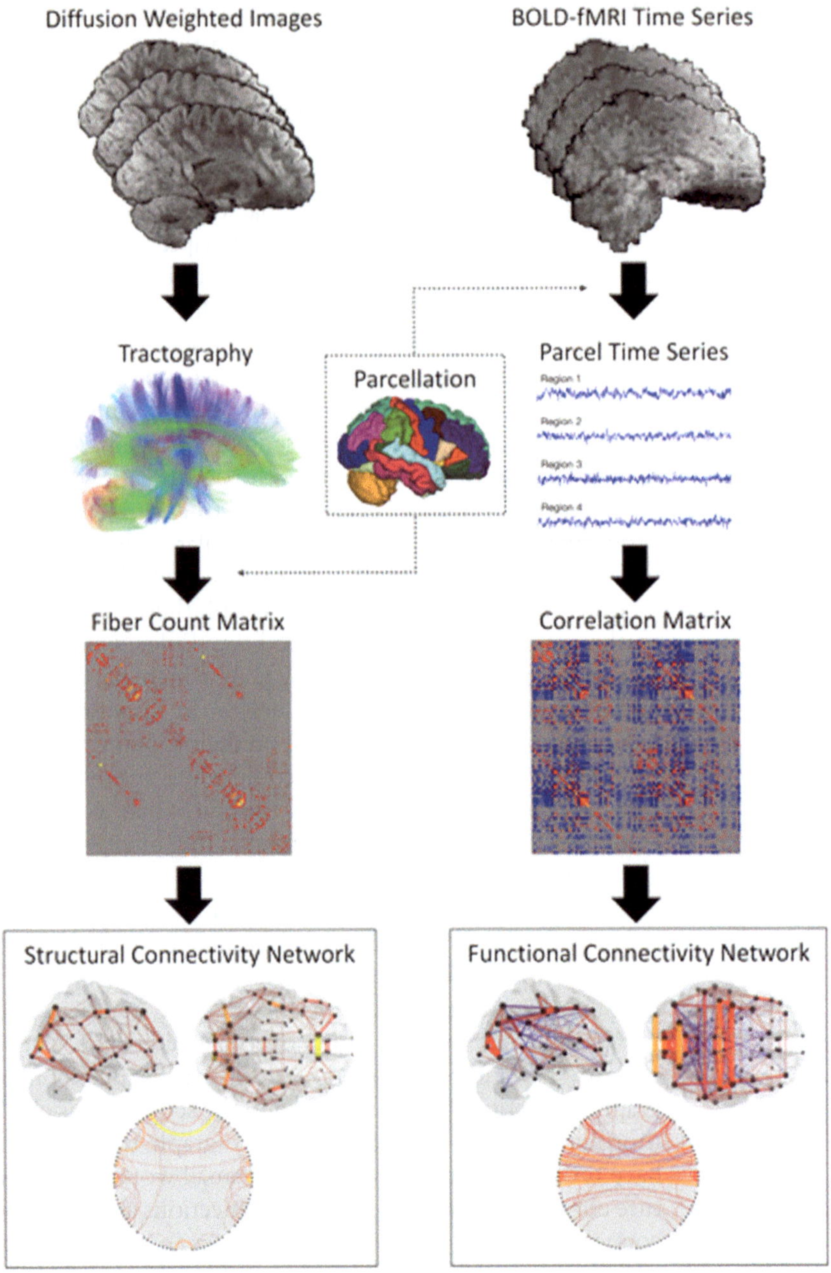

Fig. 1 Structural and functional connectome construction, from image acquisition to connectivity (adjacency) matrices, which can also be visualized with circle graphs. In a circle graph, the outer ring represents the brain regions, while the curved lines and their thickness represent the observed connections and their magnitudes

1.2 The Structural Connectome

Diffusion MRI (dMRI) allows quantification of the direction of diffusion of water molecules in the brain, which tend to travel more readily along myelinated white matter fiber bundles than across them [18]. Extraction of the orientation distribution function that represents the shape of diffusion at each image voxel allows for quantification of the directionality of the underlying fiber architecture [19]. The simplest of orientation distribution functions is that of a single tensor that can be visualized as an ellipsoid in three-dimensional space and is used to describe the underlying dMRI signal at each voxel—so-called diffusion tensor imaging (DTI) [20]. Summary statistics like fractional anisotropy (FA) and apparent diffusion coefficient (ADC) extracted from this tensor can provide a way to quantify its shape and obtain a proxy measure of the integrity of that voxel's white matter [21]. An extension of voxel-wise maps of white matter fiber orientation and integrity, tractography can estimate the strength of white matter connections between disparate brain regions. Tractography is a process wherein streamlines are constructed based on the orientation distribution function and individual anatomy, allowing approximate quantification (*see Issues and Limitations* below) of the location and strength of white matter connections between two predefined gray matter regions. An edge between region *i* and region *j* in the SC graph often represents the number of streamlines or probability of a connection between regions *i* and *j*, at times normalized by region volume or surface area to account for regional differences. Edges in the SC graph can also be constructed by averaging DTI summary statistics (e.g., FA) over voxels containing streamlines connecting region pairs.

Issues and Limitations As with any imaging modality, there are sources of noise in image acquisition and difficulties in postprocessing that can obscure the true anatomy of the white matter fibers. Particularly challenging are pathways that cross, overlap, branch, or have complex geometries—it is difficult to tease apart the diffusion signal contributing from two or more fiber bundles within one voxel [22]. Multi-shell diffusion MRI, which measures directionality of diffusion under multiple diffusion weightings, was developed to enable more accurate measurement of underlying axonal density and diameter and geometry. While tractography can reproduce accurately known connectivity, it has the disadvantage of also producing many false positive connections. One recent tractography challenge found that most of the 96 pipelines tested produced tractograms that contained 90% of the ground-truth bundles; however, these same tractograms systematically contained more invalid than valid bundles [23]. Global filtering or the use of anatomical priors has attempted to reduce these false positive effects.

1.3 The Functional Connectome

Functional connectivity is defined by correlated activity patterns measured directly or indirectly from the brain as a function of time. These activity patterns are derived from intrinsic physiological fluctuations that occur in the brain. A variety of modalities can be used to measure brain activity for functional connectivity analyses, including functional MRI (fMRI), electroencephalography (EEG), and magnetoencephalography (MEG). Of these, fMRI is the most commonly used approach for mapping the in vivo human functional connectome (FC). The nodes in the FC are again cortical or subcortical gray matter areas, while the edges represent some measure of similarity of activation over time—the most common is a correlation of the time series between region pairs. In fMRI, brain activity is estimated from blood-oxygen-level-dependent (BOLD) signal to infer patterns of regional activity with high spatial resolution [4, 24]. fMRI can be performed either while the subject is performing a task, e.g., finger-tapping or speaking, or during a task-free state. The latter version, so-called "resting-state" or task-free fMRI, was first studied by measuring the fluctuations in the BOLD signal in the motor cortex across time while the subject rested quietly in the scanner [25]. The pattern of BOLD fluctuations in the bilateral motor cortices was highly correlated, and a voxel-wise map displaying areas correlated with the motor cortex "seed" region revealed a network of structures that had previously been implicated in the motor system. Since this discovery, a number of canonical networks have been defined using resting-state functional MRI that supports specific cognitive or behavioral functions [26, 27]. This type of connectome-behavior mapping has been pushed to the forefront with the increase in availability of public, high-resolution, large-scale datasets, e.g., *the Human Connectome Project*, that has enabled data- and hypothesis-driven research into the cognitive or behavioral implications of connectome properties [28, 29].

1.4 Issues and Limitations

Functional connections represent properties of time-series data, often correlations, and not physical connections. Regions with highly correlated activity patterns may not have direct structural connections but rather result from indirect polysynaptic patterns of activation [24]. Moreover, brain regions can interact dynamically depending on the task conditions, and the complexity of these transient interactions may not be evident using a simple time-series correlation during resting conditions. fMRI specifically is not a direct measurement of neuronal activation, though many studies have shown the BOLD signal to be correlated with underlying neuronal activity [30]. There are also a number of challenges associated with processing functional connectivity MRI data, such as motion artifacts and physiological noise. Finally, the polysynaptic nature of the functional connectivity affects the interpretation of some of the graph theoretical metrics applied to functional networks. Metrics like characteristic path length and efficiency of FC networks may not be as straightforward to interpret as the same measures in SC networks, though they are widely used in the literature [31].

2 Quantifying the Lesioned Connectome

The function of each region of the brain is informed by its pattern of connections. Focal brain lesions from disease or damage, e.g., stroke, multiple sclerosis, tumor, or focal brain injury, disrupt these connections and the flow of information along the networks in which the lesioned site was embedded. A lesion's size and location will determine the impact it will have on the connectome and, therefore, the type and severity of resulting impairment. The fact that lesions have remote functional effects has been appreciated for over a century [32, 33]. Yet, one of the great challenges in neuroscience has been to understand how focal brain lesions disrupt networks to cause symptoms. There are two main approaches to mapping the hodologic effects of pathological lesions—one directly measures the connectome in patient populations, and the other uses healthy brain network maps to infer what networks were disrupted by the lesion.

2.1 Direct Measurement of the Lesioned Connectome

Imaging tools are effective in localizing the extent of the brain lesion across a variety of pathologies. For example, studies of MS [34, 35], stroke [36], and brain tumor patients show white matter lesions have altered diffusion MRI summary statistics that are hypothesized to reflect changes in myelination and axonal integrity. The location of acute ischemic lesions is identified clinically using diffusion MRI with a low number of measured directions in diffusion-weighted imaging (DWI) and the corresponding ADC maps [12]. In addition to quantifying pathology directly in the primary lesion site, there are tools available to map the hodologic effects of a lesion on the rest of the brain.

Focal brain lesions can have both structural and functional impact on areas outside the anatomical boundaries of the lesion. This notion of remote effects of a focal brain lesion was hypothesized over 100 years ago [33], and now there is conclusive supporting evidence, reviewed in [37–39]. Early studies mapping the remote functional effects of focal brain lesions, or diaschisis, showed alterations in metabolism and activation patterns in remote areas from the lesion using PET and fMRI [40]. Other studies supporting remote effects of a lesion demonstrate structural changes in remote sites in patients with lesions [41–50]. Importantly, these remote effects of the brain lesion are not at random locations but rather have been demonstrated to occur specifically at sites that are structurally or functionally connected to the lesion location [51–55].

2.2 Advantages and Disadvantages

Mapping a patient's connectome directly after the onset of a focal brain lesion provides an important method of evaluating the hodologic effects of lesion-symptom relationships. In addition, longitudinal patient studies can reveal recovery-related connectome reorganization in response to injury. Despite these obvious utilities, fMRI and dMRI studies can be difficult to implement in patients with brain lesions who may have cognitive or physical symptoms and cannot tolerate long scan times. Both approaches require advanced postprocessing expertise and additional resources for the acquisition of high-quality nonclinical imaging in patient populations. An additional drawback specific to dMRI in patients is that pathological changes to brain tissue, particularly proximal to the lesion, may introduce noise into the diffusion signal that can interfere with tractography results; *see* discussion in Crofts et al., 2011 [56]. On the fMRI side, any diseases or disorders that arise from or impact vasculature may add pathological noise to the BOLD signal that may not be straightforward to interpret, such as abnormal neurovascular coupling [57, 58]. Some of these limitations have been partially overcome through advanced preprocessing strategies [59]. Moreover, lesion-related neuronal impairment and subsequent loss of measurable activation signal preclude the elucidation of the original network organization prior to the lesion.

2.3 Lesion-Network Mapping: Inferring the Lesion-Associated Network from Normative Connectivity Data

Due to the limitations of connectome imaging in patient populations described above, there are an increasing number of studies that have utilized healthy connectome information to infer the network effects of focal brain lesions [60–64]. These techniques generally only require clinically acquired anatomical MRI (T1 or T2), for normalization purposes, and a lesion mask that indicates the area of pathology, e.g., stroke lesion, MS lesion, or tumor. The lesion masks are normalized to a common template, and estimates of the disconnectivity/connectivity profile of the lesion are generated from a large normative database of structural or functional networks.

One such tool that can map a lesion's impact on structural connectomes is the Network Modification (NeMo) Tool [65]; *see* Fig. 2. The NeMo Tool quantifies disruptions in the brain's structural network by mapping areas of damage or abnormality onto a large collection of healthy tractography data. The tool identifies streamlines that pass through the lesion mask and records the gray matter regions those streamlines connect. From this information, the tool estimates the resulting connectome changes at three different levels: changes in whole-brain network metrics (global), changes in the connectivity of gray matter regions to the rest of the network (regional), and changes in connectivity between pairs of gray matter regions (pairwise). In network terms, the regional-level measure, also called the Change in Connectivity (ChaCo) score, is akin to changes in a node's degree, while the pairwise level measure is akin to changes in network edge strength.

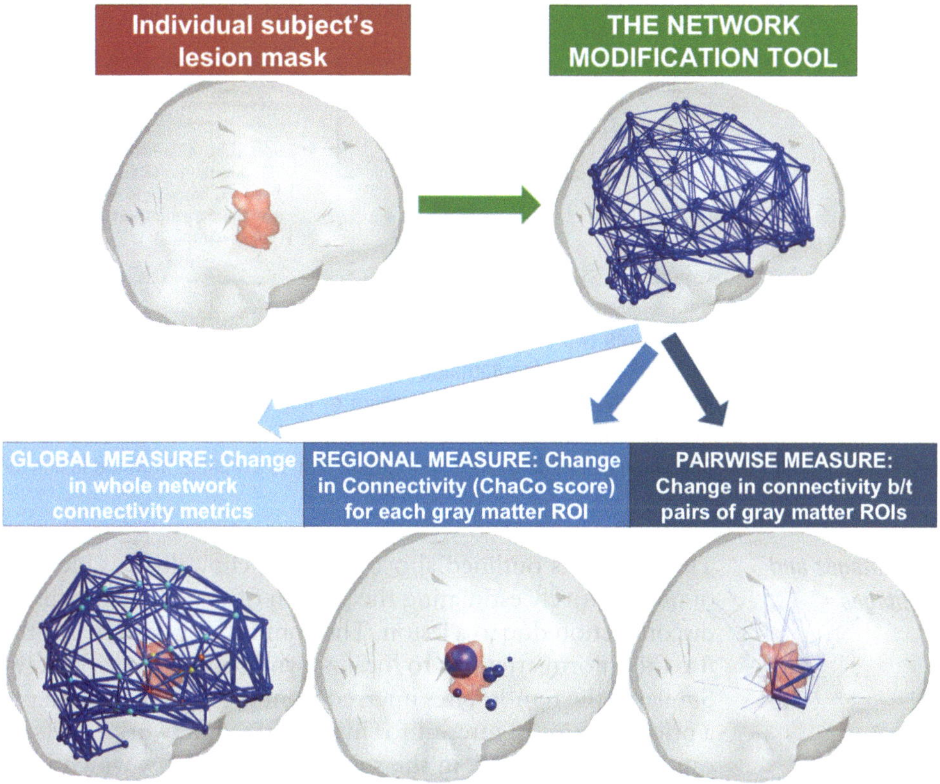

Fig. 2 The Network Modification Tool provides estimates of a lesion's impact on global, regional, and region-pair structural connectome measures by co-registering the lesion mask to a database of healthy control's tractography streamlines [9]

Another strategy to infer the functional network in which the lesion participated involves generating a functional connectivity network from the brain location affected by the lesion [60]. This process happens in two steps. First, the three-dimensional volume of a brain lesion is transferred from a structural image onto a reference brain. Next, the lesion volume is used in a resting-state fMRI (rs-fMRI) analysis using a large normative database to infer the network organization of the lesion site prior to the lesion. For lesion-network mapping, the time course of BOLD activity within the lesion volume is calculated from a large normative dataset and used to identify brain regions that are functionally connected to the lesion location in the healthy brain. As such, this approach infers the likely network organization of the brain tissue affected by a lesion based on population norms, and these inferred networks can be superimposed on the individual's brain. Lesion-network mapping has been applied to a number of lesion syndromes in recent years, ranging from hallucinations to coma [60, 66]; for a full review, *see* [62] and Chapter 10 (Fig. 3).

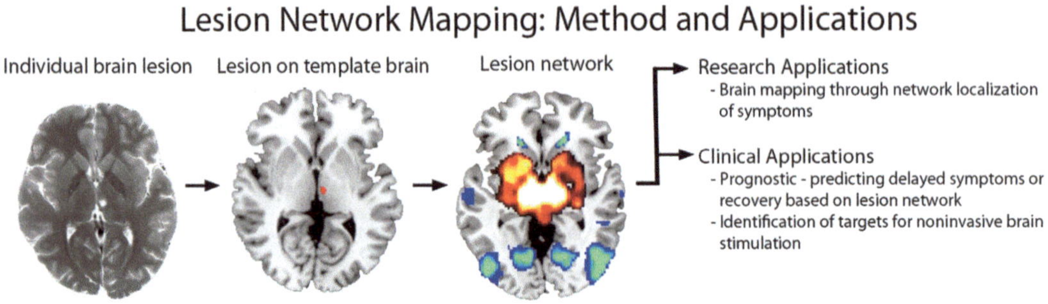

Fig. 3 Lesion-network mapping involves two steps: (1) a brain lesion from a clinical scan is mapped onto a reference brain, either traced by hand or through automated techniques (columns 1 and 2; *see* Chapters 2, 3, and 4), and (2) the lesion volume is then used as a seed region of interest for a resting-state functional connectivity MRI analysis that uses normative data (column 3). The lesion-associated networks can then be used for research or clinical applications, providing more information than is available from the lesion location alone

2.4 Advantages and Disadvantages

The two tools outlined above allow for clinically feasible methods of quantitatively estimating the pattern of structural and functional disconnection due to a lesion. The connectivity pattern can then be used to inform attempts to localize symptoms to specific networks. Similarly, the remote sites inferred from normative data can provide a priori regions of interest for hypothesized network effects, which can be tested directly in the subjects with lesions, which improves the statistical power relative to whole-brain searches. These methods relying on normative data do not require expensive, difficult-to-acquire images in the patient or require high level of expertise in image postprocessing of controls or subjects with focal brain pathology. However, there are drawbacks to these lesion-control connectome mapping methods. One is that the brain's connectomes vary greatly across the population, and these methods cannot account for an individual's particular anatomical or physiological patterns of connectivity. While attempts can be made to age and gender match the patients to the healthy connectomes used in these tools, there may be other influences on anatomy and physiology in the patient population that are not captured in the healthy connectome databases. For example, in stroke, there are other cardiovascular changes, white matter hyperintensities, and chronic cell loss that may impact the connectomes but are not directly related to the primary stroke lesion. Another drawback is that the accuracy of estimated disconnectivity patterns depends on the accurate mapping of an individual's pathological brain to a normal anatomical atlas, which can be challenging. Finally, this method does not convey information about the dynamic consequences of the lesion on networks that unfold over time as compensation occurs.

2.5 Structural-Functional Connectome Relationships and Virtual Lesions

There is converging evidence suggesting that the structural connectome constrains the functional connectome and therefore shapes patterns of spatiotemporal neuronal activity. Despite the fact that FC and SC are inexorably linked, most disconnectome studies focus on only one of the SC or FC networks' role in neurological disease. There are a few studies that incorporate both structural and functional connectome abnormalities together and even fewer that investigate the FC-SC relationship in the context of impairment or recovery [67]. One recent multimodality study showed that including information about the structural and functional connectivity network disruption due to stroke lesion could improve the prediction of aphasia severity [68]. Other recent work has focused on implementing mathematical models that formalize the relationship between SC and FC in both normal and pathological populations [69–76]. One popular method of FC-SC modeling is the use of neural mass models that describe the activation over time of large populations of neurons and are linked together according to the structural connectome. In this framework, the neural mass model equations include terms for endogenous region's properties as well as coupling parameters that allow long-range interactions between neuronal populations. One study in the context of lesions' impact on the FC-SC relationship used this neural mass modeling approach to simulate the effects of lesioning edges or nodes in the SC and quantifying the effect on the resulting FC [77]. Some of the main goals in joint structure-function modeling are to increase the accuracy of noisy connectivity measurements, identify function-specific subnetworks [76], or predict one modality from the other [73]. One recent publication with the goal of predicting function from structure used the network diffusion (ND) model [78, 79], which assumes functional activation diffuses along white matter connections. The ND model has been applied to predicting patterns of atrophy in dementia [80] and epilepsy [81] and in revealing FC-SC relationships in recovery from brain injury [82, 83].

Finally, some studies have attempted to simulate, either using noninvasive brain stimulation techniques or computational approaches, the effects of lesions on the connectomes [77]. Transcranial magnetic stimulation (TMS) can be used to suppress activation in particular brain areas and observe the resulting impairment, i.e., create a "virtual lesion" [15] (*see* Chapter 15). This approach has been used to disturb visual fields, cognitive function, and language production. Many studies have investigated the brain's ability to withstand an attack by quantifying the impact of "computational lesions," i.e., removing an edge or a node in the brain's connectome and measuring changes in the network's properties; *see* a review in [15]. One such study investigated how removing tractography streamlines passing through individual voxels effected overall SC graph theoretical measures; their resulting map combined with an individual's map of white matter integrity loss in traumatic brain injury subjects showed correlations with cognitive impairment [84].

2.6 Clinical Applications: Predicting Impairment and Recovery

Previous studies have shown correlation between the location of lesions with functional recovery in acute ischemic stroke [85, 86] and lesion size and location in motor recovery of patients with hemiplegic stroke [87]. However, knowing how lesion-induced network disruptions map to impairment and recovery may result in both deeper understanding of normal brain-behavior relationships as well as compensatory mechanisms in response to lesions [6, 9, 10, 60].

Studies that directly measure lesioned connectomes by performing imaging directly in patient populations [15, 88, 89] have shown that disruptions in the functional or structural connectomes can be mapped to baseline dysfunction and prediction of recovery [6–10, 90]. One recent study showed increased baseline functional connectivity in certain regions in stroke patients who recovered better than those who did not [10]. Connectome-based lesion-symptom mapping (CLSM; detailed in Chapter 9) focuses on measuring the structural connectome in patient populations using dMRI and tractography directly in patients. This method showed promise in identifying pairwise connections that are related to post-stroke impairments [91] and treatment response in Parkinsonism [92]. Other lesion-based patient studies show that dMRI summary statistics of white matter integrity, e.g., FA, in the cerebrospinal tract can be used to predict recovery from baseline impairment [19, 93].

The NeMo Tool's estimates of regional SC network disruption due to a lesion mask have been shown to cross-sectionally relate to processing speed in early-stage multiple sclerosis [94] and various cognitive and motor impairments after stroke [8]. Furthermore, the NeMo Tool's estimates of the lesioned connectome have been used to predict longitudinal clinical outcomes in multiple sclerosis [95] and stroke [9]. All of these papers using the NeMo Tool found that the predictions of models based on patient demographics and lesion volume did not explain as much variance as models based on the estimated SC disruption measures, providing evidence for a hodologic versus a topologic approach for accurate brain-behavior mapping and clinical applications. There was one study, however, that found adding the NeMo Tool's ChaCo scores of connectivity disruption did not improve accuracy of post-stroke language recovery prediction beyond that afforded by mapping the overlap of lesions in a gray matter atlas [96]. The NeMo Tool was also used to show that gray matter regions more structurally connected to areas of infarct at baseline tended to have worse pathology at follow-up [55]. Being able to predict regions vulnerable to secondary degeneration may enable localized therapeutic strategies aimed to mitigate the spread of damage.

In terms of lesion-network mapping with normative functional connectivity data, several studies have used lesion-network mapping to highlight regional network nodes that correspond to existing therapeutic targets for neuromodulation, such as the extra-striate visual cortex in visual hallucinations [60, 97], the motor cortex leg area for freezing of gait [98, 99], or supplementary motor area in hyperkinetic movement disorders [100, 101]. This raises the testable hypothesis that other regional nodes of networks derived from lesion-network mapping act as therapeutic targets, potentially identifying nodes of the network that are more accessible to noninvasive approaches such as transcranial magnetic stimulation.[1]

3 Conclusions and Opportunities for Future Work

One opportunity for future work lies in the validation of the estimated measures of structural or functional network disruption, i.e., comparison of extracted connectomes from individual patient data to estimated ones. A related validation step would be to show that the predictive power of the estimated measures of disruption is comparable to measures extracted from real individuals' data. Gaining a clear picture of how focal lesions result in symptoms by disrupting brain networks, either by estimated or observed disconnectivity metrics, will help to clarify the functional organization of the human brain. Moreover, this network perspective will inform our understanding of the mechanisms driving network-level reorganization that occurs with recovery of functions. This will enable the development of biomarkers for clinical use that can improve the accuracy of prognoses and aid in the development of personalized treatment plans using a precision medicine framework [102]. These personalized treatments could be based on cognitive or physical therapeutic approaches, possibly augmented by pharmacotherapy or neuromodulation. Noninvasive brain stimulation effects have been shown to depend on the integrity of white matter pathways in the SC [15] and to modify the brain's FC networks to boost recovery from stroke, depression, and TBI [103–105]. Currently, the choice of targets for brain stimulation is not well defined; many times it depends on population-level observations or heuristics. If we can better understand the role of a brain region in carrying out functional tasks within the brain's networks, and the compensatory structural and functional network reorganization that occurs following a focal brain lesion, then we may be able to identify more effective strategies to augment these plasticity mechanisms. This

[1] While showing promise for future clinical applications, to the authors' knowledge, there are currently no clinical applications of the lesion connectome mapping techniques.

method for personalized target selection could be applied in a variety of neurological disorders, improving recovery and quality of life for patients with a range of neurological diseases.

References

1. Broca PP (1861) Loss of speech, chronic softening and partial destruction of the anterior left lobe of the brain. Bull la Société Anthropol 2:235–238
2. Wernicke C (1874) Der aphasische Symptomencomplex: eine psychologische Studie auf anatomischer basis. Max Cohn & Weigert, Breslau
3. Geschwind N (1965) Disconnexion syndromes in animals and man. Brain 88. https://doi.org/10.1093/brain/88.2.237
4. Mesulam M (2005) Imaging connectivity in the human cerebral cortex: the next frontier? Ann Neurol 57:5–7. https://doi.org/10.1002/ana.20368
5. Catani M, Mesulam M (2008) What is a disconnection syndrome? Cortex 44:911–913. https://doi.org/10.1016/j.cortex.2008.05.001
6. Siegel JS, Ramsey LE, Snyder AZ et al (2016) Disruptions of network connectivity predict impairment in multiple behavioral domains after stroke. Proc Natl Acad Sci U S A 113: E4367–E4376. https://doi.org/10.1073/pnas.1521083113
7. Johansen-Berg H, Scholz J, Stagg CJ (2010) Relevance of structural brain connectivity to learning and recovery from stroke. Front Syst Neurosci 4:146. https://doi.org/10.3389/fnsys.2010.00146
8. Kuceyeski A, Navi BB, Kamel H et al (2015) Exploring the brain's structural connectome: a quantitative stroke lesion-dysfunction mapping study. Hum Brain Mapp 36:2147–2160. https://doi.org/10.1002/hbm.22761
9. Kuceyeski A, Navi BB, Kamel H et al (2016) Structural connectome disruption at baseline predicts 6-months post-stroke outcome. Hum Brain Mapp 37:2587–2601. https://doi.org/10.1002/hbm.23198
10. Puig J, Blasco G, Alberich-Bayarri A et al (2018) Resting-state functional connectivity magnetic resonance imaging and outcome after acute stroke. Stroke 49:2353–2360. https://doi.org/10.1161/STROKEAHA.118.021319
11. Hagmann P, Cammoun L, Gigandet X et al (2008) Mapping the structural core of human cerebral cortex. PLoS Biol 6:e159. https://doi.org/10.1371/journal.pbio.0060159
12. Euler L (1736) Solutio problematis ad geometriam situs pertinentis. Comment Acad Sci U Petrop 8:126–140
13. Bullmore E, Sporns O (2009) Complex brain networks: graph theoretical analysis of structural and functional systems. Nat Rev Neurosci 10:186–198. https://doi.org/10.1038/Nrn2618
14. Watts DJ, Strogatz SH (1998) Collective dynamics of "small-world" networks. Nature 393:440–442. https://doi.org/10.1038/30918
15. Aerts H, Fias W, Caeyenberghs K et al (2016) Brain networks under attack: robustness properties and the impact of lesions. Brain 77:137–185. https://doi.org/10.1093/brain/aww194
16. Bassett DS, Bullmore E (2006) Small-world brain networks. Neuroscientist 12:512–523. https://doi.org/10.1177/1073858406293182
17. van den Heuvel MP, Sporns O (2013) Network hubs in the human brain. Trends Cogn Sci 17:683–696. https://doi.org/10.1016/J.TICS.2013.09.012
18. Le Bihan D, Breton E (1985) Imagerie de diffusion in-vivo par résonance magnétique nucléaire. Comptes-Rendus de l'Académie des Sciences 93(5):27–34
19. Puig J, Pedraza S, Blasco G et al (2010) Wallerian degeneration in the corticospinal tract evaluated by diffusion tensor imaging correlates with motor deficit 30 days after middle cerebral artery ischemic stroke. AJNR Am J Neuroradiol 31:1324–1330. https://doi.org/10.3174/ajnr.A2038
20. Le Bihan D, Mangin J-F, Poupon C et al (2001) Diffusion tensor imaging: concepts and applications. J Magn Reson Imaging 13: 534–546. https://doi.org/10.1002/jmri.1076
21. Douek P, Turner R, Pekar J et al (1991) MR color mapping of myelin fiber orientation. J Comput Assist Tomogr 15:923–929
22. Jbabdi S, Johansen-Berg H (2011) Tractography: where do we go from here? Brain Connect 1:169–183. https://doi.org/10.1089/brain.2011.0033

23. Maier-Hein KH, Neher PF, Houde J-C et al (2017) The challenge of mapping the human connectome based on diffusion tractography. Nat Commun 8:1349. https://doi.org/10.1038/s41467-017-01285-x
24. Fox MD, Raichle ME (2007) Spontaneous fluctuations in brain activity observed with functional magnetic resonance imaging. Nat Rev Neurosci 8:700–711. https://doi.org/10.1038/nrn2201
25. Biswal B, Yetkin FZ, Haughton VM, Hyde JS (1995) Functional connectivity in the motor cortex of resting human brain using echo-planar MRI. Magn Reson Med 34:537–541
26. Buckner RL, Krienen FM, Yeo BTT (2013) Opportunities and limitations of intrinsic functional connectivity MRI. Nat Neurosci 16:832–837. https://doi.org/10.1038/nn.3423
27. Ferguson MA, Anderson JS, Spreng RN (2017) Fluid and flexible minds: intelligence reflects synchrony in the brain's intrinsic network architecture. Netw Neurosci 1:192–207. https://doi.org/10.1162/NETN_a_00010
28. Zimmermann J, Griffiths JD, McIntosh AR (2018) Unique mapping of structural and functional connectivity on cognition. J Neurosci 38:9658–9667. https://doi.org/10.1523/JNEUROSCI.0900-18.2018
29. Liégeois R, Li J, Kong R et al (2019) Resting brain dynamics at different timescales capture distinct aspects of human behavior. Nat Commun 10:2317. https://doi.org/10.1038/s41467-019-10317-7
30. Goense J, Whittingstall K, Logothetis NK (2012) Neural and BOLD responses across the brain. Wiley Interdiscip Rev Cogn Sci 3:75–86. https://doi.org/10.1002/wcs.153
31. Wang J, Zuo X, He Y (2010) Graph-based network analysis of resting-state functional MRI. Front Syst Neurosci 4:16. https://doi.org/10.3389/fnsys.2010.00016
32. Brown-Sequard C (1875) Seance du 18 de' cembre. C R Soc Biol 1875:424
33. Von Monakow C, Harris G (1914) Die lokalisation im grosshirn und der abbau der funktion durch kortikale herde. J. F. Bergmann, Wiesbaden
34. Sun S-W, Liang H-F, Le TQ et al (2006) Differential sensitivity of in vivo and ex vivo diffusion tensor imaging to evolving optic nerve injury in mice with retinal ischemia. NeuroImage 32:1195–1204. https://doi.org/10.1016/J.NEUROIMAGE.2006.04.212
35. Werring DJ, Clark CA, Barker GJ et al (1999) Diffusion tensor imaging of lesions and normal-appearing white matter in multiple sclerosis. Neurology 52:1626–1632
36. Werring DJ, Toosy AT, Clark CA et al (2000) Diffusion tensor imaging can detect and quantify corticospinal tract degeneration after stroke. J Neurol Neurosurg Psychiatry 69:269–272
37. Carrera E, Tononi G (2014) Diaschisis: past, present, future. Brain 137:2408–2422. https://doi.org/10.1093/brain/awu101
38. Fornito A, Zalesky A, Breakspear M (2015) The connectomics of brain disorders. Nat Rev Neurosci 16:159–172. https://doi.org/10.1038/nrn3901
39. Feeney DM, Baron JC (1986) Diaschisis. Stroke 17:817–830
40. Baron JC, D'Antona R, Serdaru M et al (1986) Cortical hypometabolism after a thalamic lesion in man: positron tomography study. Rev Neurol (Paris) 142:465–474
41. Zhang J, Meng L, Qin W et al (2014) Structural damage and functional reorganization in ipsilesional M1 in well-recovered patients with subcortical stroke. Stroke 45:788–793. https://doi.org/10.1161/STROKEAHA.113.003425
42. Kraemer M, Schormann T, Hagemann G et al (2004) Delayed shrinkage of the brain after ischemic stroke: preliminary observations with voxel-guided morphometry. J Neuroimaging 14:265–272. https://doi.org/10.1177/1051228404264950
43. Jones PW, Borich MR, Vavsour I et al (2016) Cortical thickness and metabolite concentration in chronic stroke and the relationship with motor function. Restor Neurol Neurosci 34:733–746. https://doi.org/10.3233/RNN-150623
44. Duering M, Righart R, Wollenweber FA et al (2015) Acute infarcts cause focal thinning in remote cortex via degeneration of connecting fiber tracts. Neurology 84:1685–1692. https://doi.org/10.1212/WNL.0000000000001502
45. Cheng B, Schulz R, Bönstrup M et al (2015) Structural plasticity of remote cortical brain regions is determined by connectivity to the primary lesion in subcortical stroke. J Cereb Blood Flow Metab 35:1507–1514. https://doi.org/10.1038/jcbfm.2015.74
46. Diao Q, Liu J, Wang C et al (2017) Gray matter volume changes in chronic subcortical stroke: a cross-sectional study. NeuroImage Clin 14:679–684. https://doi.org/10.1016/j.nicl.2017.01.031

47. Cai J, Ji Q, Xin R et al (2016) Contralesional cortical structural reorganization contributes to motor recovery after sub-cortical stroke: a longitudinal voxel-based morphometry study. Front Hum Neurosci 10:393. https://doi.org/10.3389/fnhum.2016.00393
48. Schaechter JD, Moore CI, Connell BD et al (2006) Structural and functional plasticity in the somatosensory cortex of chronic stroke patients. Brain 129:2722–2733. https://doi.org/10.1093/brain/awl214
49. Fan F, Zhu C, Chen H et al (2013) Dynamic brain structural changes after left hemisphere subcortical stroke. Hum Brain Mapp 34: 1872–1881. https://doi.org/10.1002/hbm.22034
50. Brodtmann A, Pardoe H, Li Q et al (2012) Changes in regional brain volume three months after stroke. J Neurol Sci 322:122–128. https://doi.org/10.1016/j.jns.2012.07.019
51. Nomura EM, Gratton C, Visser RM et al (2010) Double dissociation of two cognitive control networks in patients with focal brain lesions. Proc Natl Acad Sci U S A 107:12017–12022. https://doi.org/10.1073/pnas.1002431107
52. Gillebert CR, Mantini D, Thijs V et al (2011) Lesion evidence for the critical role of the intraparietal sulcus in spatial attention. Brain 134:1694–1709. https://doi.org/10.1093/brain/awr085
53. Lu J, Liu H, Zhang M et al (2011) Focal pontine lesions provide evidence that intrinsic functional connectivity reflects polysynaptic anatomical pathways. J Neurosci 31:15065–15071. https://doi.org/10.1523/JNEUROSCI.2364-11.2011
54. Duering M, Righart R, Csanadi E et al (2012) Incident subcortical infarcts induce focal thinning in connected cortical regions. Neurology 79:2025–2028. https://doi.org/10.1212/WNL.0b013e3182749f39
55. Kuceyeski A, Kamel H, Navi BB et al (2014) Predicting future brain tissue loss from white matter connectivity disruption in ischemic stroke. Stroke 45:717–722. https://doi.org/10.1161/STROKEAHA.113.003645
56. Crofts JJ, Higham DJ, Bosnell R et al (2011) Network analysis detects changes in the contralesional hemisphere following stroke. NeuroImage 54:161–169. https://doi.org/10.1016/j.neuroimage.2010.08.032
57. de Haan B, Rorden C, Karnath H-O (2013) Abnormal perilesional BOLD signal is not correlated with stroke patients' behavior. Front Hum Neurosci 7:669. https://doi.org/10.3389/fnhum.2013.00669
58. Veldsman M, Cumming T, Brodtmann A (2015) Beyond BOLD: optimizing functional imaging in stroke populations. Hum Brain Mapp 36:1620–1636. https://doi.org/10.1002/hbm.22711
59. Siegel JS, Snyder AZ, Ramsey L et al (2016) The effects of hemodynamic lag on functional connectivity and behavior after stroke. J Cereb Blood Flow Metab 36:2162–2176. https://doi.org/10.1177/0271678X15614846
60. Boes AD, Prasad S, Liu H et al (2015) Network localization of neurological symptoms from focal brain lesions. Brain:awv228. https://doi.org/10.1093/brain/awv228
61. Sutterer MJ, Bruss J, Boes AD et al (2016) Canceled connections: lesion-derived network mapping helps explain differences in performance on a complex decision-making task. Cortex 78:31–43
62. Fox MD (2018) Mapping symptoms to brain networks with the human connectome. N Engl J Med 379:2237–2245. https://doi.org/10.1056/NEJMra1706158
63. Raj A, Kuceyeski A, Weiner M (2012) A network diffusion model of disease progression in dementia. Neuron 73:1204–1215. https://doi.org/10.1016/j.neuron.2011.12.040
64. Zhou J, Gennatas ED, Kramer JH et al (2012) Predicting regional neurodegeneration from the healthy brain functional connectome. Neuron 73:1216–1227. https://doi.org/10.1016/j.neuron.2012.03.004
65. Kuceyeski A, Maruta J, Relkin N, Raj A (2013) The network modification (NeMo) tool: elucidating the effect of white matter integrity changes on cortical and subcortical structural connectivity. Brain Connect 3:451–463. https://doi.org/10.1089/brain.2013.0147
66. Fischer DB, Boes AD, Demertzi A et al (2016) A human brain network derived from coma-causing brainstem lesions. Neurology 87:2427–2434. https://doi.org/10.1212/WNL.0000000000003404
67. Kalinosky BT, Berrios Barillas R, Schmit BD (2017) Structurofunctional resting-state networks correlate with motor function in chronic stroke. NeuroImage Clin 16:610–623. https://doi.org/10.1016/J.NICL.2017.07.002
68. Del Gaizo J, Fridriksson J, Yourganov G et al (2017) Mapping language networks using the structural and dynamic brain connectomes.

eneuro 4:ENEURO.0204-17.2017. https://doi.org/10.1523/ENEURO.0204-17.2017
69. Cabral J, Hugues E, Sporns O, Deco G (2011) Role of local network oscillations in resting-state functional connectivity. NeuroImage 57:130–139. https://doi.org/10.1016/j.neuroimage.2011.04.010
70. Das TK, Abeyasinghe PM, Crone JS et al (2014) Highlighting the structure-function relationship of the brain with the Ising model and graph theory. Biomed Res Int 4:237898
71. Deco G, Senden M, Jirsa V (2012) How anatomy shapes dynamics: a semi-analytical study of the brain at rest by a simple spin model. Front Comput Neurosci 6:68. https://doi.org/10.3389/fncom.2012.00068
72. Fernández Galán R (2008) On how network architecture determines the dominant patterns of spontaneous neural activity. PLoS One 3:e2148. https://doi.org/10.1371/journal.pone.0002148
73. Honey CJ, Sporns O, Cammoun L et al (2009) Predicting human resting-state functional connectivity from structural connectivity. Proc Natl Acad Sci U S A 106:2035–2040. https://doi.org/10.1073/pnas.0811168106
74. Messé A, Rudrauf D, Benali H, Marrelec G (2014) Relating structure and function in the human brain: relative contributions of anatomy, stationary dynamics, and non-stationarities. PLoS Comput Biol 10:e1003530. https://doi.org/10.1371/journal.pcbi.1003530
75. Woolrich MW, Stephan KE (2013) Biophysical network models and the human connectome. NeuroImage 80:330–338. https://doi.org/10.1016/j.neuroimage.2013.03.059
76. Chu S-H, Parhi KK, Lenglet C (2018) Function-specific and enhanced brain structural connectivity mapping via joint modeling of diffusion and functional MRI. Sci Rep 8:4741. https://doi.org/10.1038/s41598-018-23051-9
77. Alstott J, Breakspear M, Hagmann P et al (2009) Modeling the impact of lesions in the human brain. PLoS Comput Biol 5:e1000408. https://doi.org/10.1371/journal.pcbi.1000408
78. Abdelnour F, Voss HU, Raj A (2014) Network diffusion accurately models the relationship between structural and functional brain connectivity networks. NeuroImage 90:335–347. https://doi.org/10.1016/j.neuroimage.2013.12.039
79. Abdelnour F, Dayan M, Devinsky O et al (2018) Functional brain connectivity is predictable from anatomic network's laplacian eigen-structure. NeuroImage 172:728–739. https://doi.org/10.1016/j.neuroimage.2018.02.016
80. Raj A, Kuceyeski A, Weiner M (2012) A network diffusion model of disease progression in dementia. Neuron 73:1204–1215. https://doi.org/10.1016/j.neuron.2011.12.040
81. Abdelnour F, Mueller S, Raj A (2015) Relating cortical atrophy in temporal lobe epilepsy with graph diffusion-based network models. PLoS Comput Biol 11:e1004564. https://doi.org/10.1371/journal.pcbi.1004564
82. Kuceyeski A, Jamison KW, Owen J et al (2019) Longitudinal increases in structural connectome segregation and functional connectome integration are associated with better recovery after mild TBI. Hum Brain Mapp. https://doi.org/10.1101/320515
83. Kuceyeski A, Shah S, Dyke JP et al (2016) The application of a mathematical model linking structural and functional connectomes in severe brain injury. NeuroImage Clin 11:635–647. https://doi.org/10.1016/j.nicl.2016.04.006
84. Kuceyeski A, Maruta J, Niogi SN et al (2011) The generation and validation of white matter connectivity importance maps. NeuroImage 58:109–121. https://doi.org/10.1016/j.neuroimage.2011.05.087
85. Nazzal ME, Saadah MA, Saadah LM, Trebinjac SM (2009) Acute ischemic stroke: relationship of brain lesion location & functional outcome. Disabil Rehabil 31:1501–1506. https://doi.org/10.1080/09638280802627702
86. Price CJ, Seghier ML, Leff AP (2010) Predicting language outcome and recovery after stroke: the PLORAS system. Nat Rev Neurol 6:202–210. https://doi.org/10.1038/nrneurol.2010.15
87. Chen CL, Tang FT, Chen HC et al (2000) Brain lesion size and location: effects on motor recovery and functional outcome in stroke patients. Arch Phys Med Rehabil 81:447–452. https://doi.org/10.1053/mr.2000.3837
88. Silasi G, Murphy TH (2014) Stroke and the connectome: how connectivity guides therapeutic intervention. Neuron 83:1354–1368. https://doi.org/10.1016/j.neuron.2014.08.052
89. Grefkes C, Fink GR (2014) Connectivity-based approaches in stroke and recovery of function. Lancet Neurol 13:206–216.

https://doi.org/10.1016/S1474-4422(13)70264-3

90. Carter AR, Astafiev SV, Lang CE et al (2010) Resting inter-hemispheric fMRI Connectivity predicts performance after stroke. Ann Neurol 67:365–375. https://doi.org/10.1002/ana.21905.Resting
91. Yourganov G, Fridriksson J, Rorden C et al (2016) Multivariate connectome-based symptom mapping in post-stroke patients: networks supporting language and speech. J Neurosci 36:6668–6679. https://doi.org/10.1523/JNEUROSCI.4396-15.2016
92. Hope TMH, Leff AP, Price CJ (2018) Predicting language outcomes after stroke: is structural disconnection a useful predictor? NeuroImage Clin 19:22–29. https://doi.org/10.1016/j.nicl.2018.03.037
93. Puig J, Blasco G, Daunis-I-Estadella J et al (2013) Decreased corticospinal tract fractional anisotropy predicts long-term motor outcome after stroke. Stroke 44:2016–2018. https://doi.org/10.1161/STROKEAHA.111.000382
94. Kuceyeski AF, Vargas W, Dayan M et al (2015) Modeling the relationship among gray matter atrophy, abnormalities in connecting white matter, and cognitive performance in early multiple sclerosis. Am J Neuroradiol 36:702–709
95. Kuceyeski A, Monohan E, Morris E et al (2018) Baseline biomarkers of connectome disruption and atrophy predict future processing speed in early multiple sclerosis. NeuroImage Clin 19:417–424. https://doi.org/10.1016/j.nicl.2018.05.003
96. Hope TMH, Leff AP, Price CJ (2018) Predicting language outcomes after stroke: is structural disconnection a useful predictor? NeuroImage Clin 19:22–29. https://doi.org/10.1016/J.NICL.2018.03.037
97. Merabet LB, Kobayashi M, Barton J, Pascual-Leone A (2003) Suppression of complex visual hallucinatory experiences by occipital transcranial magnetic stimulation: a case report. Neurocase 9:436–440. https://doi.org/10.1076/neur.9.5.436.16557
98. Kim MS, Chang WH, Cho JW et al (2015) Efficacy of cumulative high-frequency rTMS on freezing of gait in Parkinson's disease. Restor Neurol Neurosci 33:521–530. https://doi.org/10.3233/RNN-140489
99. Fasano A, Laganiere SE, Lam S, Fox MD (2017) Lesions causing freezing of gait localize to a cerebellar functional network. Ann Neurol 81:129–141. https://doi.org/10.1002/ana.24845
100. Brusa L, Versace V, Koch G et al (2005) Improvement of choreic movements by 1Hz repetitive transcranial magnetic stimulation in Huntington's disease patients. Ann Neurol 58:655–656. https://doi.org/10.1002/ana.20613
101. Laganiere S, Boes AD, Fox MD (2016) Network localization of hemichorea-hemiballismus. Neurology 86:2187–2195. https://doi.org/10.1212/WNL.0000000000002741
102. Collins FS, Varmus H (2015) A new initiative on precision medicine. N Engl J Med 372:793–795. https://doi.org/10.1056/NEJMp1500523
103. Demirtas-Tatlidede A, Vahabzadeh-Hagh AM, Bernabeu M et al (2013) Noninvasive brain stimulation in traumatic brain injury. J Head Trauma Rehabil 27:274–292. https://doi.org/10.1097/HTR.0b013e318217df55
104. Grefkes C, Fink GR (2011) Reorganization of cerebral networks after stroke: new insights from neuroimaging with connectivity approaches. Brain 134:1264–1276. https://doi.org/10.1093/brain/awr033
105. Nouri S, Cramer SC (2011) Anatomy and physiology predict response to motor cortex stimulation after stroke. Neurology 77:1076–1083. https://doi.org/10.1212/WNL.0b013e31822e1482

Chapter 9

Connectome-Based Lesion-Symptom Mapping Using Structural Brain Imaging

Ezequiel Gleichgerrcht, Janina Wilmskoetter, and Leonardo Bonilha

Abstract

Lesion-symptom mapping has been fundamental in furthering our understanding of the neurobiological basis of behavior and cognition. Since its inception, voxel-based approaches have helped establish the relationship between brain gray matter structures and behavioral function in an objective and quantifiable way. However, brain damage can extend well beyond the area of apparent gray matter injury, and functional deficits may also result from changes to the white matter tracts that provide the scaffolding for brain function. In this chapter, we discuss how connectome-based lesion-symptom mapping (CLSM) can help determine a statistical relationship between the strength of connections between brain regions across the brain and the wide variety of behavioral deficits seen in patients with different types of brain injury. We propose that CLSM can therefore provide valuable complementary information based on lesion-symptom mapping that is less constrained by cortical injury.

Key words Lesion-symptom mapping, Structural connectivity, Connectome, Diaschisis, Computational neuroimaging

1 Introduction

In Chap. 5, Baldo et al. delved into the way a technique originally introduced by Bates and collaborators [1], voxel-based lesion-symptom mapping, allowed the field of neuroscience to overcome some of the limitations previously faced by investigators who relied on inexact measurements of both brain injury and behavior. In this chapter, we will build on these ideas to introduce another approach that can further complement our understanding of the relationship between brain structure and function: connectome-based lesion-symptom mapping (CLSM).

Dorian Pustina and Daniel Mirman (eds.), *Lesion-to-Symptom Mapping: Principles and Tools*, Neuromethods, vol. 180, https://doi.org/10.1007/978-1-0716-2225-4_9,

2 The Limitations of Voxel-Based Lesion Analyses

While voxel-based lesion-symptom mapping (VLSM) has the advantage over fMRI to identify *critical* brain areas, because most lesion studies are caused by ischemic stroke, conclusions tend to be drawn about areas within the vascular constraints of the involved vessel. This is in contrast to fMRI, which partly relies on vascular constraints (i.e., structures within the same vascular territory tend to be more synchronous in their BOLD signal) but can still potentially contemplate synchronous areas anywhere in the brain. As such, lesion-behavior mapping approaches simply can neither confirm nor deny the critical role of areas outside the limits of the lesion without additional methodological techniques. Nevertheless, these techniques are a powerful tool to shed light on brain structure-function relationships *within* the confines of the lesion area. Thus, it is not surprising that VLSM has been extensively employed, especially in stroke populations, to understand different language functions [2–4], such as speech production [5–7], comprehension [1, 4], and fluency [1] as well as other behavioral impairments such as post-stroke dysphagia [8, 9], somatosensory deficits [10, 11], high-level perceptual deficits [12], and even mood disorders [13] among others.

Stroke is one of the major causes of brain damage and a commonly used disease model to identify critical brain areas associated with certain behaviors. Reasons are the abrupt and often focal damage to confined brain areas that provide opportunities for controlled observations of brain-behavior relationships. But because stroke is one the most widespread model of lesion mapping studies, one must also acknowledge the intrinsic bias that results from the organization of vascular supply in our brain. On the one hand, when vascular lesions occur in the proximal portion of a given vessel, brain injury occurs across different areas supplied by both proximal branches and distal and terminal branches, leading to an inability to statistically dissociate the contribution of these regions. On the other hand, and related to the point raised earlier, brain areas that lie outside the limits of typical perfusion territories are less susceptible to infarcts and may go unstudied because they are simply not affected by common brain insults (e.g., temporal pole, which is not commonly studied in stroke-induced aphasia because it is not perfused by the middle cerebral artery).

3 The Motivation of Connectome-Based Lesion Analyses

One key aspect when establishing associations between behavior and lesions is the fact that brain injury may extend past the area of apparent damage macroscopically seen on structural imaging. This is especially true in *disconnection* or damage to the white matter that may spread well beyond the limits of perceived necrosis (e.g., in the case of stroke) or resection (e.g., in the case of postoperative changes in cancer or epilepsy surgeries). Disconnection is important because it can present as remote dysfunction of otherwise apparently intact cortices [14–20] and is one of the key mechanisms for *diaschisis*, the dysfunction of specific brain regions due to injury or remote but connected neuronal populations [17, 21, 22].

Because of its nature, voxel-based lesion mapping can only provide an indirect and crude estimate of cortical disconnection. For example, with VLSM, one can map the location of lesioned voxels on a normative anatomical atlas of white matter tracts and calculate an indirect estimate of stroke lesion damage to certain tracts. However, white matter damage can extend beyond the confined areas of the stroke lesion and, thus, beyond the pool of lesioned voxels analyzed in VLSM approaches. One inherent limitation of lesion-behavior mapping is false negative results as a consequence of the statistical necessity that only voxels/regions in the brain can be analyzed that are lesioned in at least one participant (although a frequently used is a cutoff of 10% of participants).

It is in the context of the aforesaid limitations that a technique capable of exploring the association between different connectivity pathways and behavioral/cognitive performance may further our understanding of the relationship between damaged structure and function by means of complementary information that contemplates changes beyond the visible or apparent area of brain damage. A classic example of the importance for this comprehensive approach is not infrequently seen in post-stroke aphasia, where the clinical language profile of a given patient, that is, whether their aphasia is receptive, expressive, conductive, etc., does not necessarily match what is predicted based purely on the apparent location of the vascular injury [16, 23, 24]. As we will detail in the next sections, CLSM can leverage modern computational techniques for conventional lesion-symptom mapping by providing a comprehensive measure of network damage and residual extending beyond the visible lesion.

4 Methods of Connectome-Based Lesion-Symptom Mapping (CLSM)

The classic principle in lesion mapping stipulates that, for a given patient, injury to brain area X will lead to a loss or impairment of behavior A, and one can thus hypothesize that area X must be crucial for the execution of A. This framework can be extended to connectivity: if injury to the connection between areas Y and Z leads to loss or impairment of behavior B, then Y-Z must be crucial for execution of B. CLSM establishes the statistical relationship between injury to portions of the connectome and behavioral performance. While we will focus on structural networks in this chapter, the term "connectome" could be applied in this context to functional networks (e.g., resting state fMRI co-activation; see Chap. 10). Specifically, the structural connectome provides a roadmap of all meso- to macroscopic white matter physical connections in the brain. Modern imaging techniques combining white matter data from diffusion-weighted MRI [for review, see 25] and high-resolution gray matter maps can provide a highly reliable connectome at the individual level [26]. Importantly, the cortex is typically segmented into specific regions of interest (ROIs), usually based on predefined boundaries or standard criteria as determined by different brain-anatomical atlases. When the structural connectome is derived from a whole brain, essentially the strength of the connection between all possible pairs of ROIs is computed. The typical representation of a personalized connectome is thus by means of a two-dimensional weighted matrix, where each cell is the weight or "connection strength" between two brain regions. In this context, connection strength is not a real measure of physical white matter properties (e.g., number of axons, myelination, signal conductivity) but rather a relative concept represented by a quantified estimate of connectivity. The said connectivity strength can be estimated through a variety of methods, most commonly either by computing streamlines between pairs of ROIs (deterministic tractography) or by computing the probability of fibers connecting two specific ROIs (probabilistic tractography). In particular, connectomes reconstructed from deterministic tractography tend to have high sparsity given that a proportion of connections cannot be resolved (e.g., sites of fiber crossing); hence, deterministic connectome matrices are composed of a skewed distribution with many zeros and decaying prevalence of higher weight connections. Probabilistic connectome matrices, in turn, are also composed by skewed distributions with fewer links with higher weight, but the prevalence of zeros tends to be lower. Computational operations can also yield binary connectome matrices, for instance, by determining specific weight thresholds above which cell values will be "1" and below which cell values will be "0." It is clear, then, that individual connectomes vary in the type of information they feature depending on the approach applied to process and represent the data (step-by-step explanations of building a connectome will follow in the next section). The choice of approach should be the result of a balance between a number of variables, including the quality of the diffusion image (number of layers of diffusion data or "shells," number of diffusion directions, voxel size), the degree of confidence on the tractography algorithm to resolve complex anatomy, and the ability to process probabilistic tractography in parallel since it is a much more time-consuming approach than deterministic tractography. The connectivity values in the resulting matrix will dictate the type of statistical method that will be applied in CLSM:

binomial analyses or chi-square can be employed when dealing with binary connectomes, while parametric and nonparametric linear models can be employed when tackling weighted connectome matrices.

Our usual approach has been to use probabilistic tractography, since it yields a continuous measure of probability in terms of link weight that is amenable to linear models, including applications such as support-vector machine (SVM). Because link weight is a continuous measure, the general linear model (GLM) can be employed to establish a statistical relationship with the continuous behavioral variable in order to identify connections that are crucial for a given behavior. This approach also allows to define a weight threshold to determine the presence or absence of significant connections, yielding a binary connectome (as explained above) that can undergo GLM to compute an association with behavioral performance. A summary of the steps used to build the connectome is presented in the next section.

5 Step-by-Step CLSM

Brain damage can result in substantial neuroanatomical distortions that pose a significant challenge for neuroimaging analyses, particularly regarding spatial normalization and tissue segmentation. However, these are necessary steps during the reconstruction of the brain connectome, and, thus, we here present a step-by-step approach to construct the connectomes of individuals who survived stroke and have brain lesions secondary to post-ischemic necrosis. We opt to exemplify this with stroke, given the widespread use of this condition as a model for lesion mapping, but similar approaches could be extended to other forms of brain injury (Fig. 1).

1. File conversion: the raw magnetic resonance images, typically available in the DICOM format, must be converted to NIfTI, a popular file format among neuroscientists. It is important to choose a conversion that is able to extract and transform the diffusion gradient table [27]. An example of a freely available open-source software is dcm2niix, which converts DICOM to NIfTI (https://github.com/rordenlab/dcm2niix).
2. Lesion delineation: lesioned brain areas need to be (manually) traced to obtain masks of the area of injury that are required for the segmentation and registration process in step 3. Lesion masks are typically drawn on T2-weighted or T1-weighted images for chronic stroke and on diffusion-weighted images for acute stroke (see Chaps. 2 and 3). An example of a freely available open-source software to draw lesions is MRIcroGL, which includes dcm2niix (https://www.nitrc.org/projects/mricrogl).

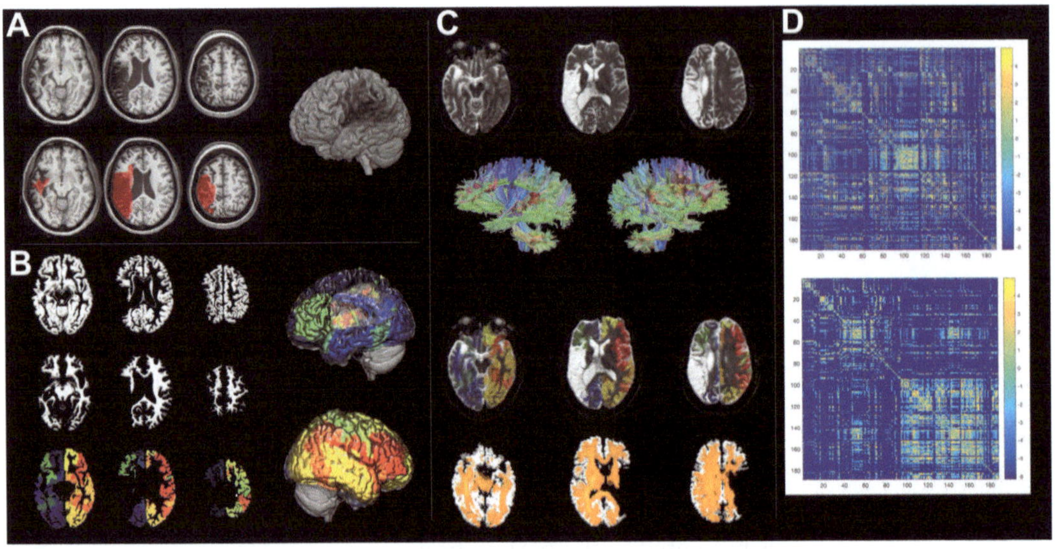

Fig. 1 The methodological steps involved in the calculation of the connectome share similarities with VLSM. First, the necrotic/gliotic image is defined on T1- or T2-weighted images as shown in Panel A. Again, here we see a 3D render of an individual patient's brain with a lesion. Subsequently, an iterative segmentation and cost function normalization approach is employed to define probabilistic maps of gray (Panel B, top row) and white (Panel B, middle row) matter. The transformation matrix from T1 to MNI space is used to transfer an anatomical atlas to T1-weighted space and segment the probabilistic gray matter into regions of interest (Panel C, bottom row). Panel C also shows the 3D renders of segmentation into regions of interest (left and right lateral views with different colors for different regions). Tractography is performed in diffusion space, so the white matter mask and the segmented gray matter maps are transferred to B0 space (Panel C), and tractography is used to assess the number of streamlines linking each possible pairs of regions. Care is taken to ensure that tractography is performed, being guided by the white matter probabilistic map, excluding the lesion site. The bottom row of Panel C shows a fiber density image in orange. Finally, a 2D matrix is generated where each entry represents the connection weight between the region in the row and column. The top matrix in Panel D shows the connectome, which is then arranged anatomically (Panel D, bottom matrix) to demonstrate the difference in the number of fibers in the left hemisphere (left upper matrix quadrant) versus the right hemisphere (right lower matrix quadrant)

3. Segmentation and normalization: the next step is to segment the brain in probabilistic gray and white matter maps, which will be used to define gray matter regions of interest and white matter masks for tractography. Conventional cortical segmentation programs have been designed for non-injured brains, and they typically lead to gross errors when processing brains with tissue damage. For this reason, it is essential to employ a tool that has been optimized for brains with injuries, particularly stroke survivors, although also applicable to surgical resections of tumors or epilepsy, discrete trauma, etc. The "Clinical Toolbox" is an extension of the Statistical Parametric Mapping (SPM) software, which was developed by our group [28] in order to optimize the segmentation and registration of brains with distorted anatomy due to large lesions. The "Clinical

Toolbox" is freely available (https://www.nitrc.org/projects/clinicaltbx/). The lesion mask (step 2) is used to minimize the impact of the lesion on the normalization estimates, either via explicit masking [29] or by substituting healthy tissue for homologous regions of the intact hemisphere [30] (see Chap. 4 for further explanation). This yields transformation matrices for normalization into standard stereotaxic space (MNI space) and vice versa. Normalization and tissue segmentation are performed iteratively, generating probabilistic tissue maps in native T1 and standard MNI space. Once transformation matrices are obtained, a neuroanatomical atlas, e.g., Automated Anatomical Labeling [31, 32], can be nonlinearly registered onto the probabilistic gray matter map (in native T1 space) and used to divide the gray matter into ROIs. Visual inspection of the probabilistic maps is important to ensure anatomically accurate segmentation. Subsequently, the ROIs generated from the previous step, as well as the white matter map, are moved to B0 space. To achieve this, the T1-weighted image can be registered to an inverted fractional anisotropy (FA) map. The motivation for using an inverted FA is to better match the gray and white matter gray levels and thus maximize the similarity between images. Alternatively, if T2-weighted image is available, it can be linearly co-registered onto the native T1 image and then linearly transformed to the B0 image. The transformation matrices are subsequently applied to the map of segmented ROIs and to the white matter probabilistic tissue map, yielding cortical ROIs and white matter maps in DWI space. The segmentation and normalization can be computed with our in-house developed freely available MATLAB scripts (https://github.com/neurolabusc/nii_preprocess).

4. Tractography: with all data in diffusion space, either a probabilistic or a deterministic diffusion tractography algorithm can now be applied to determine the strength of all pairwise connections between all ROIs set by the atlas. As stated earlier, this strength is either the probability that two ROIs are connected (probabilistic approach) or the number of streamlines with a correction that accounts for distance travelled and the volumes of the ROIs involved (deterministic approach). Probabilistic tractography can be computed with our in-house developed freely available MATLAB scripts (https://github.com/neurolabusc/nii_preprocess).

5. Statistical analyses: with a matrix of link strengths, CLSM is performed by assessing link or node-wise statistics. The statistical analyses applied in CLSM are similar to those applied in VLSM. To draw a parallel, in many ways, each cell in the connectome matrix in CLSM is treated like each voxel in

VLSM. The main difference is that in VLSM one assesses the relationship between lesioned voxels and impaired behavior and in CLSM one assesses the relationship between preserved connections and preserved behavior. This is an important difference when interpreting the results and also one of the main advantages of CLSM. In contrast to VLSM, where only brain areas can be assessed, which are lesioned in at least one participant, with CLSM, potentially any connection can be assessed because the variability in the independent MRI variable (connection strength in CLSM, lesion in VLSM) does not depend on the presence of a lesion. Much like in VLSM, correction for multiple comparisons is crucial (e.g., with permutation thresholding, false discovery rate, or Bonferroni correction among others). The number of segmented brain areas will dictate the number of connections and, thus, the number of multiple comparisons. CLSM can be performed using the freely available MATLAB scripts NiiStat (https://www.nitrc.org/projects/niistat/).

6. Additional analyses: furthermore, graph theory measures can be derived from this matrix to characterize global and regional aspects of the patient's connectome. The freely available MATLAB Brain Connectivity Toolbox (http://www.brain-connectivity-toolbox.net/) provides scripts to compute graph theory measures for structural and functional brain networks.

6 Applications of CLSM

CLSM can be applied in different ways to understand the relationship between white matter structure and function. A whole-brain connectome approach has the potential to identify key connections for specific behaviors without any a priori assumptions. For instance, our group [33] has previously demonstrated that CLSM can reveal connections between brain regions that contribute to specific language functions otherwise not predicted by VLSM. Specifically, we unveiled connections involving parietal regions that contribute to auditory comprehension as well as a link between the *pars orbitalis* and the dorsal part of the middle frontal gyrus crucial for speech repetition. This was achieved by employing a commonly used clinical assessment tool, the Western Aphasia Battery [34], which measures performance across different language functions. We have also applied this approach to more specific language outcomes, such as syntactic processing: damage to a ventral connection between temporal and frontal cortices was associated with impaired sentence comprehension and production, particularly of non-canonical sentences (i.e., syntactically more complex) [35]. Here, the behavioral measure was the

Northwestern Assessment of Verbs and Sentences, a task specifically designed to probe syntactic processing [36]. But behavioral measures in CLSM do not need to be confined to pen and paper or cognitive tasks. Any measurable outcome can potentially be studied with this approach. For example, physical therapist observations of upper and lower extremity motor function in individuals with chronic stroke were investigated using this approach, revealing that damage to the ipsilateral connections between lesioned primary motor cortex and cerebral peduncle, thalamus, and red nucleus all contributed to variability in contralateral limb locomotion [37].

CLSM can also help explain clinical outcomes (e.g., success of epilepsy surgery or likelihood of improving post-stroke anomia with speech therapy) that may be dependent on the integrity of white matter tracts beyond cortical sparing [38, 39]. The flexible nature of what "behavioral data" are used for CLSM (e.g., test performance, or clinical diagnosis, or outcome, or any other quantifiable variable) gives this approach the potential to be applicable across a wide array of neurological and psychiatric diseases as well as cognitive functions [e.g., 40] and behavioral domains.

The connectome lesion mapping approach can be further complemented by other computational methods, such as graph theory, in order to characterize the global and regional properties of the networks. For example, one can examine the network's efficiency and its organization into modules or motifs and even determine the influence that specific regions exert on the whole network. Exploring these measures and how they relate to behavioral outcomes has the potential to provide valuable clinical and theoretical information. For instance, if CLSM reveals that the connection between brain regions Y and Z is critical for behavior A, understanding the influence that Y and Z exert on the network by means of different graph theory measures such as "betweenness centrality" or "rich-hubness" or "modularity", among others, can be important for understanding structural-functional relations as well as for clinical prognosis and development of tailored rehabilitation interventions in the future. For example, Taylor and collaborators [41] showed in patients with temporal lobe epilepsy that 15 specific connections seemed to be useful in predicting seizure control after epilepsy surgery, again by relating a connectivity measure (connection strength) with a clinical measure (postoperative seizure outcome).

There are other approaches related to CLSM that are worth mentioning. Instead of using diffusion data from the participants themselves to perform individual tractography, another approach is to calculate virtual lesions of normative tractography. Here, normative white matter connectivity maps (e.g., calculated based on tractography from healthy individuals) are used in conjunction with the lesion map to estimate which tracts are damaged by the lesion [42–45] (see also the description of this method in Chap. 8).

For individual tractography (as outlined in this chapter), the whole-brain approach essentially focuses on fiber tractography between all ROIs defined by a cortical (i.e., gray matter) atlas, but a number of methods have been developed that rely directly on white matter atlases [46, 47]. For example, white matter bundle analysis relies on a predefined tractography atlas based on a priori anatomical knowledge of white matter pathways [48], such as the arcuate fasciculus, the superior longitudinal fasciculus, and so forth [e.g., see 49–51]. Similar methods identify pathways based on DTI metrics (e.g., variations in functional anisotropy), either by still requiring predefined known white matter bundles [e.g., 52] or by aligning DTI from multiple patients onto a single common space to infer a tract "skeleton" that represents the center of all tracts common to the pooled group [53]. These types of approach require a homogeneous distribution of links across subjects, and they can both underestimate link weights (e.g., by not including fibers running in parallel) and overestimate link weights (e.g., by counting fibers that happen to overlap with a specific predefined tract location on their way from one structure to another). This is likely prone to error in cases of damaged brains, with large variability in residual white matter connections. In favor of this approach – specifically in the context of injured brains – however, is the fact that probabilistic white matter maps can constrain tractography, thus limiting errors and demanding less time and computational resources. One must also consider that while CLSM estimates, for each connection, the fibers going between two specific ROIs (deterministic) or the probability of these two ROIs to be connected (probabilistic), white matter bundle approaches can possibly provide information about pathways with uncertain origins (i.e., with clear approximate white matter location but without knowing exactly what cortical regions those tracts are connecting), which can prove useful in several clinical and behavioral measures [e.g., [48, 54].

7 Limitations of CLSM

Like every neuroimaging method, CLSM has its limitations, and results should be interpreted with caution, especially in the context of the caveats intrinsic to tractography methods. Currently, there is no agreement on the best approaches to perform probabilistic or deterministic tractography leading to a lack of comparability between studies. No matter which approach is used, special attention should be placed to maximize the sensitivity and specificity of fiber tracking. This applies to healthy normal subjects and on an even higher level to subjects with brain damage where tractography can be exceptionally challenging. We recommend to avoid tracking in areas of gliosis and liquefactive necrosis (i.e., within the stroke lesion) and to further enhance the resolution of areas of fiber crossing.

Another limitation of CLSM is the segmentation of the brain into smaller parcels that are used as start and end points (seeds) for fiber tracking. As for any other approach that relies on brain segmentation, there are endless options on how to segment regions, for example, in terms of their size or where borders are set between regions. Because streamlines are estimated between two regions, the size and location of regions will impact the detection (or failed detection) of connections.

CLSM only assesses direct links between two regions. While direct links are crucial, brain network analyses have commonly demonstrated that brain signaling occurs in a complex network involving direct as well as indirect connections as means of communication between regions. Recently, we demonstrated that while direct connections between two areas can be lost due to a stroke lesion, indirect connections through a third (or more) region (s) might still be preserved, and those indirect connections independently predicted treated aphasia recovery [55]. Therefore, CLSM sheds light on one part of network organization but does not assess the entirety of network connections and communication between brain regions.

8 Conclusions

Lesion-behavior mapping is a very powerful approach in furthering our understanding of the neurobiological basis of behavior and can reveal a more direct relationship between structure and function than activation techniques. The emergence of voxel-based lesion mapping initially eliminated the need to predefine regions of interest or classify patients into groups based on arbitrary cutoff scores. It also introduced the ability to statistically evaluate the relationship between brain structure and behavioral function in an objective and quantifiable way. Yet, brain damage, as argued above, may extend well beyond the area of apparent gray matter injury, and functional deficits may result, among others, from changes to the white matter tracts that provide the scaffolding for brain function. CLSM is a whole-brain approach that aims to determine a statistical relationship between the strength of connections between all regions of the brain and the wide variety of behavioral performance seen in patients with different types of brain injury. CLSM can therefore provide valuable complementary information based on lesion-symptom mapping less constrained by cortical injury. This chapter serves as an introduction to a methodological approach with objective advantages for the study of brain structure-function relationships, because CLSM can identify connections crucial for behavior independently of structural presumptions. Nevertheless, for all lesion-behavior mapping approaches (VLSM, CLSM), confirming a statistical association between an imaging marker and a clinical symptom or behavior is critical in order to validate the biophysical relevance and the utility of the tool.

References

1. Bates E, Wilson SM, Saygin AP, Dick F, Sereno MI, Knight RT et al (2003) Voxel-based lesion-symptom mapping. Nat Neurosci 6(5): 448–450. https://doi.org/10.1038/nn1050
2. Fridriksson J, Yourganov G, Bonilha L, Basilakos A, Den Ouden DB, Rorden C (2016) Revealing the dual streams of speech processing. Proc Natl Acad Sci U S A 113(52):15108–15113. https://doi.org/10.1073/pnas.1614038114
3. Mirman D, Chen Q, Zhang Y, Wang Z, Faseyitan OK, Coslett HB et al (2015) Neural organization of spoken language revealed by lesion-symptom mapping. Nat Commun 6: 6762. https://doi.org/10.1038/ncomms7762
4. Dronkers NF, Wilkins DP, Van Valin RD, Jr., Redfern BB, Jaeger JJ. (2004) Lesion analysis of the brain areas involved in language comprehension. Cognition 92(1–2):145–177. https://doi.org/10.1016/j.cognition.2003.11.002
5. Borovsky A, Saygin AP, Bates E, Dronkers N (2007) Lesion correlates of conversational speech production deficits. Neuropsychologia 45(11):2525–2533. https://doi.org/10.1016/j.neuropsychologia.2007.03.023
6. Basilakos A, Rorden C, Bonilha L, Moser D, Fridriksson J (2015) Patterns of poststroke brain damage that predict speech production errors in apraxia of speech and aphasia dissociate. Stroke 46(6):1561–1566. https://doi.org/10.1161/strokeaha.115.009211
7. Fridriksson J, Kjartansson O, Morgan PS, Hjaltason H, Magnusdottir S, Bonilha L et al (2010) Impaired speech repetition and left parietal lobe damage. J Neurosci 30(33): 11057–11061. https://doi.org/10.1523/jneurosci.1120-10.2010
8. Galovic M, Leisi N, Pastore-Wapp M, Zbinden M, Vos SB, Mueller M et al (2017) Diverging lesion and connectivity patterns influence early and late swallowing recovery after hemispheric stroke. Hum Brain Mapp. https://doi.org/10.1002/hbm.23511
9. Wilmskoetter J, Bonilha L, Martin-Harris B, Elm JJ, Horn J, Bonilha HS. Mapping acute lesion locations to physiological swallow impairments after stroke. NeuroImag Clin 2019;22:101685. https://doi.org/10.1016/j.nicl.2019.101685
10. Meyer S, Kessner SS, Cheng B, Bonstrup M, Schulz R, Hummel FC et al (2016) Voxel-based lesion-symptom mapping of stroke lesions underlying somatosensory deficits. NeuroImag Clin 10:257–266. https://doi.org/10.1016/j.nicl.2015.12.005
11. Preusser S, Thiel SD, Rook C, Roggenhofer E, Kosatschek A, Draganski B et al (2015) The perception of touch and the ventral somatosensory pathway. Brain J Neurol 138(Pt 3): 540–548. https://doi.org/10.1093/brain/awu370
12. Karnath HO, Fruhmann Berger M, Kuker W, Rorden C (2004) The anatomy of spatial neglect based on voxelwise statistical analysis: a study of 140 patients. Cereb Cortex 14(10): 1164–1172. https://doi.org/10.1093/cercor/bhh076
13. Kim NY, Lee SC, Shin JC, Park JE, Kim YW (2017) Voxel-based lesion symptom mapping analysis of depressive mood in patients with isolated cerebellar stroke: a pilot study. NeuroImag Clin 13:39–45. https://doi.org/10.1016/j.nicl.2016.11.011
14. Carrera E, Tononi G (2014) Diaschisis: past, present, future. Brain : a journal of neurology 137(Pt 9):2408–2422. https://doi.org/10.1093/brain/awu101
15. Mukherjee P (2005) Diffusion tensor imaging and fiber tractography in acute stroke. Neuroimaging Clin N Am 15(3):655–665., xii. https://doi.org/10.1016/j.nic.2005.08.010
16. Fridriksson J, Bonilha L, Rorden C (2007) Severe Broca's aphasia without Broca's area damage. Behav Neurol 18(4):237–238
17. Bonilha L, Nesland T, Rorden C, Fillmore P, Ratnayake RP, Fridriksson J (2014) Mapping remote subcortical ramifications of injury after ischemic strokes. Behav Neurol 2014:215380. https://doi.org/10.1155/2014/215380
18. Bonilha L, Rorden C, Fridriksson J (2014) Assessing the clinical effect of residual cortical disconnection after ischemic strokes. Stroke 45(4):988–993. https://doi.org/10.1161/STROKEAHA.113.004137
19. Bonilha L, Fridriksson J (2009) Subcortical damage and white matter disconnection

associated with non-fluent speech. Brain 132 (Pt 6):e108. https://doi.org/10.1093/brain/awn200

20. Catani M, Mesulam M (2008) What is a disconnection syndrome? Cortex J Devoted Study Nerv Syst Behav 44(8):911–913. https://doi.org/10.1016/j.cortex.2008.05.001
21. Catani M, ffytche DH. (2005) The rises and falls of disconnection syndromes. Brain J Neurol 128(Pt 10):2224–2239. https://doi.org/10.1093/brain/awh622
22. Catani M, Dell'acqua F, Bizzi A, Forkel SJ, Williams SC, Simmons A et al (2012) Beyond cortical localization in clinico-anatomical correlation. Cortex J Devoted Study Nerv Syst Behav 48(10):1262–1287. https://doi.org/10.1016/j.cortex.2012.07.001
23. Croquelois A, Bogousslavsky J (2011) Stroke aphasia: 1,500 consecutive cases. Cerebrovasc Dis 31(4):392–399. https://doi.org/10.1159/000323217
24. Dronkers NF (2000) The pursuit of brain-language relationships. Brain Lang 71(1):59–61. https://doi.org/10.1006/brln.1999.2212
25. Assaf Y, Johansen-Berg H, Thiebaut de Schotten M (2017) The role of diffusion MRI in neuroscience. NMR Biomed. https://doi.org/10.1002/nbm.3762
26. Bonilha L, Gleichgerrcht E, Fridriksson J, Rorden C, Breedlove JL, Nesland T et al (2015) Reproducibility of the structural brain connectome derived from diffusion tensor imaging. PLoS One 10(8):e0135247. https://doi.org/10.1371/journal.pone.0135247
27. Li X, Morgan PS, Ashburner J, Smith J, Rorden C (2016) The first step for neuroimaging data analysis: DICOM to NIfTI conversion. J Neurosci Methods 264:47–56. https://doi.org/10.1016/j.jneumeth.2016.03.001
28. Rorden C, Bonilha L, Fridriksson J, Bender B, Karnath HO (2012) Age-specific CT and MRI templates for spatial normalization. NeuroImage 61(4):957–965. https://doi.org/10.1016/j.neuroimage.2012.03.020
29. Brett M, Leff AP, Rorden C, Ashburner J (2001) Spatial normalization of brain images with focal lesions using cost function masking. NeuroImage 14(2):486–500. https://doi.org/10.1006/nimg.2001.0845
30. Nachev P, Coulthard E, Jager HR, Kennard C, Husain M (2008) Enantiomorphic normalization of focally lesioned brains. NeuroImage 39(3):1215–1226. https://doi.org/10.1016/j.neuroimage.2007.10.002
31. Joliot M, Jobard G, Naveau M, Delcroix N, Petit L, Zago L et al (2015) AICHA: an atlas of intrinsic connectivity of homotopic areas. J Neurosci Methods 254:46–59. https://doi.org/10.1016/j.jneumeth.2015.07.013
32. Tzourio-Mazoyer N, Landeau B, Papathanassiou D, Crivello F, Etard O, Delcroix N et al (2002) Automated anatomical labeling of activations in SPM using a macroscopic anatomical parcellation of the MNI MRI single-subject brain. NeuroImage 15(1):273–289. https://doi.org/10.1006/nimg.2001.0978
33. Yourganov G, Fridriksson J, Rorden C, Gleichgerrcht E, Bonilha L (2016) Multivariate connectome-based symptom mapping in post-stroke patients: networks supporting language and speech. J Neurosci 36(25):6668–6679. https://doi.org/10.1523/jneurosci.4396-15.2016
34. Kertesz A (2007) The Western aphasia battery - revised. Grune & Stratton, New York
35. den Ouden DB, Malyutina S, Basilakos A, Bonilha L, Gleichgerrcht E, Yourganov G et al (2019) Cortical and structural-connectivity damage correlated with impaired syntactic processing in aphasia. Hum Brain Mapp 40(7):2153–2173. https://doi.org/10.1002/hbm.24514
36. Thompson CK (2011) Northwestern assessment of verbs and sentences. Northwestern University, Evanston
37. Peters DM, Fridriksson J, Stewart JC, Richardson JD, Rorden C, Bonilha L et al (2018) Cortical disconnection of the ipsilesional primary motor cortex is associated with gait speed and upper extremity motor impairment in chronic left hemispheric stroke. Hum Brain Mapp 39(1):120–132. https://doi.org/10.1002/hbm.23829
38. Gleichgerrcht E, Kocher M, Bonilha L (2015) Connectomics and graph theory analyses: novel insights into network abnormalities in epilepsy. Epilepsia 56(11):1660–1668. https://doi.org/10.1111/epi.13133
39. Bonilha L, Gleichgerrcht E, Nesland T, Rorden C, Fridriksson J (2016) Success of anomia treatment in aphasia is associated with preserved architecture of global and left temporal lobe structural networks. Neurorehabil Neural Repair 30(3):266–279. https://doi.org/10.1177/1545968315593808
40. Gleichgerrcht E, Fridriksson J, Rorden C, Nesland T, Desai R, Bonilha L (2016) Separate neural systems support representations for actions and objects during narrative speech in post-stroke aphasia. Neuroimag Clin 10:140–

145. https://doi.org/10.1016/j.nicl.2015.11.013

41. Taylor PN, Sinha N, Wang Y, Vos SB, de Tisi J, Miserocchi A et al (2018) The impact of epilepsy surgery on the structural connectome and its relation to outcome. Neuroimag Clin 18: 202–214. https://doi.org/10.1016/j.nicl.2018.01.028
42. Kuceyeski A, Maruta J, Niogi SN, Ghajar J, Raj A (2011) The generation and validation of white matter connectivity importance maps. NeuroImage 58(1):109–121. https://doi.org/10.1016/j.neuroimage.2011.05.087
43. Kuceyeski A, Navi BB, Kamel H, Raj A, Relkin N, Toglia J et al (2016) Structural connectome disruption at baseline predicts 6-months post-stroke outcome. Hum Brain Mapp 37(7):2587–2601. https://doi.org/10.1002/hbm.23198
44. Kuceyeski A, Maruta J, Relkin N, Raj A (2013) The network modification (NeMo) tool: elucidating the effect of white matter integrity changes on cortical and subcortical structural connectivity. Brain Connect 3(5):451–463. https://doi.org/10.1089/brain.2013.0147
45. Pustina D, Coslett HB, Ungar L, Faseyitan OK, Medaglia JD, Avants B et al (2017) Enhanced estimations of post-stroke aphasia severity using stacked multimodal predictions. Hum Brain Mapp 38(11):5603–5615. https://doi.org/10.1002/hbm.23752
46. Thiebaut de Schotten M, Kinkingnehun S, Delmaire C, Lehericy S, Duffau H, Thivard L et al (2008) Visualization of disconnection syndromes in humans. Cortex J Devoted Study Nervous Syst Behav 44(8):1097–1103. https://doi.org/10.1016/j.cortex.2008.02.003
47. Thiebaut de Schotten M, Dell'Acqua F, Ratiu P, Leslie A, Howells H, Cabanis E et al (2015) From Phineas Gage and Monsieur Leborgne to H.M.: revisiting disconnection syndromes. Cereb Cortex 25(12):4812–4827. https://doi.org/10.1093/cercor/bhv173
48. Catani M, Mesulam M (2008) The arcuate fasciculus and the disconnection theme in language and aphasia: history and current state. Cortex 44(8):953–961. https://doi.org/10.1016/j.cortex.2008.04.002
49. Catani M, Mesulam MM, Jakobsen E, Malik F, Martersteck A, Wieneke C et al (2013) A novel frontal pathway underlies verbal fluency in primary progressive aphasia. Brain 136(Pt 8): 2619–2628. https://doi.org/10.1093/brain/awt163
50. Catani M (2008) Thiebaut de Schotten M. A diffusion tensor imaging tractography atlas for virtual in vivo dissections. Cortex J Devoted Study Nervous Syst Behav 44(8):1105–1132. https://doi.org/10.1016/j.cortex.2008.05.004
51. Craig MC, Catani M, Deeley Q, Latham R, Daly E, Kanaan R et al (2009) Altered connections on the road to psychopathy. Mol Psychiatry 14(10):946–953. https://doi.org/10.1038/mp.2009.40
52. Ivanova MV, Isaev DY, Dragoy OV, Akinina YS, Petrushevskiy AG, Fedina ON et al (2016) Diffusion-tensor imaging of major white matter tracts and their role in language processing in aphasia. Cortex 85:165–181. https://doi.org/10.1016/j.cortex.2016.04.019
53. Smith SM, Jenkinson M, Johansen-Berg H, Rueckert D, Nichols TE, Mackay CE et al (2006) Tract-based spatial statistics: voxelwise analysis of multi-subject diffusion data. NeuroImage 31(4):1487–1505. https://doi.org/10.1016/j.neuroimage.2006.02.024
54. Agosta F, Henry RG, Migliaccio R, Neuhaus J, Miller BL, Dronkers NF et al (2010) Language networks in semantic dementia. Brain J Neurol 133(Pt 1):286–299. https://doi.org/10.1093/brain/awp233
55. Wilmskoetter J, Fridriksson J, Basilakos A, Phillip Johnson L, Marebwa BK, Rorden C, et al (2019) Propagation speed within the ventral stream predicts treatment response in chronic post-stroke aphasia. 11th annual meeting of the Society for the Neurobiology of Language (SNL). Helsinki, Finland

Chapter 10

Lesion Network Mapping Using Resting-State Functional Connectivity MRI

Juho Joutsa, R. Ryan Darby, and Michael D. Fox

Abstract

Brain lesions can allow for causal links between symptoms and human neuroanatomy. However, lesions causing the same symptom often fail to overlap a single brain region, leaving the localization unclear. Resting-state functional connectivity MRI is a powerful tool for mapping human brain networks. Using resting-state functional connectivity, one can test whether lesions causing the same symptom map to a functionally connected brain network rather than a single brain region. This approach, termed "lesion network mapping," has proven useful for mapping a wide variety of lesion-induced neurological and psychiatric symptoms to brain networks. These lesion network mapping results are reproducible across independent datasets and show promise for identifying therapeutic targets for neuromodulation. Here, we review the methodology for lesion network mapping using functional connectivity MRI.

Key words rs-fcMRI, Brain networks, Connectome, Diaschisis, Localization

1 Functional Lesion Network Mapping

1.1 General Principle

Brain lesions enable causal links between the location of brain damage and neurological and psychiatric symptoms [1–3]. However, a challenge of traditional lesion analyses has been that the lesions causing a single symptom or symptom complex often do not overlap. There are potentially multiple reasons for this lack of overlap, including a phenomena called *diaschisis*, in which damaged brain tissue has functional effects on remote but connected brain regions [4]. These remote functional effects can occur across multiple synapses and may be responsible for many lesion-induced symptoms. Up until very recently, there have not been tools to account for these remote effects in lesion analyses. However, new brain imaging techniques and large publicly available datasets have produced detailed maps of human brain connectivity that allow one to identify the network of remote brain regions to which a lesion location is connected [3].

Dorian Pustina and Daniel Mirman (eds.), *Lesion-to-Symptom Mapping: Principles and Tools*, Neuromethods, vol. 180,
https://doi.org/10.1007/978-1-0716-2225-4_10, © Springer Science+Business Media, LLC, part of Springer Nature 2022

Resting-state functional connectivity MRI (rs-fcMRI) is based on spontaneous fluctuations in brain activity that produce spontaneous fluctuations in blood oxygenation that can be measured with MRI. In rs-fcMRI scanning, subjects lay still in the scanner without performing any task. Brain regions that show synchronous activity, i.e., simultaneous increases or decreases in spontaneous brain activity, are considered to be functionally connected. Functional connections in rs-fcMRI correspond to activation patterns in task-based fMRI and follow the anatomical connections in the brain [5]. One of the important advantages of rs-fcMRI is that it can be used to define neural networks with polysynaptic connections and not just direct monosynaptic white matter tracts.

Lesion network mapping is similar to traditional lesion mapping in that both techniques seek to map lesion locations causing the same symptom to a common neuroanatomical substrate [6]. The main difference is that traditional lesion mapping seeks to map lesions to a common brain region, while lesion network mapping seeks to map lesions to a common brain circuit or network. As such, lesion network mapping allows for lesions causing the same symptom to occur in multiple different brain regions and then tests whether these lesion locations are part of a common functionally connected brain network. If all lesion locations causing a symptom are connected to the same brain region, then connectivity with this brain region defines a brain network that encompasses the lesion locations causing this symptom [3, 6], in other words, lesions causing the symptom map to a single brain network.

Resources needed for lesion network mapping include the lesion location from each individual patient and a functional connectome, generally from a large cohort of healthy subjects, which we will discuss more specifically in the next paragraphs. A major advantage of this technique is that no functional imaging data is needed from the lesion patients themselves. In fact, if one were to acquire rs-fcMRI data in these lesion patients, it would not be possible to compute connectivity with the lesion location as the neuronal tissue at the lesion location is lost. In addition, there can be functional changes and reorganization of brain connectivity after lesion has occurred that can complicate the interpretation of rs-fcMRI in patients with the brain lesion [7, 8].

Lesion network mapping proceeds in three steps (illustrated in Fig. 1): First, the lesion locations are mapped to a standard atlas brain. Second, functional connectivity between each lesion location and all other brain voxels in the normative population is computed and thresholded to define brain regions connected with each lesion location (i.e., lesion networks). Third, the resulting lesion networks are overlaid, and regions connected to the greatest number of lesion locations are identified. One of the advantages of this approach is that connectivity can be computed to all brain voxels from each lesion, resulting in whole-brain coverage in the analyses

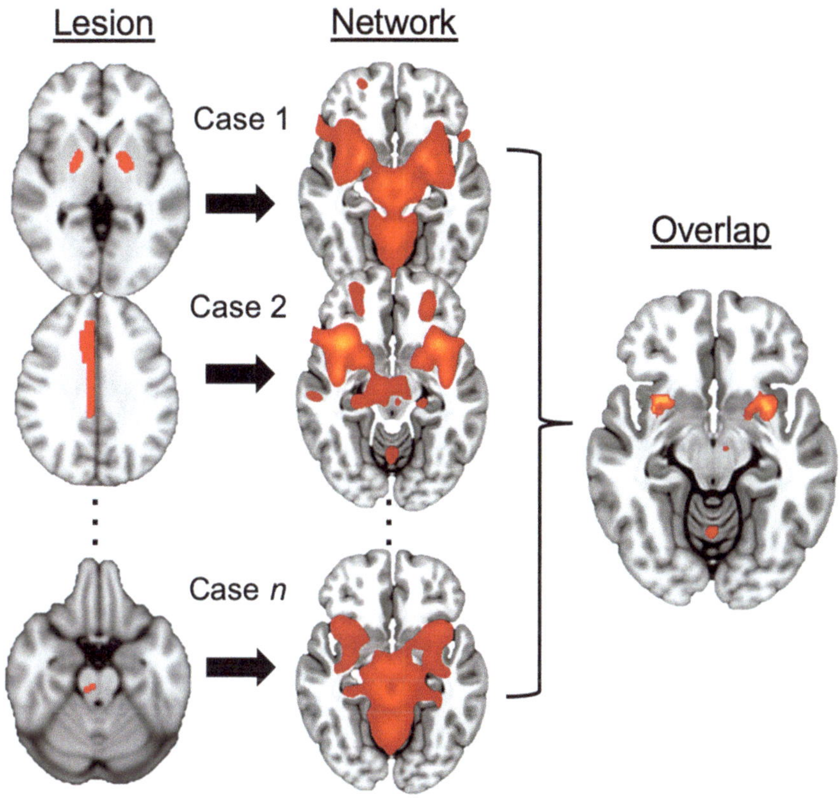

Fig. 1 Lesion network mapping principle
Lesions are first traced onto a standard brain template. Next, whole-brain connectivity of each lesion location is calculated using the functional connectome data and thresholded to identify brain regions functionally connected with the lesion location. Finally, the thresholded lesion networks are overlaid to identify common brain regions

in contrast to traditional VSLM where voxels affected only in few patients (typically <10% of the sample) need to be excluded from the statistical analyses.

1.2 Lesions

With functional lesion network mapping, lesions can be identified either from published images from the literature or using the original anatomical imaging data from the patients. It should be noted that using published images of the lesions results in 2D slices of the lesions, which does not fully represent the original lesion and couldn't be used in traditional lesion mapping analyses (i.e., investigating overlap in lesion location). Individual anatomical imaging data showing the full lesions from the patients is always preferable if available. However, collecting this data is expensive and time-consuming, and may not be feasible for rare lesion-induced symptoms.

The seed signal for rs-fcMRI analysis is calculated as the average of the voxels within the lesion. This allows for determining the lesion connectivity maps using representative sections of the lesions when the full lesion is not available. In cases where the original imaging data was available, the connectivity maps between the full lesion and the 2D representation have shown high spatial correlation when analyzed across all brain voxels ($r \sim 0.9$), supporting the validity of 2D lesion approach [6, 9]. The use of 2D lesion data, however, still is likely to result in decreased power compared to the use of full 3D lesions. This limitation is likely to increase noise in the data and bias the results against finding a common substrate for the symptom.

Although there are not yet studies formally investigating the effect of lesion type, lesion type is likely to have an impact on the accuracy of the localization of the original lesion. For example, the connectivity patterns of lesion from patients with parkinsonism or freezing of gait caused by central nervous system lymphoma did not line up with lesion networks of all other patients, suggesting that this type of lesion may be unreliable for lesion network mapping analyses [10, 11]. An ideal case (for any type of lesion analysis) is a patient with a single lesion with clearly identifiable borders and highly plausible causal relationship with the symptom in question without confounding other symptoms. In clinical practice, these cases rarely exist. Lesions frequently cause several symptoms, and some of the symptoms may occur with long delays, although the symptoms are still considered very likely to have been caused by the lesion. For example, lesion-induced dystonia may develop months or years after the lesion has occurred [12].

Functional lesion network mapping analyses published to date have used different strategies to collect lesions, depending on the symptom in question and availability of data. The inclusion criteria regarding lesion etiology (e.g., stroke, tumor, trauma), delay between the lesion and symptom onset, patient characteristics, and comorbidities vary from study to study. Although detailed inclusion and exclusion criteria likely need to be decided case by case depending on the research question and available data, there are certain general recommendations that should usually be considered and reported:

1. Patients' age.
   Ideally, the connectome data used to compute a lesion network should match the age of the patient with the brain lesion. This is likely to be particularly important when investigating symptoms caused by lesions in children. However, the value of an age-matched connectome must be weighed against the accessibility and sample size of publicly available connectomes. For example, a connectome derived from 1000 healthy subjects aged 20–30 may provide a better approximation of an

individual 60-year-old patient's connectivity than an age-matched connectome derived from 20 subjects. Prior work using an age-matched connectome for older adults showed little difference in lesion network mapping results [6]. Further work is needed to determine the value of connectomes better matched to the age or other characteristics of the lesion patient.

2. Preexisting brain disorders.
   Patients with preexisting brain disorders may have abnormal brain function and anatomy, and lesions in these patients may have different neurobiological effects as compared to healthy individuals. However, it is worth noting that changes in brain connectivity associated with brain disease are often small relative to the overall connectivity pattern. If region A is connected to region B and region C, this connectivity may be slightly stronger or weaker in a particular brain disease, but region A will still be connected to regions B and C. Prior work using disease-matched connectomes to investigate connectivity of brain stimulation sites (which can be considered as virtual lesions) showed little difference compared to results derived from a large normative connectome [13–15].
3. Delay between the lesion and resulting symptoms.
   Neurological and psychiatric symptoms may appear immediately after the damage to the brain tissue but sometimes occur months or even years later. The delay may depend on the brain tissue damage or symptom (e.g., hemiparesis is usually evident immediately after the lesion, but dystonia usually occurs after a long delay). The reasons for these differences are still under investigation but may involve slowly developing neuroplastic changes in brain function. Whether lesion network mapping can identify areas most likely to undergo such plastic changes remains unclear, but the technique has proven valuable in delayed lesion syndromes such as post-stroke pain [6] and dystonia [12]. In general, the tighter the temporal relationship between lesion onset and the resulting symptoms, the tighter the causal link, and the more likely that lesion network mapping (or lesion mapping in general) is likely to identify the causal neuroanatomical substrate for the symptom. Therefore, acute lesions cases with immediate symptom onset could be considered ideal for lesion network mapping. However, it should be kept in mind that limiting the lesion dataset, for example, to strokes only, can lead to poor spatial coverage, spatial bias due to vascular anatomy, and decreased sample sizes.
4. Lesion type.
   Although any lesion type can be used for functional lesion network mapping, lesions that produce rapid-onset symptoms

with clearly definable lesion borders are optimal. Focal ischemic strokes without significant surrounding edema tend to fit these criteria best. Slowly progressive infiltrative or compressive central nervous system tumors are not ideal because the identification of the exact areas of damaged brain tissue is difficult. Areas of encephalomalacia following traumatic brain injury are likely even less accurate. Errors in defining the damaged brain regions will result in increased noise in the data and can lead to false negatives. However, results with different types of lesion seem to be replicable [16], and recent research is expanding on what can be considered as a lesion [13] (*see* paragraph 3).

1.3 Connectivity Analyses and Connectomes

The term "functional connectome" is often used to describe the average functional connectivity between different brain regions derived from a large database of rs-fcMRI collected from many individuals. There are several connectomes constructed from different datasets with different characteristics, e.g., healthy volunteers, or disease-specific connectomes. These datasets include the Human Connectome Project [17] and Brain Genomics Superstruct Project that include thousands of healthy volunteers [18]. There are publicly available tools for voxel-based or seed-based connectivity analyses, such as Neurosynth (https://neurosynth.org/) [19], Lead-DBS (https://www.lead-dbs.org/) [20], or BCBtoolkit (http://toolkit.bcblab.com/).

Different connectomes can differ in image acquisition (e.g., resolution, scanner strength, scan duration) and in the data preprocessing (e.g., global signal regression, motion correction strategy). Although the basic principles of rs-fcMRI data analysis are fairly standard, there is no consensus on the optimal preprocessing and analysis strategy, and new methods and tools are constantly being developed [5, 21–24]. Therefore, we will refrain from making solid recommendations regarding the optimal image acquisition and preprocessing of connectome data for lesion network mapping. In some cases, we have directly compared different connectome processing pipelines and found lesion network mapping results to be similar [6, 25].

Briefly, rs-fMRI data is recorded, while the subjects lay in the scanner awake and not performing any task. Following data collection, preprocessing usually includes distortion correction, registration to an anatomical image, motion correction, and regressing out non-neuronal noise and signals that are common to all brain voxels. This latter regressor, termed the "global signal," has been particularly controversial, and there are many approaches for trying to correct for it in rs-fcMRI datasets [21]. We recommend some type of global signal correction, as this allows for visualization of negative correlations (aka anticorrelations), which we have found to be important in linking lesion locations to symptoms [6, 9, 12, 25–

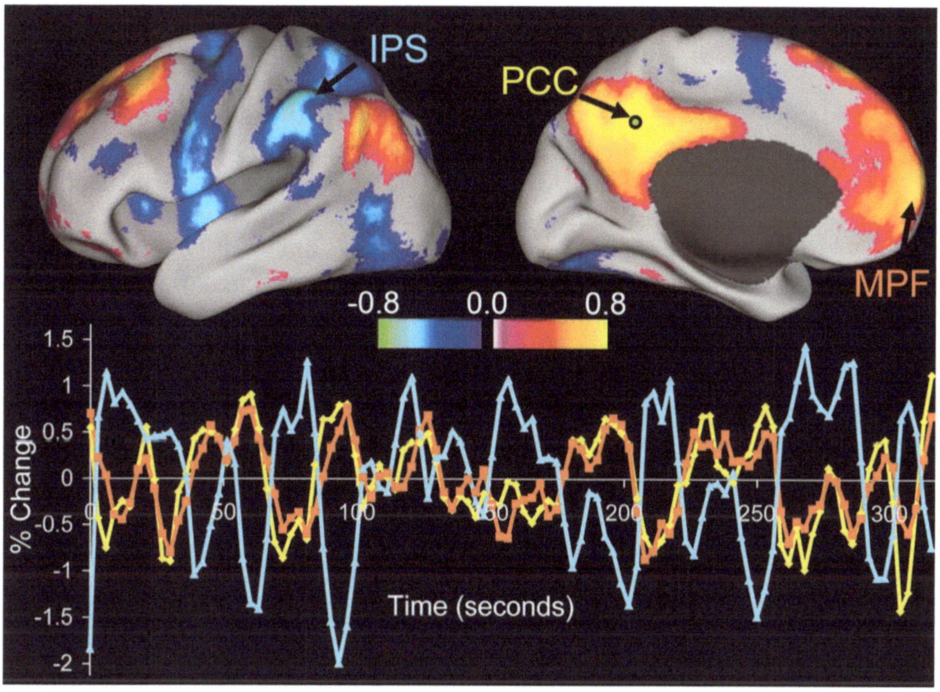

Fig. 2 Resting-state functional connectivity MRI
Intrinsic correlations of BOLD activity in the posterior cingulate cortex (PCC) seed region and all other voxels in the brain for a single subject during rest [30]. Copyright (2005) National Academy of Sciences

27]. For more information about rs-fMRI data collection and analysis, please see recent reviews about the methodology [28, 29].

Lesion location connectivity is calculated by using seed-based correlation analysis [5]. The rs-fcMRI signal is averaged across all voxels within the lesion; then correlations between this signal and all other brain voxels are computed to create a rs-fcMRI map for each individual. These individual maps are then combined across all subjects in the connectome dataset, resulting in a single map which reflects the connectivity of the lesion location in the "average/healthy" human brain (Fig. 2). For lesion network mapping of new-onset neurological and psychiatric symptoms in otherwise healthy individuals, this map provides a good approximation of the connectivity profile in the lesion patient, as the connections in these patients were presumably normal before the occurrence of the lesion. However, studying connectivity from lesions in patients with underlying brain disorders may require the use of disease-specific connectomes, although the major outlines of brain connectivity are still likely to be quite similar.

1.4 Lesion Network Mapping Technique

Lesion network mapping starts with identifying a set of lesion cases that share a specific symptom of interest. The more homogeneous the symptom and the more heterogeneous the lesion locations, the more suitable the dataset is for lesion network mapping. Note that each individual case is likely to also have other lesion-induced symptoms, but the only symptom they should share across all cases is the symptom of interest. After identifying suitable cases with either a published image of the lesion or original imaging data sufficient to identify the lesion, lesion network mapping itself involves four steps: (1) lesion segmentation and registration to template space (*see* Chaps. 2, 3, and 4), (2) defining the lesion networks, (3) overlapping these networks to identify common regions connected to all or most of the lesions, and (4) identifying lesion network components that are specific to the symptom or behavior of interest. Next we will go through these steps and discuss methodological issues with each of the steps.

1.4.1 Transferring the Lesions

The target space of the lesions depends on the connectome, because lesion data should match with the anatomical template of that connectome as closely as possible. The most commonly used space of the connectomes is MNI152 with several different anatomical templates. The most straightforward technique to transfer the lesions to the MNI152 space is to trace them by hand using one of the many available software developed for visualizing neuroimaging data. This technique can be used for both published 2D lesion images and individual anatomical 3D imaging data. Note that being able to use 2D lesion data differs from standard VLSM analyses, which require full 3D lesions. Drawing of the lesions requires a certain level of knowledge on human neuroanatomy and anatomical imaging (*see* Chaps. 1 and 2). The challenges include recognizing the correct anatomical slices and structures, which may not always be straightforward from individual patient data and accurate identification of lesion borders. The latter is especially challenging with infiltrative lesions such as malignant tumors and lesions with surrounding edema.

An alternative method for delineating the lesions is to use the individual neuroimaging data if it is available. As all lesion analyses, lesion network mapping relies on the accurate transformation of the anatomical image to the target space, which is usually performed by using computational algorithms implemented in neuroimaging software packages. Calculating transformation for imaging data from patients with brain lesions however is a challenge, because by definition the brain anatomy is not normal (*see* Chap. 4). This issue is less problematic in patients with small lesions, which cause smaller changes in the gross brain anatomy compared to large lesions. Linear transformations are usually less sensitive for errors caused by lesions compared to nonlinear transformations, but linear transformations are also less accurate for matching in space the gyri and sulci. The lesion may be drawn directly on the individual anatomical image and then transferred to the target space with the transformation parameters derived from the actual brain image (recommended approach) or may be drawn directly in the target space after the anatomical image has been transformed. Whichever the method is, quality to control for the resulting standard space lesions is important.

1.4.2 Lesion Networks

Lesion location masks transferred to the standard brain are used as seeds for rs-fcMRI seed-based correlation analysis. The seed signal is calculated as the average of all voxels within the lesion mask. The seed signal is then correlated with the signal in all brain voxels resulting in statistical map reflecting the strength of correlation in each voxel. Early functional lesion network mapping studies used data from 98 healthy volunteers [6], but more recent studies have included 1000 healthy volunteers [9], which is likely to reduce the noise in the resulting connectivity networks. Seed-based correlation analysis results in whole-brain connectivity maps reflecting the strength of connectivity of the seed (i.e., the lesion) with each voxel in the brain. The whole-brain connectivity map for each lesion location is referred to as a lesion network.

For simplicity, the individual lesion connectivity maps are usually binarized using a threshold to define voxels "connected" with the lesion location. This binarization is motivated by traditional lesion analyses, in which voxels are either "lesioned" or not. However, the threshold for defining a region as "connected" is somewhat arbitrary and can vary with the connectome used. In general, it is good to ensure that results are independent of the specific threshold used [26, 31, 32] and to complement any threshold-dependent analyses with analyses that do not depend on threshold (*see* specificity below).

rs-fcMRI analysis can identify both positively and negatively connected brain regions [30]. Although negatively correlated regions are connected to the seed region, they may have different biological interpretation compared to positively correlated regions. Brain activation and deactivation pattern during tasks seem to correlate with positive and negative connectivity patterns, respectively. Lesion network mapping studies have usually separated positively and negatively connected regions and highlighted some interesting differences between the two. For example, lesions causing cervical dystonia are positively connected to the cerebellum, a brain region that shows atrophy, and negatively connected to the primary sensory cortex, which may become hypertrophied in patients with cervical dystonia [12, 33, 34].

In the last step, the binarized lesion networks are overlaid, and the number of overlapping networks in each voxel is calculated. These overlap maps are used to identify brain regions that are connected to all or most of the lesion locations. Brain regions with high overlap of lesion networks are considered to be *sensitive* but not necessarily *specific* for the studied symptom.

1.4.3 Specificity

Although a high lesion network overlap indicates that a brain region is connected to all or most lesion locations that cause the symptom, the same region might also be connected to lesion locations that do not cause the symptom. As such, simple lesion network overlap can be biased toward identifying "hub" regions that are simply connected to the greatest number of other brain areas. As such, lesion network overlap findings should be complemented by a specificity analysis, comparing results to lesion locations not associated with the symptoms of interest, similarly as with standard VLSM analysis (*see* Chap. 5).

In the published studies using lesion network mapping, control lesion datasets have included randomly generated artificial lesions, lesions causing nonspecific or wide variety of different symptoms, lesion causing specific symptoms that are unrelated to the symptom of interest, or lesions causing symptoms that more closely match the studied symptom (e.g., parkinsonism maps have been compared to other movement disorder maps) [10]. However, lesion network mapping can also be used to analyze symptom severity on a continuous scale, again paralleling traditional VLSM [16].

Specificity analyses can be conducted using binary lesion networks (i.e., connected or not) or using continuous lesion connectivity maps (i.e., strength of the connection). The former is more similar to traditional lesion mapping analyses and can be performed with VLSM tools, while the latter is more similar to functional imaging analyses. Many studies have performed both to ensure that results are independent of the statistical approach. Brain regions that are both sensitive (i.e., identified by lesion network overlap) and specific (i.e., more connected to lesions associated with the studied symptom than with other lesions) are likely to be attributable to the studied symptom.

2 Using Lesion Network Mapping to Localize Neuropsychiatric Symptoms

2.1 Localization of Symptoms

By the time of writing this chapter, lesion network mapping has been used to localize several neurological and psychiatric symptoms (*see* Table 1). In the original validation of the method, Boes et al. (2015) investigated four different neurological symptoms caused by brain lesions, i.e., peduncular hallucinosis (visual hallucinations following brainstem or thalamic stroke), auditory hallucinations, central pain, and aphasia [6]. These four symptoms were chosen because there were clear a priori hypotheses about where each symptom should localize. They showed that lesions causing peduncular hallucinosis are negatively connected to the extrastriate visual cortex. Similarly, lesions causing the three other symptoms were connected to brain regions that have previously been linked with the corresponding brain function, for example, auditory hallucinations to the superior temporal gyrus where the primary

Table 1
Symptoms studied with lesion network mapping

Symptom	References
Akinetic mutism	Darby et al. PNAS 2018a [35]
Alien limb	Darby et al. PNAS 2018a [35]
Anosognosia	Klingbeil et al. Neuroimage 2019 [36]
Aphasia	Boes et al. Brain 2015 [6]
Bodily self-consciousness	Wawrzyniak et al. Neuroimage 2018 [37]
Cervical dystonia	Corp et al. Brain 2019 [12]
Coma	Fischer et al. Neurology 2016 [38]
Consciousness	Snider et al. Hum Brain Mapp 2020 [25]
Criminality	Darby et al. PNAS 2018b [9]
Delusions	Darby et al. Brain 2017 [27]
Depression	Padmanabhan et al. Biol Psychiatry 2019 [16]
Freezing of gait	Fasano et al. Ann Neurol. 2017 [11]
Hallucinations	Boes et al. Brain 2015 [6] Kim et al. Mol Psych 2021 [39]
Hemichorea	Laganiere et al. Neurology 2016 [40]
Holmes' tremor	Joutsa et al. Ann Neurol 2019 [41]
Impaired decision-making	Sutterer et al. Cortex 2016 [42]
Mania	Lee et al. J affect Disord. 2019 [43]
Memory	Ferguson et al. Nat Commun 2019 [44]
Pain	Boes et al. Brain 2015 [6]
Parkinsonism	Joutsa et al. Brain 2018a [10]
Prosopagnosia	Cohen et al. Brain 2019 [45]

auditory cortex is located. These findings suggested that the technique could be used to localize different neurological symptoms (*see* Fig. 3).

Following the validation of the technique, lesion network mapping has been applied to a number of different neurological and psychiatric conditions, ranging from motor symptoms to complex neuropsychiatric entities, such as criminal behavior. In some of the cases, there was a single hub attributable for the symptom (e.g., the posterior putamen in hemichorea or the claustrum in parkinsonism) [10, 40]. Interestingly, lesions causing certain other symptoms such as cervical dystonia showed both a region where the lesions were positively connected (cerebellum) and a region where

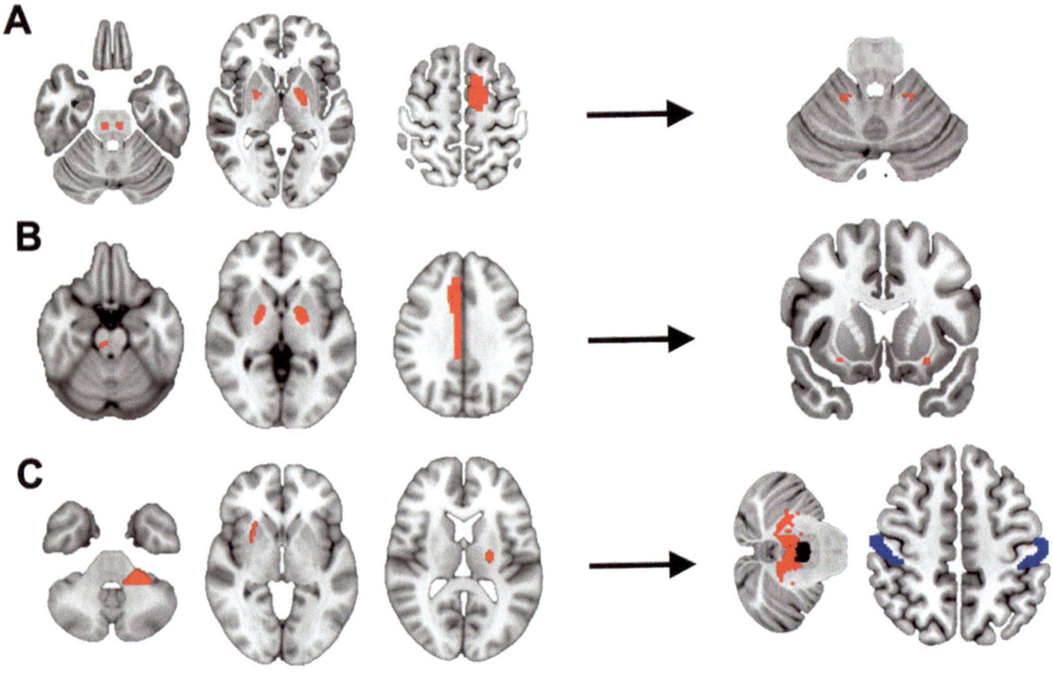

Fig. 3 Lesion network mapping in neurological symptoms
Examples of lesions and connected brain regions identified by functional lesion network mapping. (**a**) Freezing of gait [11]. (**b**) Parkinsonism [10]. (**c**) Cervical dystonia [12]

the lesions were negatively connected (primary sensory cortex) [12]. Similar findings were made with complex psychiatric symptoms, such as delusions and criminality, consistent with two-hit models of symptom generation, where more than one affected brain circuit is required for the symptoms to appear [9, 12].

2.2 Relevance across Etiologies

Lesion network mapping can be applied only to symptoms caused by brain lesions. However, many symptoms are rarely caused by lesions, and vast majority of the patients suffering from these symptoms have some other etiology than a focal brain lesion, such as primary psychiatric disease. This raises the question if the findings from lesion network mapping can be generalized across different etiologies of the same symptom.

There are many examples that suggest that symptom localization derived from lesion network mapping does generalize. Lesions causing hemichorea are connected to the posterolateral putamen, an area that appears abnormal on MRI when hemichorea is caused by hyperglycemia [40]. Lesions causing parkinsonism predict brain atrophy patterns in patients with primary parkinsonism syndromes, including Parkinson's disease [10]. Network mapping of brain atrophy patterns localizes clinical dementia syndromes to syndrome-specific brain networks in patients with Alzheimer's disease, behavioral variant frontotemporal dementia, corticobasal syndrome, and primary progressive non-fluent aphasia [31, 46]. Further, atrophy patterns in dementia patients with specific symptoms like delusions or hallucinations localize to the same brain networks identified using lesion network mapping [31, 47]. These studies demonstrate not only that lesion network mapping results are relevant to other diseases but also that the network mapping approach can be extended to understanding symptom localization in other neurological disorders, like dementia.

Lesion network mapping may also be useful in guiding the search for abnormalities in patients with similar symptoms that are not caused by a lesion. For example, lesion network mapping of mania identifies brain regions that are abnormal in non-lesioned patients with primary mania [43], while lesion network mapping of disordered agency and volition identifies brain regions that are abnormal in non-lesioned patients with catatonia and functional neurological disorders [35]. Lesion network mapping of dystonia identifies regions that are abnormal in non-lesioned patients with primary dystonia [12]. In this latter example, the regions identified with lesion network mapping (cerebellum and sensory cortex) were more abnormal in patients with primary dystonia than multiple other regions. This supports the view that lesion network mapping might identify brain regions causing similar symptoms in idiopathic conditions and other disorders, although further work is needed to validate these findings.

3 Mapping Treatment Targets

Several lines of evidence suggest that lesion network mapping-based localization may have relevance for treatment. In Parkinson's disease, clinical efficacy of DBS treatment was correlated with connectivity of the stimulation site to the claustrum, which is the most sensitive and specific brain region to which lesion locations causing parkinsonism are connected [10]. In cervical dystonia, DBS electrode locations providing good symptom reduction were positively connected to the cerebellum and negatively connected to the somatosensory cortex, matching lesion network mapping findings [12]. Lesion locations causing Holmes' tremor were connected to DBS sites that provided therapeutic benefit [48]. Finally, lesion locations associated with depression were connected to effective transcranial magnetic stimulation (TMS) sites in the dorsolateral prefrontal cortex [16].

Rarely, brain lesions can provide symptomatic benefit instead of functional deficits, a phenomenon termed "paradoxical functional facilitation" [49]. For example, a patient with essential tremor can have a stroke that relieves their tremor and provides overall improvement in the limb function. Ideally, such cases would

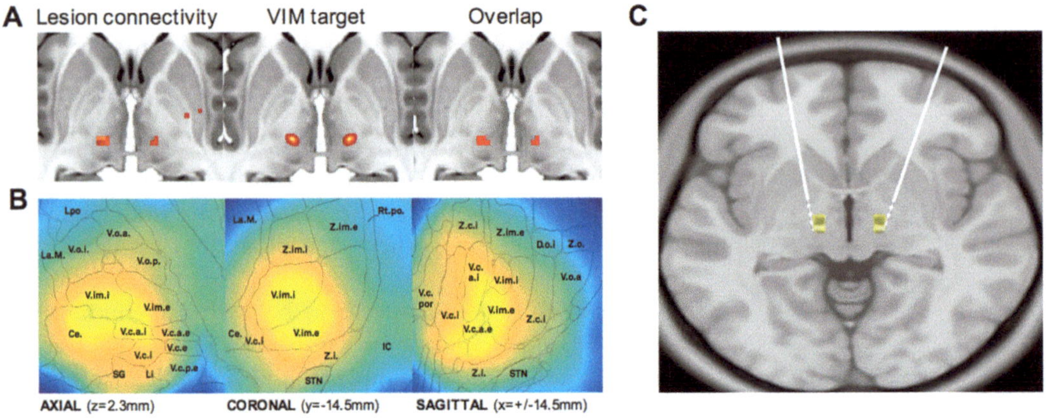

Fig. 4 Lesion network mapping of beneficial brain lesions identifies effective neurosurgical target in essential tremor
(**a**) Lesion locations improving tremor in patients with essential tremor are connected to the ventral intermediate nucleus of the thalamus (VIM), which is the main neurosurgical target in essential tremor. (**b**) Lesion network overlap overlaid on a high-resolution thalamic atlas. (**c**) Lead trajectories in a patient with essential tremor who got a good therapeutic response running immediately next to the lesion network overlap hotspot. Original publication of the figure [50].

guide surgical intervention to improve tremor in other patients. However, the lesion locations that improve tremor are so heterogeneous that the surgical target was unclear. Joutsa et al. collected studied case reports of paradoxical functional facilitation in essential tremor in a proof-of-concept study to investigate if lesion network mapping could be used to identify treatment targets [50]. Although only less than 30% of the lesions included the thalamus, all of them were functionally connected to the ventral intermediate nucleus of the thalamus, which is the current neurosurgical treatment target in essential tremor (Fig. 4). In fact, the voxel most strongly predictive of tremor relief was nearly identical to the thalamic coordinate currently targeted with deep brain stimulation. These findings provide the most direct evidence to date that lesion network mapping can identify effective treatment targets. Further work will show if this approach leads to identification of novel neuroanatomical targets for brain disorders without established surgical or neuromodulation treatment protocols.

It is worth noting that the same connectivity method described here for lesion locations can also be applied to brain stimulation sites. Connectivity with the DBS site can be used to predict improvement in Parkinson's symptoms and tremor [13, 51]. Connectivity with TMS sites can predict improvement in depression symptoms [14, 15]. DBS and TMS sites effective for the same symptom tend to be connected to the same circuit [52]. Future work using connectivity to combine lesion locations and brain stimulation sites that modulate the same symptom may be particularly valuable in isolating therapeutic targets.

4 Limitations and Future Directions

Because BOLD signal is driven mainly by gray matter, functional connectivity MRI may not be optimal for investigating the connections of lesion locations mainly within the white matter. White matter has been shown to have sufficient BOLD signal for functional connectivity imaging [53], and lesion network mapping has successfully been applied to lesions located mainly in the white matter [16]. However, techniques such as diffusion tensor imaging that are designed for investigating connectivity with white matter may complement the functional connectivity lesion network mapping method described here (*see* Chap. 9). As described earlier in this chapter, there are different techniques to study lesion connectivity. Each of these methods have different strengths and are likely to provide complementary information, and combining techniques is likely to be valuable and provides more detailed information about the underlying biology. As the methodology for connectivity imaging improves and larger publicly available datasets are made available, lesion network mapping is likely to become more accurate.

When lesions causing a specific symptom overlap in a single brain region, damage to this brain region is likely to be causal to the symptoms. However, when lesions causing a specific symptom overlap in a single network, causal inference applies on the network but not necessarily for any single brain region within this network [3]. Many of the symptoms studied with lesion network mapping to date are relatively rare consequences of brain lesions compared to idiopathic disorders causing these symptoms. Several findings suggest that studying lesion connectivity can inform about the neural origin of symptoms in idiopathic neurological and psychiatric disorders, but this remains to be thoroughly investigated and may not apply for all symptoms [10, 12]. Lesion network mapping probably works best for well-phenotyped symptoms, such as motor symptoms, but not so well for more heterogeneous and loosely characterized symptoms, such as depression, which have required larger sample sizes [10–12, 16].

Finally, it is important to note that there has yet to be a prospective clinical trial testing the efficacy of a therapeutic target derived from lesion network mapping. As described above, there are many retrospective associations that suggest that lesion network mapping results align with previously identified therapeutic targets. However, the true test of the technique will depend on whether it can identify new targets for symptoms that lack effective treatments.

References

1. Karnath HO, Sperber C, Rorden C (2017) Mapping human brain lesions and their functional consequences. NeuroImage 165:180–189. https://doi.org/10.1016/j.neuroimage.2017.10.028
2. Adolphs R (2016) Human lesion studies in the 21st century. Neuron 90(6):1151–1153. https://doi.org/10.1016/j.neuron.2016.05.014
3. Fox MD (2018) Mapping symptoms to brain networks with the human connectome. N Engl J Med 379(23):2237–2245. https://doi.org/10.1056/NEJMra1706158
4. Monakow C (1914) Die Lokalisation im Grosshirn: und der Abbau der Funktion durch kortikale Herde. Verlag von J.F. Bergmann, Wiesbaden
5. Fox MD, Raichle ME (2007) Spontaneous fluctuations in brain activity observed with functional magnetic resonance imaging. Nat Rev Neurosci 8(9):700–711. https://doi.org/10.1038/nrn2201
6. Boes AD, Prasad S, Liu H, Liu Q, Pascual-Leone A, Caviness VS, Fox MD (2015) Network localization of neurological symptoms from focal brain lesions. Brain 138(Pt 10):3061–3075. https://doi.org/10.1093/brain/awv228
7. de Haan B, Rorden C, Karnath HO (2013) Abnormal perilesional BOLD signal is not correlated with stroke patients' behavior. Front Hum Neurosci 7:669. https://doi.org/10.3389/fnhum.2013.00669
8. Veldsman M, Cumming T, Brodtmann A (2015) Beyond BOLD: optimizing functional imaging in stroke populations. Hum Brain Mapp 36(4):1620–1636. https://doi.org/10.1002/hbm.22711
9. Darby RR, Horn A, Cushman F, Fox MD (2018) Lesion network localization of criminal behavior. Proc Natl Acad Sci U S A 115(3):601–606. https://doi.org/10.1073/pnas.1706587115
10. Joutsa J, Horn A, Hsu J, Fox MD (2018) Localizing parkinsonism based on focal brain lesions. Brain 141(8):2445–2456. https://doi.org/10.1093/brain/awy161
11. Fasano A, Laganiere SE, Lam S, Fox MD (2017) Lesions causing freezing of gait localize to a cerebellar functional network. Ann Neurol 81(1):129–141. https://doi.org/10.1002/ana.24845
12. Corp DT, Joutsa J, Darby RR, Delnooz CCS, van de Warrenburg BPC, Cooke D, Prudente CN, Ren J, Reich MM, Batla A, Bhatia KP, Jinnah HA, Liu H, Fox MD (2019) Network localization of cervical dystonia based on causal brain lesions. Brain 142(6):1660–1674. https://doi.org/10.1093/brain/awz112
13. Horn A, Reich M, Vorwerk J, Li N, Wenzel G, Fang Q, Schmitz-Hübsch T, Nickl R, Kupsch A, Volkmann J, Kühn AA, Fox MD (2017) Connectivity predicts deep brain stimulation outcome in Parkinson's disease. Ann Neurol. https://doi.org/10.1002/ana.24974
14. Weigand A, Horn A, Caballero R, Cooke D, Stern AP, Taylor SF, Press D, Pascual-Leone A, Fox MD (2018) Prospective validation that Subgenual connectivity predicts antidepressant efficacy of transcranial magnetic stimulation sites. Biol Psychiatry 84(1):28–37. https://doi.org/10.1016/j.biopsych.2017.10.028
15. Fox MD, Buckner RL, White MP, Greicius MD, Pascual-Leone A (2012) Efficacy of transcranial magnetic stimulation targets for depression is related to intrinsic functional connectivity with the subgenual cingulate. Biol Psychiatry 72(7):595–603. https://doi.org/10.1016/j.biopsych.2012.04.028
16. Padmanabhan JL, Cooke D, Joutsa J, Siddiqi SH, Ferguson M, Darby RR, Soussand L, Horn A, Kim NY, Voss JL, Naidech AM, Brodtmann A, Egorova N, Gozzi S, Phan TG, Corbetta M, Grafman J, Fox MD (2019) A human depression circuit derived from focal brain lesions. Biol Psychiatry 86(10):749–758. https://doi.org/10.1016/j.biopsych.2019.07.023
17. Van Essen DC, Smith SM, Barch DM, Behrens TE, Yacoub E, Ugurbil K, Consortium W-MH (2013) The WU-Minn human connectome project: an overview. NeuroImage 80:62–79. https://doi.org/10.1016/j.neuroimage.2013.05.041
18. Holmes AJ, Hollinshead MO, O'Keefe TM, Petrov VI, Fariello GR, Wald LL, Fischl B, Rosen BR, Mair RW, Roffman JL, Smoller JW, Buckner RL (2015) Brain genomics Superstruct project initial data release with structural, functional, and behavioral measures. Sci Data 2:150031. https://doi.org/10.1038/sdata.2015.31
19. Yarkoni T, Poldrack RA, Nichols TE, Van Essen DC, Wager TD (2011) Large-scale automated synthesis of human functional neuroimaging data. Nat Methods 8(8):665–670. https://doi.org/10.1038/nmeth.1635
20. Horn A, Kühn AA (2015) Lead-DBS: a toolbox for deep brain stimulation electrode localizations and visualizations. NeuroImage 107:

127–135. https://doi.org/10.1016/j.neuroimage.2014.12.002

21. Murphy K, Fox MD (2017) Towards a consensus regarding global signal regression for resting state functional connectivity MRI. NeuroImage 154:169–173. https://doi.org/10.1016/j.neuroimage.2016.11.052
22. Snyder AZ, Raichle ME (2012) A brief history of the resting state: the Washington university perspective. NeuroImage 62(2):902–910. https://doi.org/10.1016/j.neuroimage.2012.01.044
23. Bijsterbosch J (2017) Introduction to resting state fMRI functional connectivity. Oxford neuroimaging primers, 1st edn. Oxford University Press, Oxford/New York
24. Esteban O, Markiewicz CJ, Blair RW, Moodie CA, Isik AI, Erramuzpe A, Kent JD, Goncalves M, DuPre E, Snyder M, Oya H, Ghosh SS, Wright J, Durnez J, Poldrack RA, Gorgolewski KJ (2019) fMRIPrep: a robust preprocessing pipeline for functional MRI. Nat Methods 16(1):111–116. https://doi.org/10.1038/s41592-018-0235-4
25. Snider SB, Hsu J, Darby RR, Cooke D, Fischer D, Cohen AL, Grafman JH, Fox MD (2020) Cortical lesions causing loss of consciousness are anticorrelated with the dorsal brainstem. Hum Brain Mapp. https://doi.org/10.1002/hbm.24892
26. Kim NY, Hsu J, Talmasov D, Joutsa J, Soussand L, Wu O, Rost NS, Morenas-Rodríguez E, Martí-Fàbregas J, Pascual-Leone A, Corlett PR, Fox MD (2019) Lesions causing hallucinations localize to one common brain network. Mol Psychiatry. https://doi.org/10.1038/s41380-019-0565-3
27. Darby RR, Laganiere S, Pascual-Leone A, Prasad S, Fox MD (2017) Finding the imposter: brain connectivity of lesions causing delusional misidentifications. Brain 140(Pt 2):497–507. https://doi.org/10.1093/brain/aww288
28. Lv H, Wang Z, Tong E, Williams LM, Zaharchuk G, Zeineh M, Goldstein-Piekarski AN, Ball TM, Liao C, Wintermark M (2018) Resting-state functional MRI: everything that nonexperts have always wanted to know. AJNR Am J Neuroradiol 39(8):1390–1399. https://doi.org/10.3174/ajnr.A5527
29. Chen K, Azeez A, Chen DY, Biswal BB (2020) Resting-state functional connectivity: signal origins and analytic methods. Neuroimaging Clin N Am 30(1):15–23. https://doi.org/10.1016/j.nic.2019.09.012
30. Fox MD, Snyder AZ, Vincent JL, Corbetta M, Van Essen DC, Raichle ME (2005) The human brain is intrinsically organized into dynamic, anticorrelated functional networks. Proc Natl Acad Sci U S A 102(27):9673–9678. https://doi.org/10.1073/pnas.0504136102
31. Darby RR, Joutsa J, Fox MD (2019) Network localization of heterogeneous neuroimaging findings. Brain 142(1):70–79. https://doi.org/10.1093/brain/awy292
32. Cohen AL, Soussand L, Corrow SL, Martinaud O, Barton JJS, Fox MD (2019) Looking beyond the face area: lesion network mapping of prosopagnosia. Brain 142(12):3975–3990. https://doi.org/10.1093/brain/awz332
33. Zheng Z, Pan P, Wang W, Shang H (2012) Neural network of primary focal dystonia by an anatomic likelihood estimation meta-analysis of gray matter abnormalities. J Neurol Sci 316(1–2):51–55. https://doi.org/10.1016/j.jns.2012.01.032
34. Batla A, Sánchez MC, Erro R, Ganos C, Stamelou M, Balint B, Brugger F, Antelmi E, Bhatia KP (2015) The role of cerebellum in patients with late onset cervical/segmental dystonia?--evidence from the clinic. Parkinsonism Relat Disord 21(11):1317–1322. https://doi.org/10.1016/j.parkreldis.2015.09.013
35. Darby RR, Joutsa J, Burke MJ, Fox MD (2018) Lesion network localization of free will. Proc Natl Acad Sci U S A 115(42):10792–10797. https://doi.org/10.1073/pnas.1814117115
36. Klingbeil J, Wawrzyniak M, Stockert A, Karnath HO, Saur D (2020) Hippocampal diaschisis contributes to anosognosia for hemiplegia: evidence from lesion network-symptom-mapping. NeuroImage 208:116485. https://doi.org/10.1016/j.neuroimage.2019.116485
37. Wawrzyniak M, Klingbeil J, Zeller D, Saur D, Classen J (2018) The neuronal network involved in self-attribution of an artificial hand: a lesion network-symptom-mapping study. NeuroImage 166:317–324. https://doi.org/10.1016/j.neuroimage.2017.11.011
38. Fischer DB, Boes AD, Demertzi A, Evrard HC, Laureys S, Edlow BL, Liu H, Saper CB, Pascual-Leone A, Fox MD, Geerling JC (2016) A human brain network derived from coma-causing brainstem lesions. Neurology 87(23):2427–2434. https://doi.org/10.1212/WNL.0000000000003404
39. Kim NY, Hsu J, Talmasov D, Joutsa J, Soussand L, Wu O, Rost NS, Morenas-Rodríguez E, Martí-Fàbregas J, Pascual-Leone A, Corlett PR, Fox MD (2021) Lesions causing hallucinations localize to one common

brain network. Mol Psychiatry 26(4): 1299–1309. https://doi.org/10.1038/s41380-019-0565-3

40. Laganiere S, Boes AD, Fox MD (2016) Network localization of hemichorea-hemiballismus. Neurology 86(23): 2187–2195. https://doi.org/10.1212/WNL.0000000000002741

41. Joutsa J, Shih LC, Fox MD (2019) Mapping holmes tremor circuit using the human brain connectome. Ann Neurol 86(6):812–820. https://doi.org/10.1002/ana.25618

42. Sutterer MJ, Bruss J, Boes AD, Voss MW, Bechara A, Tranel D (2016) Canceled connections: lesion-derived network mapping helps explain differences in performance on a complex decision-making task. Cortex 78:31–43. https://doi.org/10.1016/j.cortex.2016.02.002

43. Lee I, Nielsen K, Nawaz U, Hall MH, Öngür D, Keshavan M, Brady R (2019) Diverse pathophysiological processes converge on network disruption in mania. J Affect Disord 244:115–123. https://doi.org/10.1016/j.jad.2018.10.087

44. Ferguson MA, Lim C, Cooke D, Darby RR, Wu O, Rost NS, Corbetta M, Grafman J, Fox MD (2019) A human memory circuit derived from brain lesions causing amnesia. Nat Commun 10(1):3497. https://doi.org/10.1038/s41467-019-11353-z

45. Cohen AL, Ferguson MA, Fox MD (2021) Lesion network mapping predicts post-stroke behavioural deficits and improves localization. Brain 144(4):e35. https://doi.org/10.1093/brain/awab002

46. Seeley WW, Crawford RK, Zhou J, Miller BL, Greicius MD (2009) Neurodegenerative diseases target large-scale human brain networks. Neuron 62(1):42–52. https://doi.org/10.1016/j.neuron.2009.03.024

47. Weil RS, Hsu JK, Darby RR, Soussand L, Fox MD (2019) Neuroimaging in Parkinson's disease dementia: connecting the dots. Brain Commun 1(1):fcz006. https://doi.org/10.1093/braincomms/fcz006

48. Joutsa J, Shih LC, Fox MD (2019) Mapping holmes tremor circuit using the human brain connectome. Ann Neurol. https://doi.org/10.1002/ana.25618

49. Kapur N (1996) Paradoxical functional facilitation in brain-behaviour research: a critical review. Brain 119(Pt 5):1775–1790

50. Joutsa J, Shih LC, Horn A, Reich MM, Wu O, Rost NS, Fox MD (2018) Identifying therapeutic targets from spontaneous beneficial brain lesions. Ann Neurol 84(1):153–157. https://doi.org/10.1002/ana.25285

51. Al-Fatly B, Ewert S, Kübler D, Kroneberg D, Horn A, Kühn AA (2019) Connectivity profile of thalamic deep brain stimulation to effectively treat essential tremor. Brain 142(10): 3086–3098. https://doi.org/10.1093/brain/awz236

52. Fox MD, Buckner RL, Liu H, Chakravarty MM, Lozano AM, Pascual-Leone A (2014) Resting-state networks link invasive and noninvasive brain stimulation across diverse psychiatric and neurological diseases. Proc Natl Acad Sci U S A 111(41):E4367–E4375. https://doi.org/10.1073/pnas.1405003111

53. Ding Z, Huang Y, Bailey SK, Gao Y, Cutting LE, Rogers BP, Newton AT, Gore JC (2018) Detection of synchronous brain activity in white matter tracts at rest and under functional loading. Proc Natl Acad Sci U S A 115(3): 595–600. https://doi.org/10.1073/pnas.1711567115

Chapter 11

Multivariate Lesion-Behavior Mapping

Yee-Haur Mah, Ashwani Jha, Tianbo Xu, and Parashkev Nachev

Abstract

Lesion-behavior mapping is commonly used to infer the macroscopic functional organization of the brain from the behavioral consequences of anatomically defined focal brain injury. It is a task conventionally assumed to be tractable with mass-univariate methods by analogy with spatial inference from functional and morphometric data. Here we demonstrate that this assumption is critically violated by the fundamental nature of pathological lesions and their causal relationship with the neural substrate. This cardinal fault does not merely limit mass-univariate inference to compositionally simple distributed patterns of neural dependence; it distorts it across all plausible patterns of dependence to a degree determined by the distributed structure of the lesioning process that mass-univariate inference definitionally simplifies or ignores. We argue that high-dimensional multivariate inference is essential to robust lesion-behavior mapping and provide an overview of the approach to creating a multivariate inferential framework optimized for the task, with attention to aspects of experimental design, algorithmic development, and quantification of fidelity.

Key words Lesion-behavior mapping, Lesion-deficit mapping, Multivariate modeling, Spatial inference, Focal brain injury

1 The Problem

The principal difficulty with lesion-behavior mapping is that it appears so deceptively simple. We already know from functional imaging that the brain's functional anatomy generalizes across the population—at least at some spatial scales—so it would seem our task is simply to fit enough focally lesioned instances to reveal the underlying map, much as one might fit pieces in a jigsaw puzzle. The lesion mapper's primary preoccupations, dwelt on at length in the preceding chapters, are then analogous to the puzzle solver's concern with finding the pieces (lesion segmentation), identifying their interlocking teeth (minimal voxel resolution), locating them on the map (lesion registration), and determining the minimal number needed to reveal the picture (coverage and power). The necessary equipment—intellectual and technical—seems correspondingly simple: as long as our imaging is of good quality,

Dorian Pustina and Daniel Mirman (eds.), *Lesion-to-Symptom Mapping: Principles and Tools*, Neuromethods, vol. 180, https://doi.org/10.1007/978-1-0716-2225-4_11,

inter-subject registration is adequate, and we segment the lesions crisply and assemble a dataset of sufficient size, coverage, and phenotypic variance, our job is more or less done. Or so, at least, one might infer from the confident tone in which the anatomical localizations thereby derived are widely reported in the literature.

The task is in reality one of the hardest in neuroscience. The reasons are many, but we need only dwell on the three of greatest importance: the arrow of causality, the index of fidelity, and the interaction between the anatomies of functions and lesions. We shall see reflection on these reasons can only lead one to conclude that the approach to lesion-behavior mapping requires wholesale revision, for it rests on quicksand nothing could securely be built on. High-dimensional multivariate modeling is not just an option here; it is the only viable solution.

2 The Arrow of Causality

The BOLD response is not caused by behavior but by the distributed neural activity on which cognition and behavior depend (Fig. 1, left). It certainly does not itself cause behavior, unless the subject is artificially observing it through a closed feedback loop. So when a given anatomical locus is activated during a task, its statistical association with the behavior is *not* dependent on the observation of activation of any other locus. Of course, the statistics here are indifferent to causality, and any plausible causal model of how the underlying neural mechanism—as revealed by BOLD signals—supports behavior ought to involve consideration of all other activated loci across the brain, among other distributed factors. But such mechanistic inference is downstream of the spatial inference, which is here merely correlative. Equally, regional gray matter concentrations in volumetric brain morphometry do not cause behavior, but reflect developmental processes and plasticity in response to long-term patterns of neural activity. The statistical association of greater or lesser concentration at any one voxel with an ability is also determinable without reference to other voxels. For example, under idealized, noise-free conditions, the association of $voxel_1$ in Fig. 1 with a given behavior is not altered by the state of $voxel_2$ or any other $voxel_n$ because each is independently conditional on the underlying neural substrate, given the same observed behavior. A model of $voxel_1$ alone is a valid form of spatial inference, even if the causal interpretation of the operation of the brain as a whole will naturally require integration across all material loci.

An inferential framework that ignores the relation between different loci other than modeling spatial uncertainty under an assumption of unstructured, "blob-like" local coherence is therefore perfectly licit here—for the purpose of spatial inference—even if it carries no causal weight. Indeed, SPM—the mass-univariate

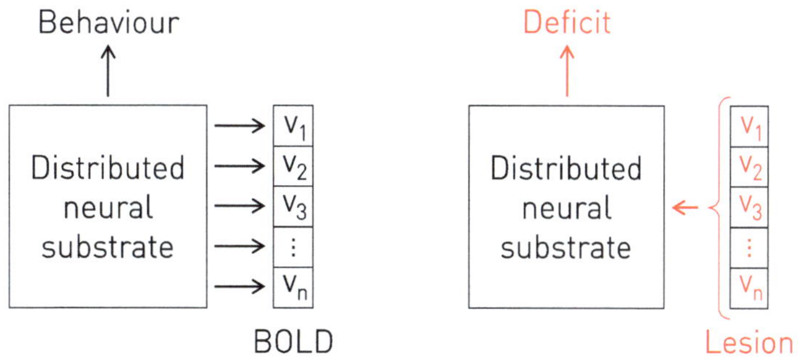

Fig. 1 A comparison of the causal relations in functional imaging (left) vs lesion-deficit mapping (right). Behavior is causally dependent on the distributed neural substrate of the brain, whose spatially organized activity is (in part) revealed by BOLD changes across a set of voxels 1. . .n. Under idealized, noise-free conditions, the association of a given voxel with the state of the brain underlying a given behavior is not conditional on the state of any other voxel, just as the association with pastoral scenes of blue-colored pieces in the top left corner of a set of picture puzzles is not dependent on the color of pieces in the bottom right. A conclusion about the topic of the puzzle *as a whole* is of course dependent on *all* of the pieces, but that is something to be settled downstream of the spatial inference. By contrast, the association of a given voxel with a given behavioral deficit is conditional on *all* voxels, for the deficit is caused by the lesion *as a whole*. The picture puzzle analogy here would be trying to associate the state of a single spatial location with the recognizability of the entire depicted scene when necessarily many contiguous pieces are simultaneously changed

approach to functional imaging in general—is successful precisely because it leaves causal interpretation downstream of the fundamental statistical framework.

By stark contrast, in lesion deficit mapping, causality and spatial inference are inextricably intertwined. The vanishingly rare case of focal excitotoxicity aside, lesion cannot be caused by the underlying distributed neural substrate. Equally, again niche cases aside, it would be absurd to conclude that the observed deficit caused the lesion. Generally, the only possible association between a lesion and a deficit is causation of the latter by the former: reversing the arrow of causality between the state of a set of voxels and the underlying neural substrate (Fig. 1, right). Crucially, the deficit is caused by the lesion *as a whole*—typically spanning many voxels—and is almost invariably caused only by lesions of spatial extent far greater than a single voxel. The statistical association of a given $voxel_1$ with the deficit is now conditional on every other affected voxel, for it is all lesioned voxels taken together that interact with the underlying neural substrate to cause the observed deficit. A single part of a lesion therefore *cannot* be associated with a deficit without taking into consideration all other loci jointly affected, for the only possible association is a causal one, and causation here must be distributed beyond any single voxel.

Surely, one might be tempted to object, when we assemble a set of lesions differing in their spatial characteristics and assess their voxel-wise differences with some kind of mass-univariate test [1], identifying the relevant voxels is precisely what we do? No, since it is the lesion *as a whole*, not any one voxel, that causes the change in behavior, it could be *any* set of features of the lesion correlated with the test voxel that are material here. Indeed, lesion-behavior mappers widely concede this point by modeling total lesion volume as an obvious confounder [2, 3]. But if that global confound is admitted, so should any other lesion morphological characteristic of potential causal impact. And unless the distribution of lesion characteristics is random—which it is abundantly clear it is not [4–7]—there can be no expectation that these confounds will be eliminated at the limit of infinite data: on the contrary, they can only be entrenched by replication.

Consider the proposition from another angle. An effective way of testing the coherence of a statistical approach is to ask oneself what the null hypothesis is. In SPM, the null hypothesis is that the BOLD signal of a given voxel alone is not associated with the neural activity underlying the behavior under study: this makes perfect sense. In mass-univariate lesion-deficit mapping, the null hypothesis is that the integrity of a voxel is not necessary for a behavior. But across most of the brain, no *single* voxel could ever be necessary for a behavior so both null and its alternative are essentially nonsensical.

Again, one might object: conventional statistics are indifferent to causality, so why is the distinction decisive here? It is because, unlike BOLD, a lesion *cannot* be merely correlated; it *must* be causal if it is associated with the deficit at all, and if causation is necessarily distributed across all lesion characteristics, the only valid form of statistical association is one that takes all of them jointly into account.

No map drawn from mass-univariate can be trusted, for the underlying statistics are invalid at the most basic level. We need to model the relationships *between* voxels, within a high-dimensional multivariate framework. This is inevitably much harder than any mass-univariate approach and is complicated by two further characteristics of the problem.

3 The Index of Fidelity

How do we know our map is correct? This is not a question functional imagers generally ask themselves, for correctness is implied by the foundational statistical framework. When you reject the null hypothesis at a given voxel, the alternative hypothesis is the only one left standing, and its acceptance is widely argued to be inevitable, even if that is not what the statistics strictly speaking imply (at least for the frequentist). But when our model is multivariate, and especially when high-dimensional, there is a vast space of possible models to survey. Here it means nothing to reject one or even a tractably large array of models, for there will always be many alternatives left. Our task becomes one of *model comparison*, of finding the best possible model within a space that will rarely, if ever, be comprehensively surveyable. It ceases to make sense to talk of inference here: one could never say of a solution that it *must* be the right one, only that it might be more or less accurately descriptive or predictive, within the limited space the data—and our computational abilities—allow us to traverse.

Now predictive fidelity is easy enough to measure: we can quantify the model's ability to predict the patient's behavior from the lesion together with nuisance covariates of one kind or another. Naturally, the model needs to be tested on out-of-sample data it has not "seen," for a high-capacity model can trivially memorize the data it has been exposed to. But generalizable predictive fidelity merely tells us the model has captured the underlying functional anatomy sufficiently well to make good predictions *from lesion data*. Given that lesion scales are typically much larger than functional anatomical scales, and lesions themselves exhibit a characteristic structure (as we shall see further on), the decision boundaries identified by a discriminative model may be only a partial guide to the functional anatomical landscape. Under pressure to keep the number of parameters low, a model will tend to ignore voxels that are irrelevant to the decision, even if they are very much relevant to function [8, 9].

How can we tell which voxels will be ignored in this way? The simple answer is that we *cannot* without knowing *both* the functional anatomical architecture *and* the lesion architecture [10]. This means that the behavior alone does not give us an easy ground truth for the anatomy, for it conflates two complex anatomical patterns. Not only do we not possess an anatomical index of fidelity here; a three-dimensional volume of weights cannot possibly provide an adequate description, for the relationships between loci might depart from the simple conjunction the natural reader of functional anatomical maps tends to infer. There may be complex partial or complete disjunctions here, including exclusive ones, that no scalar volumetric map could conceivably capture. We cannot reasonably insist on representing the ground truth as a simple map, even if we could derive one.

4 Functional and Lesional Anatomies

What little we know about BOLD can reasonably reassure us its physiological drivers operate below the spatial scale of the macroscopic functional organization of the brain. Functional imagers therefore rarely dwell on its elemental spatial structure, though its temporal characteristics are a matter of great interest. Lesions, by contrast, are almost always on a scale that is either comparable or—more commonly—greater than the functional anatomy they are used to illuminate. Lesion anatomy therefore matters at least as much—indeed arguably more—for it is the "lens" through which functional anatomy is "seen."

To illustrate the point, consider a simpler topological system—a two-dimensional flat surface—where the criticality for a given hypothetical function is defined as distributed in the pattern of the symbol "?". We do not know the pattern and must infer it from a set of hypothetical lesions distributed across the plane that disrupt function if, and only if, one or more critical locus falls within their boundaries. If the lesions are contiguous, the acuity of the inferred picture is a simple function of their size. If the lesions overlap, acuity is the outcome of a complex relation between the size of the lesions, their number, and the geometry of their overlap, exactly analogous to the scenario in image super-resolution. It should be obvious that a large set of small, finely overlapping lesions will yield the best acuity and a small set of large, coarsely overlapping lesions the worst (Fig. 2).

Now consider the maximum fidelity achievable with infinite data. With contiguous lesions, the minimum lesion size imposes a hard ceiling on the achievable resolution at less the Nyquist limit of the sampling spatial frequency thereby defined, less because the sampling signal is nonlinear and non-stationary. For lesions detected by their clinical manifestations—by far the commonest—this will almost invariably be many times the typical voxel resolution. With uniformly distributed, consistently overlapping lesions, "sub-lesion" resolution beyond the Nyquist limit is theoretically achievable but cannot be analytically prescribed where the sampling variation is as complex as it is here.

Now if the lesions are simple blobs, the retrieved anatomy will be degraded—more so the larger the lesions and the lesser the overlap—but otherwise undistorted. The distribution of degradation across the field will be as uniform as the distribution of the lesions. But what if the lesions exhibit a characteristic, non-random structure that more or less systematically varies with location? Far from a hypothetical, the existence of such a structure is rendered inevitable by the etiology of common pathological lesions. In stroke, for example—by far the commonest source in lesion-function studies—lesions are naturally the joint outcome of the topology of the vascular tree, the mechanisms of vascular occlusion, and the clinical eloquence of the affected brain [7].

In such circumstances, the retrieved picture will not only be degraded; it may be distorted beyond all recognition in a manner that is *not* remedied by including more data but *entrenched* by it. Figure 3 shows the impact of this within another synthetic

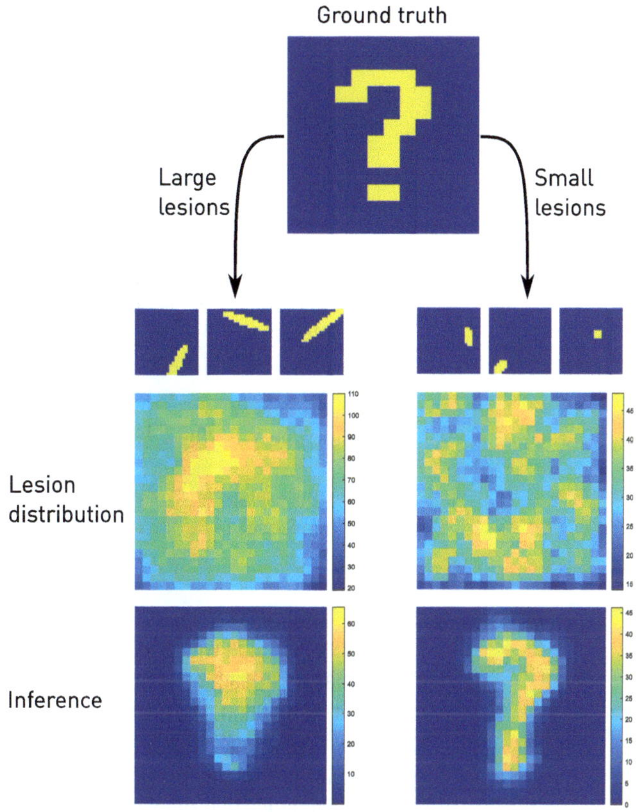

Fig. 2 Lesion-behavior mapping as nonlinear, stochastic filtering. The task is to infer an unknown functional anatomical map (here illustrated in two dimensional 24 × 24 variable space by the symbol "?") with the aid of a series of natural lesions, each associated with a behavioral label, that act as distorting filters through which the true map is perceived. Among other characteristics, the size of the lesions will limit the precision and accuracy of the inferred map, here drawn with mass-univariate Fisher's exact tests

two-dimensional model: adding even relatively simple structure to the lesion distribution may fatally corrupt the inferred picture (second row).

Lesion anatomy is then a vital concern here, independently from functional anatomy, and *prior* to it. The aspects that concern us are *not* merely the extent and uniformity of coverage, dwelt on at length in the past, but the complex patterns of inter-voxel dependence across lesions. Their specification is irreducible to the scalar volume maps we are so accustomed to, for they occupy a space of voxel covariances with as many dimensions as there are voxels. Characterizing lesion anatomy is not a simple problem, and it must be solved before lesions are used to understand the anatomical organization of brain function. What approach should we take?

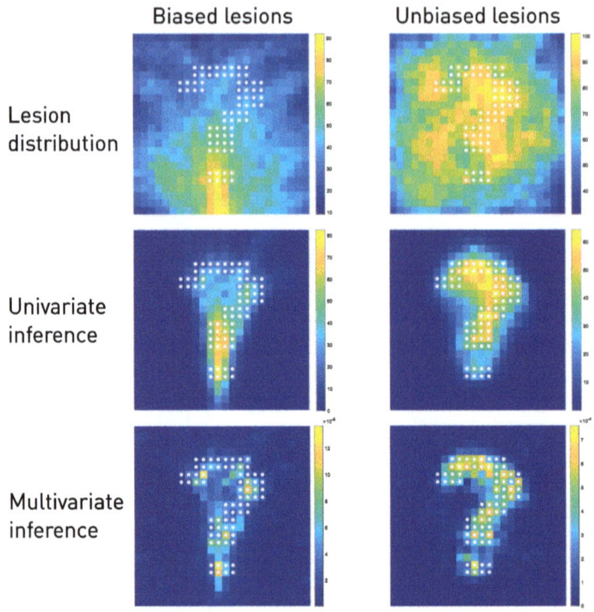

Fig. 3 Demonstration with a two-dimensional (24 × 24 variable) synthetic model of the potential impact on spatial inference of complex patterns of covariance in the lesion distribution. For each of two patterns of simulated damage ("Biased," left, and "Unbiased," right), we created 1000 single, uniformly distributed synthetic lesions in the form of ellipses (*see* Fig. 2 for examples) with a major to minor diameter ratio ranging between 0.1 and 0.2. In the unbiased case, the orientation of the lesions was uniformly random; in the biased case, the range of randomly chosen orientations varied systematically with location. The resultant lesion distributions are shown in the first row. A synthetic ground truth in the shape of a question mark (white glyphs) was then used to label each lesion as generating a binary deficit where an overlap of four or more pixels was present. This "deficit vector" was applied in a pixel-wise Fisher's exact test to generate a field of *p* values, shown as the negative log in the second row. Note the marked distortion of the inferred map in the biased condition. The last row shows the feature importances from a high-dimensional multivariate model based on a sampling/boosting tree-based ensemble algorithm [11]. Note the inferred picture shows comparatively less distortion

5 Empirically Informed Simulation

We have seen that any lesion-function study will retrieve a map that convolves two unknown complex anatomical patterns: the lesion architecture and the true functional architecture. Any method that seeks to isolate the latter must at the very least quantify the effect of the former. This can only be done by setting an *artificial*, synthetic ground truth for function and attempting to retrieve it with *organic*, real lesions representative of the lesion architecture. That such an exercise is in silico does not make it merely theoretical; on the contrary, it is the *only* way of obtaining empirical evidence for the fidelity of any lesion-function mapping approach. We are otherwise left with the impossible task of estimating the value of a variable known only from its product with another unknown variable.

What ground truths should we posit? If we assume nothing about the functional organization of the brain other than what functional imaging data has already suggested (and even that is arguably question-begging), we need to evaluate a vast space of loci of variable size interacting conjunctively, disjunctively, and any partial combination of the two [7, 12]. This implies testing not one candidate functional map, but an array of maps so large an evaluation is likely to be tractable only when focused on a relatively narrow set of possibilities, focused on restricted sets of functions within relatively constrained anatomical territories.

In any event, it makes sense to approach the problem hierarchically, beginning with the simplest imaginable pattern of functional dependence: single, independent voxels. If a given method fails to achieve adequate retrieval with this simplest of models, escalation to more complex models is unwarranted for they can only be worse. If it succeeds, the next stage reasonably consists of models of larger areas, including their interactions, guided by orthogonal functional and structural data such as co-activation and connectivity. Here, too, success at retrieving one set of relations—simple pairwise conjunctions, for example—does not warrant generalization to disjunctions or more complex conjunctions: the power of the method in vivo is always limited to what has been demonstrated in silico.

This is the approach we take in demonstrating the distortion mass-univariate methods inevitably introduce [7]. Here a family of single-voxel dependence models was evaluated across the entire brain, for each model labeling a stack of 581 lesions as "affected" or "unaffected" dependent on their overlap with the index voxel. A voxel-wise Fisher's exact test applied in mass-univariate fashion then yielded, for each model, a Bonferroni-corrected cluster, to whose centroid we derived a vector quantifying the direction and magnitude of the spatial error (Fig. 4).

The average mis-localization incurred by the mass-univariate approach was 15.7 mm (standard deviation 9.2 mm), a distance sufficient to move from gray to white matter, between Brodmann areas and across cerebral lobes.

Quantifying the error in this way may be thought to provide a means of eliminating it within the mass-univariate framework, for example, by applying the inverse of the mis-localization vector field to the inferred map [13]. But the pattern of error is *specific* to the underlying lesion-deficit model, which is by definition unknown and is here hypothetically posited only to demonstrate the error. It is not just that a single-voxel dependence model is biologically implausible, any model—simple or complex—will produce its

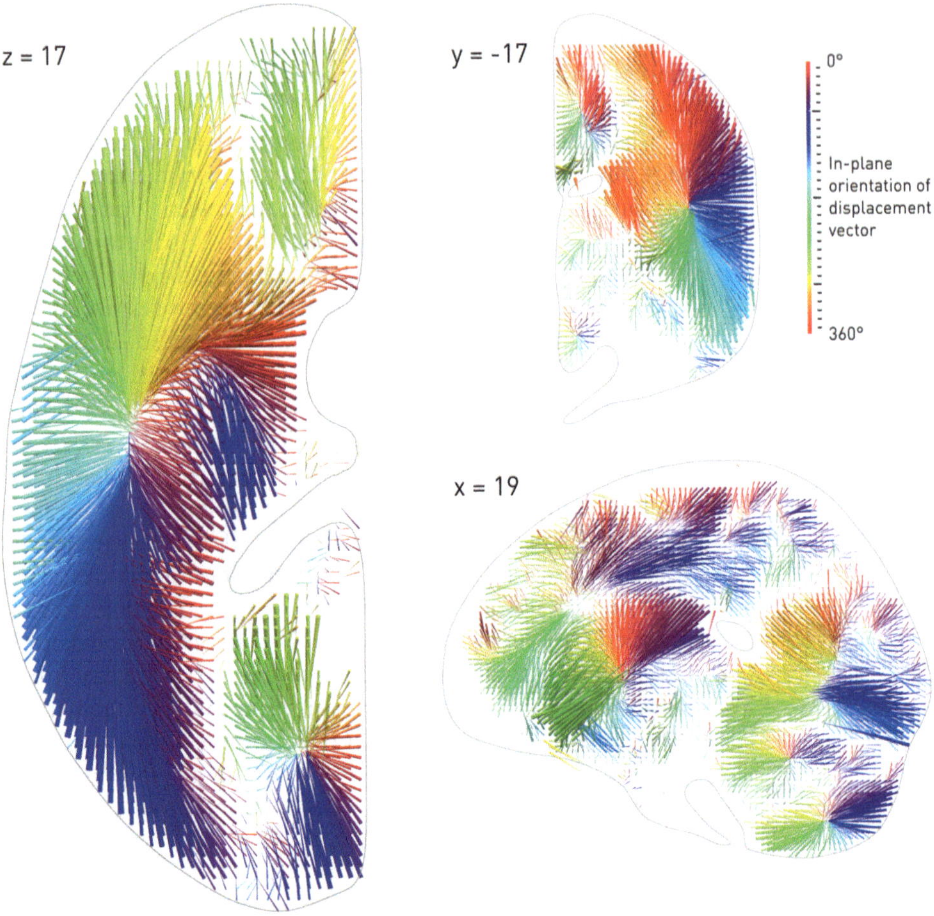

Fig. 4 Three-dimensional vector plot of the direction (color map) and magnitude (length of arrow) of mis-localization at adequately sampled voxels within 3 representative planes (left axial $z = 17$, top coronal $y = -17$, and bottom sagittal $x = 19$), based on a sample of 581 acute stroke lesions, normalized into standard stereotactic space, and mirrored onto one hemisphere. The value at each voxel was calculated by labeling the stack of 581 lesioned volumes as being "affected" or "unaffected" depending on whether or not that voxel fell within the lesion in each volume, running a standard voxel-wise Fisher's exact test-based mass-univariate analysis on the two groups, and identifying the center of the resultant significant cluster, identified by the asymptotic p value thresholded at a Bonferroni-corrected $p < 0.01$. This procedure was performed at all voxels hit more than 3 times in the dataset. Note that the mis-localization tends to follow the organization of the vascular tree, with clusters corresponding to the branches of the middle cerebral, anterior cerebral, and posterior circulations (dotted rings). (*See* Ref. [7] for details)

own peculiar mis-localization field that cannot be transferrable to any other through a simple voxel-wise operation.

That the mis-localization arises from the collision of two complex anatomical patterns means that no simple scalar adjustment could conceivably correct it. For example, adding lesion volume as a covariate will bias the spatial inference against loci more commonly affected in large lesions [10]. Since such loci tend to fall far from the vascular tree—the reason they correlate with lesion size—they tend to be prone to greater distortion, and downgrading them artifactually reduces the mean error overall. But the problem is not solved: to one hidden bias, another is simply added, in a different pattern.

Equally, we cannot, as has been suggested by some [13], arbitrarily censor voxels at some proportional threshold—e.g., where a voxel is involved in less than 5% of all lesions in any one study. Lesions are differentially distorted by such censoring, to a degree and in a pattern that unpredictably interacts with their comparative frequency. The use of a proportional criterion implies that in a dataset of (say) 10 000 lesions, we would be compelled to ignore voxels sampled as many as 500 times. The only permissible way of bringing rare voxels back into the picture would then be to remove lesions confined to common areas, further distorting the lesion patterns fed into the model. Of course, rarely hit voxels tend to be further away from the vascular tree, and so pruning them trivially reduces the mean localization error. But to accept this is to assume that the function of areas of the brain varies with their vascular position, indeed that they might simply be ignored on that account alone. This is justifiable neither statistically nor neuroanatomically.

The point of in silico modeling here is to give us a lower bound on the error, the best it could conceivably be under circumstances of perfect lesion-deficit fidelity and a given, usually simple, anatomical dependence. It is crucial to appreciate that simple models such as single-voxel dependence set a much *lower* standard than biological plausibility requires and are used merely to identify methods, such as VLSM, that are obviously inadequate. Pace Sperber and Karnath, it is *never* sufficient to show that a method appears to work within a synthetic model beneath the threshold of plausibility: only models of the complexity likely to obtain in reality may constitute a persuasive test. And as long as the underlying pathological process determining the lesion is neither simple nor random [5], no straightforward correction could release a mass-univariate approach from the risk of spatial distortion.

6 Anatomical Inference in the High-Dimensional Realm

Simulations, then, provide us with a robust framework for validating the performance of any lesion-function mapping method. Before we come to the central question of what method we should use, let us examine what is required of the input data and the output metrics and what it is that we are fundamentally modeling.

6.1 Data

We are dealing with a complex, multivariate problem whose intrinsic dimensionality is unknown. There is no viable mechanism here for estimating power and no reason to expect performance not to continue to improve with adding data beyond the typical limit of practical feasibility: $>10^4$ cases [14]. If outcome prediction were our aim, we would want the data to be as representative of the population of interest as possible. But, since our aim here is anatomical inference, our sole concern is with the aspects of the lesions that maximize the fidelity of spatial retrieval. This makes a "minimal behaviourally eloquent lesion" the ideal [15] one that is of the smallest size and with the simplest spatial pattern of dependence of its constituent voxels. The comparative rarity of lesions does not generally give us the luxury of choice, but as imaging collections become larger and more comprehensive, selection will be increasingly feasible. What should be the balance between data volume and quality? This can only be ascertained in simulations with fixed ground truths as already discussed.

Though it is natural to be preoccupied with the quality of the lesion parameterization, the associated behavior is equally important. The desiderata of behavioral characterization in neuroscience apply here indeed are amplified by the difficulty of the problem. It is better to use psychometric functions than scalar scores; adaptive rather than fixed difficulty; continuous rather than discrete measures; plausibly stationary rather than dynamic characteristics; and many, decorrelated rather than few, correlated dimensions of performance. It is common to cite patient tolerability as an excuse, but it is striking how few studies even consider the efficiency, the "sweat factor" of the paradigms they deploy, let alone try to optimize it. Techniques matured in the field of psychophysics are readily applicable here [16, 17].

Although the volume and diversity of imaging data with matched behavioral labels are paramount, we should bear in mind unlabeled imaging data—often far more abundant—is far from useless here. Aside from enabling simulations, it allows us to derive optimally surveyable representations of lesions that facilitate the subsequent discriminative tasks. Where, as here, complexity is extreme, we need all the help we can get, and semi-supervised inferential frameworks tend to make the best use of the data we have.

6.2 Metrics

We have seen that the space of hypothetical anatomical dependents is infeasibly large here. No frequentist framework is applicable, and no criterial voxel-wise *p* values could be meaningfully derived. What alternatives do we have?

First, our trust in the model's estimated parameters depends on its predictive performance, quantified wholly out-of-sample. A model with high—generalizable—predictive fidelity can be trusted to have at least *part* of the picture right; a poorly predictive model is useless, even if the poor performance is caused by confounding from extra-anatomical factors and the spatial fidelity is actually adequate. There is nothing but prediction here to guide our inference: if it is poor, everything downstream of it will be impoverished too. Even where prediction is excellent, the solution may be grounded in only a small number of loci, if that is all it takes to achieve good performance. We might trust the specificity but not the sensitivity of the result.

Second, the weights of the model are readily interpretable as a voxel-wise map only if the model is linear. Within a complex nonlinear model, a voxel-wise projection is an approximation that may be both crude and misleading and cannot be assumed to generalize across either functions or anatomies. Here we must *specify* the observed relationships between critical loci rather than just the loci themselves, leaving the output naturally intelligible only as a graph. It is obviously difficult to render such a graph intuitive at a fine anatomical resolution.

So the best we can do is to evaluate a modeling architecture's capacity to retrieve sets of *hypothetical* functional anatomical ground truths, instantiating specified forms of multifocal relationships, and to trust the weights estimated from real outcome data—projected into a space as simple as the relationships allow—only as far as the synthetic models justify and only for the relational complexity proven to be tractable with the method at hand. Such inference is naturally bound to be hesitant, not because there is a better method, but because the problem is far too complex to allow strong conclusions.

6.3 Forms of Description

Though our ambition is to retrieve the underlying functional anatomy, we may be content to specify it in the form or "vocabulary" of lesions. Rather than asking what is the critical neural substrate of a function, here we may ask what are the spatial characteristics of the lesions that tend to disrupt it. The answer might be intelligible as an archetypal lesion, but it might equally be a *family* of otherwise unrelated lesions. At the limit of infinite data, the critical neural substrate will fall within the inferred boundaries, but will also include non-critical regions that are invariably collaterally damaged, reflecting a limitation of the data rather than the inferential method. It is an approach likely to dissatisfy the neuroscientist, but it might be fine for the neurologist, for whom localization is of interest only in the setting of pathology. Note both are forms of spatial inference; the difference is only that the latter makes no attempt to deconvolve functional and lesional anatomies.

One might also choose to assume a more or less quantized vocabulary of functional anatomy, relying on structural or other indices of local homogeneity to parcellate the brain into regions much larger than typical acquisition voxels, greatly simplifying subsequent analysis, especially when complex interactions are modeled. Naturally, the assumption of homogeneity places a ceiling on the fidelity of the inferred picture and may well be misguided anyway given the mounting evidence of continuous anatomical patterns in the brain. Such discretization also makes it harder to remove the distorting effects of the lesion architecture, for they are then buried beneath the spatial scale of discretization [18]. But it also facilitates the derivation of network models of dependence, potentially capturing complex non-monotonic relations.

Finally, one might wish to treat gray and white matter differently, a localization with the latter being naturally interpretable as a *disconnection* of the former [19]. Rather than mapping white and gray matter anatomically, it is reasonable to project the damage solely onto gray matter, the components arising from white matter appropriately weighted by strength of gray matter connectivity. One can think of this as an "anatomical support vector machine," transforming white matter damage into a more widely distributed anatomical space where functionally distinct lesions are nonetheless more easily separable. Such disconnectome analyses will usually model only a first order of connection, but that is not an intrinsic limitation of the technique. Where interactions between distributed areas are modeled, the structural connectivity between nodes is a plausible weighting for every edge, and a disconnectome is a perfectly logical addition. As with cortical anatomy, an assumption of inter-subject homogeneity is implicit, but that seems no less reasonable here than there.

6.4 Modeling Approaches

We should confess at the outset that no wholly satisfactory method for high-dimensional multivariate lesion-function mapping exists. Even if it did, the field of complex modeling is evolving so rapidly; here we can give only tentative guidance, hardened mostly where we know, conceptually know, that a particular avenue is invalid or likely to produce biased results. We assume the brains are already in reasonable register and that lesion segmentation is adequate: these aspects are naturally common to all lesion-deficit mapping and need no special treatment here.

6.5 Prior Feature Selection

Where the objective is prediction rather than inference, it is natural to seek to limit the number of input dimensions (characteristics or "features" of the lesion) through a prior feature selection step, eliminating those that are irrelevant to that task either because they bear no identifiable relation to the outcome or because they are duplicated by others. Since our objective is anatomical inference rather than prediction, it is natural to be liberal here, erring on the side of inclusivity. In particular, whereas for the purposes of prediction one might drop one of two redundant features, both remain relevant to lesion-deficit mapping: our interest here is not merely in the decision boundaries but the constitution of the anatomical territories on both sides. Nonetheless, a degree of prior selection is sensible given how taxing the downstream inferential problem is with the scale of data commonly available and the degree of biological heterogeneity likely to obtain in reality. It is also sensible to favor Bayesian methods, which are by their very nature are less liable to be misled by irrelevant features.

How feature selection is done otherwise requires no special commentary: the full range of techniques may theoretically be applied, even mass-univariate statistics if extracted from independent data at highly liberal thresholds. But careful consideration must be given to the downstream consequences. Feature selection insensitive to nonlinear, widely distributed effects may strip them from the data, rendering them undetectable by the lesion-deficit model. Equally, it may conceal distortions arising from stereotyped patterns of collateral damage that we have seen fatally undermine direct mass-univariate inference.

6.6 Linear Methods

Those approaching the problem from the world of functional imaging are naturally attracted to methods that operate directly in voxel space. Since the size of voxels needs not be congruent with the underlying functional anatomy, this will over-parameterize some regions and under-parameterize others. Where operating at the acquisition resolution, or close to it, there is nothing to be done about under-parameterization. But over-parameterization needlessly expands the dimensionality of the model and introduces collinearity that causes numerical instability. This invites two pre-processing steps: down-sampling to a more tractable resolution and/or selecting voxels through prior feature selection. Each is obviously lossy, and the distortion introduced by the latter will tend to correlate with the degree of dimensionality reduction.

The resultant dimensionality will nonetheless be typically high in proportion to the number of cases, leaving the model underdetermined by the data. In Bayesian penalized logistic regression, this is done by applying a prior scheme that promotes a sparse solution (*see* [20] for a state-of-the art implementation and [21] for an applied example). There is a presumption of economy here, the ubiquitous Occam's razor, that need not obtain in the brain, but it is the best we can do. As with any methods here, one could explore the effect of different priors in simulations and choose the approach that best balances spatial accuracy with the efficiency needed in real-life studies. The simplicity of formulation makes the interpretation of the inferred anatomy simple enough: as with conventional regression, we have beta coefficients to plot, here qualified by their estimated distributional parameters to yield Bayesian credibility intervals for their plausible values. Model statistics are interpretable here—of course qualified by measures of the captured variance—and the Bayesian framework makes principled model comparison easy.

Though commonly used for classification, support vector machines [22] offer a highly flexible framework for modeling, of particular relevance to our domain being regression, multiclass classification, and multi-label classification—corresponding to continuous, exclusive categorial, and non-exclusive categorial deficits—and the use of unlabeled data through semi-supervised learning [23] (find implementations at https://www.csie.ntu.edu.tw/~cjlin/libsvm/). When a linear kernel is used, they yield voxel-wise weights interpretable as measures of relevance. Such measures are not, however, interpretable along the lines of conventional model statistics, nor can be licitly rendered interpretable by grafting conventional statistics on top of the kernel machinery. For example, pace Zhang and his colleagues [24], permutation-derived statistics on the variation in weights across a training or validation run produce p values on the null that the weights *of the given model* are random, not that the neural dependents of the behavior under study are random. The weights of a model with no capacity to learn the underlying lesion-deficit association and zero predictive power might well be stably non-random, for the optimization may have become stuck in some local minimum.

Those relaxed about the vocabulary of the input can project the voxel data into a more compact latent space, reducing collinearity and the number of irrelevant features upstream of the main modeling architecture. While this is traditionally done by principal component analysis, the incoherence of a negative lesion makes other methods such as logistic principal component analysis [25] or non-negative matrix factorization more appropriate [26]. A further advantage of the latter is its tendency to yield a naturally parts-based decomposition of the input, increasing the localization of the subsequently inferred picture. Where the target behavior is multidimensional, dimensionality reduction at the input can be naturally replicated at the output, through either partial least squares or canonical correlation analysis.

Any of these methods is open to a Bayesian formulation (*see* [21] for an example application). Of the many attractions a Bayesian framework offers, of particular relevance to our application is the graceful handling of missing data and resistance to overfitting with relatively compact datasets, i.e., efficiency.

6.7 Nonlinear Methods

One might increase the expressive power of the models outlined above by transforming the inputs, creating additional nonlinearly transformed terms and interactions just as one might with conventional regression. This leaves each term easily interpretable, but inevitably within limits of expressiveness. The natural evolution from Bayesian penalized regression is to Gaussian processes [27], and though they have seen applications in modeling structural changes in the brain (e.g., [28]), the implicit redescription of the problem in functional form complicates the anatomical interpretation. Similarly, when support vector machines are deployed with nonlinear kernels, there is no easy way of relating the inferred anatomy to its voxel-wise coordinates without making questionable simplifications it is the very point of complicating the model to escape in the first place. Though tree-based algorithms, especially those operating in ensemble mode such as gradient boosting machines [29] (implemented here https://xgboost.readthedocs.io/en/latest/), offer natural measures of feature importance, they will tend to produce a very sparse picture, strongly amplifying features closest to the decision boundary.

There is generally an inevitable trade-off between model fit and interpretability, on which intelligibility in the form of a 3D map of voxel-wise weights may be too tight a constraint. An alternative is to confine the modeling of complex relations to a small subset of target features or to operate with regions-of-interest on a sufficiently large scale for their combinatorial interactions to be surveyable [8, 30]. The former requires another means of identifying the candidate areas; the latter assumes functional homogeneity across implausibly large anatomical territories. But it is hard to see how very complex relations such as non-monotonic interactions can be captured with the scale of data commonly available.

Given their proven capacity to approximate any function [31], even in relatively shallow architectures, artificial neural nets are an obvious tool to consider here, interpreting neuronal activation maps as measures of anatomical dependence (e.g., with implementations here https://www.cs.toronto.edu/~radford/fbm.software.html). But the complexity of lesion-function relations, certainly at high anatomical resolutions, is too high to be tractable with fully connected nets at the scale of commonly available labeled data. Though convolutional neural nets save parameters by translating the input into a compact visual "dictionary," they are not well suited to the binary data of lesion masks. If nets make an impact here, it will be downstream of learning compact representations of the space of possible lesions through deep generative models of lesions, contextualized by non-imaging variables of interest. The quality of such models is currently limited by the scale of available data and the lack of an efficient scheme for decomposing lesions.

7 A Perspective

That the approaches we have outlined are so hesitantly sketched, so bare of the usual accoutrements of statistical modeling, is a reflection of the fundamental difficulty of the problem and the lack of sufficient constraint on the space of possible solutions. Progress here is likely to depend on three things: first, a reality-induced adjustment of the lesion-mappers ambitions and aspirations; second, careful attention to the sensing instrument here—lesions themselves—and third, refinement in application of the few methods flexible enough to be trusted, given sufficient data. Let us take each of these elements in turn.

7.1 The Limits of Lesion-Function Mapping

Though causally strong, lesion-function mapping is spatially impaired by the pathological nature of lesions. Even at the limits of infinite data and elementary functional anatomy, it can only yield maps with a constrained bandwidth of representation far narrower than the acquisition parameters of the image. We need to adjust to the idea that reporting and interpreting localizations with voxel precision is meaningless here. Equally, the space of conceivable patterns of functional organization is so great, we can only hope to discriminate between candidate subsets defined by other methods, limiting conclusions to preference between models, not general inferences about the brain. Moreover, to race ahead to inferential models reckless of their predictive power is to reach a finishing line no one could care about, for the parameters of a poorly predictive model are of no consequence where its alternatives remain unexplored. We need to focus on prediction first and accept grander ambitions are likely to be merely grandiose.

7.2 Lesion-Induced Bias

We have seen the distribution of lesions, at least of vascular lesions, is complex. But unlike functional anatomy, we have a direct view of it and at the relatively modest expense of imaging alone are able to model it independently of the questions it is used to address. Moreover, we may explicitly determine its capacity to reveal functional-anatomical relations, operating with synthetic ground truths that enable us to quantify fidelity robustly. Attention to reformulating lesions in ways that simplify the task downstream is likely to be rewarding. Such reformulation may be topological, a matter of maximally disentangling the constituent spatial components so the picture they reveal is minimally distorted. But it might also be guided by known anatomical features of the brain, such as the differential behavior of white and gray matter. In any event, the fidelity of any lesion-function map will always be gated by the instrument used to derive it—natural lesions—so that ought to be the overriding focus of attention.

7.3 Complex Modeling Methods

The success of deep nets across complex modeling domains—from machine vision to language translation—is no accident: they are excellent universal function approximators [32]. But their capacities will always be gated by two things: the availability of data of sufficient scale and quality and the computational burden of optimization. Though the importance of data is now widely recognized, few appreciate the expense of arriving at an optimal architecture, not just parameter tuning but the fundamental architectural framework itself. Methodological innovation here is driven by data of a very different type—typically two-dimensional, three-channel photographic images—available at grand scale and dominated by broadly affine relations between material elements. The transition to volumes is not merely incremental, but creates categorially different problems, and the nonlinearities of brains are not replicated by any other kind of widely studied data. Were research in machine learning preoccupied with classifying porridge rather than distinguishing cats from dogs in holiday scenes, our circumstances might be different. But as it is—relatively simple tasks such as lesion segmentation aside—the theoretical benefits of deep learning will take some time to realize.

References

1. Bates E, Wilson SM, Saygin AP, Dick F, Sereno MI, Knight RT, Dronkers NF (2003) Voxel-based lesion-symptom mapping. Nat Neurosci 6:448–449
2. Karnath HO, Fruhmann BM, Kuker W, Rorden C (2004) The anatomy of spatial neglect based on voxelwise statistical analysis: a study of 140 patients. Cereb Cortex 14:1164–1172
3. Schwartz MF, Faseyitan O, Kim J, Coslett HB (2012) The dorsal stream contribution to phonological retrieval in object naming. Brain 135: 3799–3814
4. Phan TG, Donnan GA, Wright PM, Reutens DC (2005) A digital map of middle cerebral artery infarcts associated with middle cerebral artery trunk and branch occlusion. Stroke 36: 986–991
5. Dronkers NF, Colliot O, Iba-Zizen MT, Cabanis EA (2007) Paul Broca's historic cases: high resolution MR imaging of the brains of Leborgne and Lelong. Brain J Neurol 130: 1432–1441
6. Lee E, Kang D-W, Kwon SU, Kim JS (2009) Posterior cerebral artery infarction: diffusion-weighted MRI analysis of 205 patients. Cerebrovasc Dis 28:298–305
7. Mah Y-H, Husain M, Rees G, Nachev P (2014) Human brain lesion-deficit inference remapped. Brain J Neurol 137:2522–2531
8. Smith DV, Clithero JA, Rorden C, Karnath H-O (2013) Decoding the anatomical network of spatial attention. Proc Natl Acad Sci USA 110:1518–1523
9. Sperber C, Karnath H-O (2017) Impact of correction factors in human brain lesion-behavior inference: validity of human brain lesion-behavior inference. Hum Brain Mapp 38:1692–1701
10. Xu T, Jha A, Nachev P (2018) The dimensionalities of lesion-deficit mapping. Neuropsychologia 115:134–141
11. Seiffert C, Khoshgoftaar TM, Van Hulse J, Napolitano A (2009) RUSBoost: a hybrid approach to alleviating class imbalance. IEEE Trans Syst Man Cybern Part Syst Hum 40: 185–197
12. Pustina D, Avants B, Faseyitan OK, Medaglia JD, Coslett HB (2018) Improved accuracy of lesion to symptom mapping with multivariate sparse canonical correlations. Neuropsychologia 115:154–166
13. Sperber C, Karnath H-O (2017) Impact of correction factors in human brain lesion-behavior inference. Hum Brain Mapp 38: 1692–1701
14. Schulz M-A, Yeo T, Vogelstein J, Mourao-Miranada J, Kather J, Kording K, Richards BA, Bzdok D (2019) Deep learning for brains?: Different linear and nonlinear scaling in UK Biobank brain images vs. machine-learning datasets. bioRxiv:757054
15. Husain M, Parton A, Hodgson TL, Mort D, Rees G (2003) Self-control during response conflict by human supplementary eye field. Nat Neurosci 6:117
16. Wichmann FA, Hill NJ (2001) The psychometric function: I. Fitting, sampling, and goodness of fit. Percept Psychophys 63: 1293–1313
17. Kuss M, Jäkel F, Wichmann FA (2005) Bayesian inference for psychometric functions. J Vis 5:8–8
18. Nachev P (2015) The first step in modern lesion-deficit analysis. Brain 138:e354
19. Thiebaut de Schotten M, Foulon C (2017) The rise of a new associationist school for lesion-symptom mapping. Brain 141:2–4
20. Makalic E, Schmidt DF (2016) High-dimensional Bayesian regularised regression with the BayesReg package. ArXiv161106649 Stat. Available at: http://arxiv.org/abs/1611.06649
21. Jha A, Teotonio R, Smith A-L, Bomanji J, Dickson J, Diehl B, Duncan JS, Nachev P (2020) Metabolic lesion-deficit mapping of human cognition. Brain 143:877–890
22. Cortes C, Vapnik V (1995) Support-vector networks. Mach. Learn. 20:273–297

23. Xu T, Rolf Jäger H, Husain M, Rees G, Nachev P (2018) High-dimensional therapeutic inference in the focally damaged human brain. Brain 141:48–54
24. Zhang Y, Kimberg DY, Coslett HB, Schwartz MF, Wang Z (2014) Multivariate lesion-symptom mapping using support vector regression. Hum Brain Mapp 35:5861–5876
25. Tipping M (2004) Bayesian Inference: An Introduction to Principles and Practice in Machine Learning. Adv. Lect. Mach. Learn. 3176:41–62
26. Lee DD, Seung HS (1999) Learning the parts of objects by non-negative matrix factorization. Nature 401:788
27. Rasmussen CE (2003) Gaussian processes in machine learning. In: Summer school on machine learning. Springer, Berlin, pp 63–71
28. Gutierrez Becker B, Klein T, Wachinger C (2018) Gaussian process uncertainty in age estimation as a measure of brain abnormality. NeuroImage 175:246–258
29. Friedman JH (2001) Greedy function approximation: a gradient boosting machine. Ann. Stat.:1189–1232
30. Toba MN, Zavaglia M, Rastelli F, Valabrégue R, Pradat-Diehl P, Valero-Cabré A, Hilgetag CC (2017) Game theoretical mapping of causal interactions underlying visuo-spatial attention in the human brain based on stroke lesions. Hum Brain Mapp 38: 3454–3471
31. Park J, Sandberg IW (1991) Universal approximation using radial-basis-function networks. Neural Comput 3:246–257
32. Poggio T, Banburski A, Liao Q (2020) Theoretical issues in deep networks. Proc Natl Acad Sci. https://www.pnas.org/content/early/2020/06/08/1907369117 Accessed 26 Aug 2020

Chapter 12

Lesion-Based Prediction and Predictive Inference

Daniel Mirman and Melissa Thye

Abstract

The successes of lesion-symptom mapping (LSM) have brought interest in predicting symptoms or deficit patterns from lesion features. This chapter covers two forms of lesion-based prediction. The first is predictive inference: a set of statistical procedures (such as k-fold cross-validation) that can be used to infer brain-behavior relationships. This improves the rigor and generalizability of claims about brain-behavior relationships compared to standard associationist inferences, which tend to be overly optimistic about the strength of lesion-symptom relationships. Predictive inference is fairly straightforward to implement, and some LSM methods—especially those based on machine learning—already use it. The second is prospective or longitudinal prediction: using lesion data to predict the degree of deficit at some future point in time or to predict the degree of response to a particular treatment. This requires collecting longitudinal data and accounting for structural and functional changes that occur over this period. Although these issues are challenging, they also present important opportunities for progress in both basic and translational neuroscience.

Key words Prediction, Cross-validation, Statistical inference, Longitudinal data analysis, Lesion-symptom mapping, Functional reorganization

1 Introduction

The recent dramatic increase in use of lesion-symptom mapping (LSM) methods has brought with it increased interest in lesion-based symptom prediction (LSP). That is, in addition to studying lesion-symptom associations using LSM, there is both scientific and clinical value in being able to predict symptoms or deficit patterns from lesion features. In this chapter, we discuss two distinct senses for the phrase "lesion-based prediction of deficits." One use of the term "prediction" is in the prospective or longitudinal sense: using lesion data to predict the degree of deficit at some future point in time or, perhaps even more usefully, to predict the degree of response to a particular treatment. This kind of prediction has tremendous translational significance for personalized treatment approaches. "Prediction" can also be used in the inferential sense: referring to a set of statistical procedures that can be used to infer

Dorian Pustina and Daniel Mirman (eds.), *Lesion-to-Symptom Mapping: Principles and Tools*, Neuromethods, vol. 180, https://doi.org/10.1007/978-1-0716-2225-4_12, © Springer Science+Business Media, LLC, part of Springer Nature 2022

brain-behavior relationships. For predictive inference, the lesion and behavioral data can be acquired at (approximately) the same time, which is typical for LSM studies, but any longitudinal claims from these cross-sectional data are weaker and require additional assumptions. Of these two forms of prediction, predictive inference is substantially easier to implement and is important for improving the rigor of claims about brain-behavior relationships. It is also a valuable precursor to the longitudinal sense of prediction, so we begin by discussing predictive inference and then proceed to additional challenges for prospective lesion-behavior prediction.

2 Predictive Inference

To understand the concept of predictive inference, it is useful to distinguish it from standard associationist inference. Imagine fitting a regression model for data from 50 participants that contain an outcome variable and 2 predictor variables. If statistically significant effects were observed, they would typically be described as "predicting" the outcome, but there was no prediction here. It would be more accurate to say that these variables were "associated" with the outcome. Now imagine that data were collected from an additional 50 participants; it would be possible to evaluate the model that was fit using the first 50 participants by testing how well it *predicts* the outcome in these new 50 participants. This is the crux of predictive inference: fitting statistical models on one set of data and evaluating how well they describe an independent set of data. It is usually possible to construct this kind of analysis by fitting a model on a subset of the data and evaluating it against the remaining (withheld) subset (more on this below).

This approach is valuable because the full-sample model fits are usually overly optimistic about prediction accuracy. This is a type of overfitting [1]: any data sample will contain spurious associations that only exist in that sample, as well as (possibly) associations that would generalize to other samples. When a full-sample model is fit, these two kinds of associations are merged and indistinguishable, producing an overly optimistic fit to the observed data that may not generalize to other data sets. As Yarkoni and Westfall discuss, the difference can be quite extreme when sample sizes are modest and predictors are not particularly strong. In their example, for $N = 50$, a model with 20 predictors that each have a $r = 0.1$ correlation with the outcome will have in-sample $R^2 = 0.45$ (i.e., will account for 45% of the variance in the sample, which is quite impressive) but out-of-sample $R^2 = 0.02$ (i.e., will be virtually useless for predicting outcomes for new participants).

To provide a concrete neuropsychological example, we used data from the Moss Aphasia Psycholinguistics Project Database [2]. In particular, we focused on the issue of sample size because neuropsychological samples are typically limited by practical

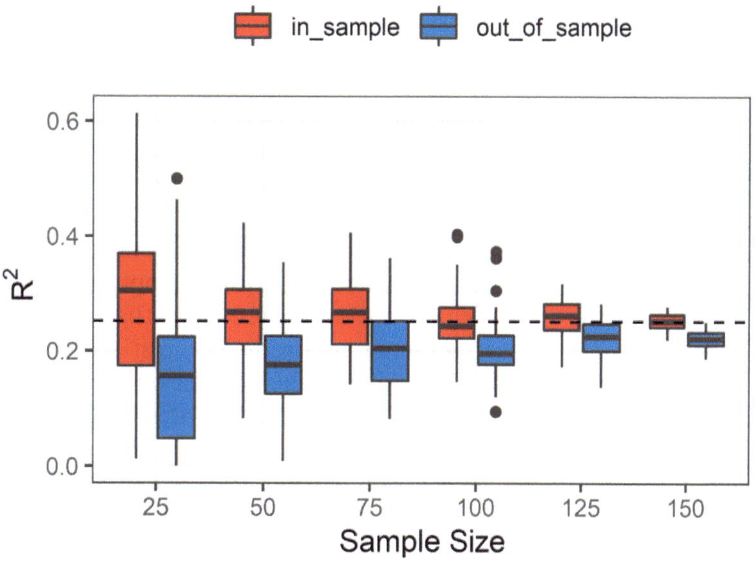

Fig. 1 In-sample association strength and out-of-sample prediction accuracy (R^2) at different sample sizes

constraints. We used a model with two predictors, for which the full sample ($N = 167$) $R^2 = 0.252$ and conducted analyses using 50 random subsamples at different sample sizes (Fig. 1). As sample size decreases, the in-sample R^2 tends to inflate (overfitting the sample), and, naturally, the variability of R^2 also increases (i.e., for smaller subsamples the model can appear much better or much worse than it does for the full sample). In contrast, the out-of-sample R^2 (leave-one-out cross-validation, described in more detail below) is less optimistic and tends to decrease as sample size decreases. That is, as samples get smaller, the in-sample model becomes more prone to false positives and becomes less and less generalizable to new data (out-of-sample R^2).

This example demonstrates that predictive inference is a fundamental statistical challenge, especially for smaller samples. It is not specific to LSM, but most LSM situations face a particularly difficult version of this problem: sample sizes are limited, many voxels can contribute to the deficit, and each voxel only contributes weakly. As a consequence, an LSM result may be statistically significant, but the out-of-sample prediction could be poor. For example, we have found that a speech production deficit composite score is reliably associated with lesions in the "dorsal stream" [3–5], which is a very sensible result that is consistent with a large literature [6]. However, for an individual out-of-sample participant, the overlap between their lesion and this LSM-defined "critical region"[1] is not a

[1] It is common to treat all voxels within the region as equivalent and aggregate them into a region of interest, though a more sophisticated model that used differential voxel weights would likely perform somewhat better.

particularly good predictor of speech production deficit: $R^2 = 0.160$ (and adding overall lesion size to the model is only a tiny, not statistically significant improvement: $R^2 = 0.166$; for more detailed discussion, *see* Thye and Mirman [5]).

The goal of generalizable inference is one of the cornerstones of study design, informing everything from sample selection to stimulus and task design, and it should also inform statistical analysis methods. The usual associationist methods quantify the strength of the evidence for within-sample inferences (lesions in that brain region are associated with that deficit *in this sample*), but this represents an overly optimistic estimate of the strength of the relationship in the population (or in other samples from the same population). Predictive inference methods provide a way to estimate how well a lesion-symptom relationship can be expected to generalize beyond the given sample. In the next section, we describe cross-validation methods for predictive inference.

2.1 Cross-Validation

The general principle of cross-validation is to divide a sample of data into a "training" set, which will be used to fit (train[2]) the model, and a "testing" set, which will be used to evaluate the model. Critically, although the researcher might start out with one large data set, for the purposes of the analysis, the training and testing set will be independent, providing an estimate of out-of-sample prediction accuracy.

A general framework called *k-fold cross-validation* encompasses a continuum of specific ways of implementing the general cross-validation principle, varying in the relative size of the training and testing set. At $k = 2$, this is the "split-half" approach: randomly divide the data set in half so there are two equal subsets, A and B; fit (train) the model on the A half; then predict performance for the B half. The critical measure is the prediction accuracy (e.g., correlation) for the B set. Then repeat the procedure with reversed groups: fit the model on the B half and evaluate it on the A half. The individual models will be based on half-samples, but the overall evaluation will be based on the full sample.

More generally, k can be any integer ≥ 2 and $\leq N$; then, for each fold, fit/train the model withholding that fold; and evaluate the model using that withheld test set. For example, for five-fold cross-validation, the data set would be divided into five equal, random subsets. There would be five models, each trained on 80% of the data and tested on the remaining 20%. Then the prediction accuracy of each model would be combined to calculate the overall prediction accuracy. If $k = N$, then this is a special case known as "leave-one-out" (LOO) cross-validation, where each data point

[2] The cross-validation approach has roots in machine learning where models are typically "trained" rather than "fit," hence the somewhat different, but equivalent, terminology.

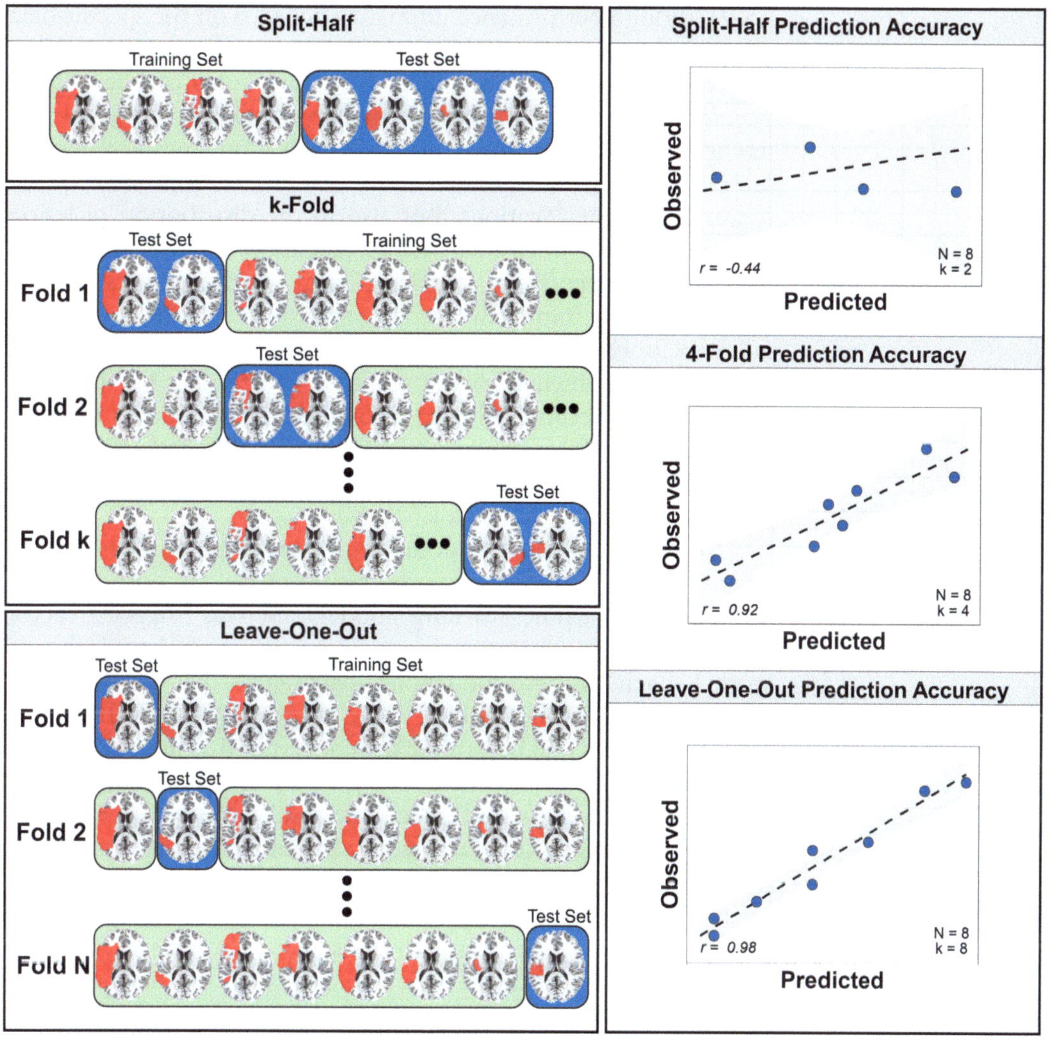

Fig. 2 Schematic illustrations of three versions of cross-validation applied to prediction-based lesion-symptom inference

(participant) is a fold, and, for each participant, the model is fit on all but that one participant's data and then tested on that participant's data. This approach estimates how well the model would describe the outcome for a hypothetical next participant to complete the study. These variants of the general cross-validation approach are illustrated in Fig. 2.

There is no one correct value for *k*. For smaller data sets, two-fold (split-half) cross-validation may be impractical if the training set is too small to fit a stable model. Leave-one-out is a widely used method, but is generally not recommended because the trained model will systematically underfit the test set (the single excluded data point). In practice, *k* values of 4–10 are typically used and have been recommended based on several simulation studies

[7–9], though performance ultimately depends on the specific data set. Because randomly dividing the sample can create groups that are very different on a variable of interest, folds can be created *pseudo*-randomly, ensuring some degree of balance across the different folds. In LSM, the situation is even more complex because samples vary in the distribution of outcome (symptom severity) scores and lesion locations, but also in the distribution of lesion size and amount of overlap. Cross-validation is likely to perform somewhat differently for samples of large overlapping lesions and samples of small minimally overlapping lesions. To our knowledge, no simulation studies have evaluated these types of data situations. It may be worthwhile to test multiple values of k to verify the stability of the results. It is also recommended to conduct multiple cross-validation runs in order to evaluate (and report) the stability of the model and to reduce the variance that results from randomly splitting the sample into folds.

Note that any kind of model can be used in a predictive inference framework, from a basic regression model to more complex multivariate machine learning models such as support vector machines, random forests, or neural networks (including deep networks). Each of these models has parameters that are estimated during model fitting/training (e.g., regression coefficients) and hyperparameters that determine the context of that fitting/training (e.g., the terms included in the regression model). If these vary simultaneously across folds, then the results from different folds may not be directly comparable. On the other hand, setting fixed hyperparameters based on a full-sample model may overfit the cross-validation space (for the same reasons as discussed above). Therefore, it may be useful to construct hierarchical or nested cross-validation in order to tune and evaluate both the parameters and hyperparameters.

2.2 Predictive Inference and LSM

Predictive inference has become more common in LSM research, especially as machine learning techniques have started to be incorporated into LSM analyses. For example, LSM based on support vector regression [10, 11] and sparse canonical correlations [12] use cross-validation to optimize parameters and evaluate model fit (for more on these multivariate LSM methods, *see* Chapter 11). These methods have also been used for multimodal integration [13] and for distinguishing the role of overall lesion size from lesion location [5]. Associationist inference describes a particular sample, which is an optimistic estimate of the effect in the population; predictive inference will more accurately estimate how an observed pattern is likely to generalize (with standard considerations of sampling bias, etc.). Therefore, if our scientific goal is to produce generalizable knowledge about lesion-symptom relationships, then predictive inference is the more effective strategy.

3 Prospective Prediction

Predicting deficits and symptoms based on the pattern of brain damage has tremendous value for prognosis and personalized medicine. Accurate predictions about natural recovery could help clinicians set expectations for their patients and the patients' families and caregivers. Predictions about response to different treatment strategies would be particularly useful for developing personalized treatment plans. In addition to the generalization challenges that are faced with predictive inference, prospective prediction entails several further challenges, though, as we will discuss below, these challenges are also opportunities for addressing important scientific questions.

3.1 Challenges and Opportunities

First, and most obviously, prospective prediction requires collecting longitudinal data. Participants must be recruited for an initial test session that must minimally include the neuroimaging that will be used for prediction and a second session some non-trivial amount of time later to collect the behavioral data that will be predicted. Ideally, both neuroimaging and behavioral data would be collected at each time point in order to longitudinally track changes in both brain and behavior. Recruiting and testing participants with brain damage can be challenging and requires substantial human and financial resources, especially when neuroimaging is involved. Collecting longitudinal data is more common in research on neurodegenerative diseases where tracking the course of neural and behavioral degeneration is central, so perhaps insights and procedures from that domain can help. In the case of abrupt-onset brain injury such as stroke or TBI, it may be particularly useful for the initial test session to be during the acute or subacute stage, shortly after the brain injury occurred, but this is more difficult because the priority during this stage is on the medical stability of the patient and there may be limited time or opportunity for research procedures. There may also be ethical or regulatory considerations if the patient's cognitive status during this stage impairs their ability to provide informed consent. Dealing with these considerations can induce selection biases if, for example, only milder cases are included or only cases with particular deficit profiles, which will then limit the scope of the prediction model.

Second, brain lesions change over time. This is obviously true for neurodegenerative diseases like frontotemporal lobar degeneration and Alzheimer's disease, but it is also true for stroke (for a more detailed discussion of lesions, *see* Chapter 1). In the acute stage (first 24 h), in addition to the infarcted tissue, there is hypometabolism in regions outside the infarct (diaschisis) and hypoperfusion in the ischemic penumbra (perilesional tissue), and there can be structural shifts due to swelling (edema). In the subacute stage (days-to-weeks after stroke), the brain undergoes substantial natural recovery including synaptogenesis and resolving edema. As the swelling goes down and the hypoperfusion (partially) resolves, some of the behavioral deficits also resolve. By the chronic stage (about 3 months post-stroke), most brain changes have subsided, but even after the first year, lesions continue to expand at an estimated rate of 1.59 cc or 6.8% per year [14], and there is whole-brain atrophy, especially in the lesioned hemisphere. Further, Wallerian degeneration, which is the secondary degeneration of white matter fibers distant to the primary lesion, occurs from 2 months to many years post-stroke. Lesions can also change because plasticity occurs throughout the life span, especially in response to experience. Experience-dependent structural changes post-stroke include bilateral increases in white matter volume [15, 16] and gray matter volume [17–21].

These changes mean that the specific longitudinal time window—the timing of the initial scan and of the deficit that is being predicted—is critical and the prediction model might be substantially different for different time windows. For example, a prediction of symptoms 6 months after stroke based on an acute-stage scan will likely be very different from using a scan at 6 months post-stroke to predict long-term outcomes 2 or 5 years later. These considerations refer to actual biological changes in the lesion, but as described in Chapters 2 and 3 [22], delineating the lesioned tissue from brain scans is itself an inexact process, so there will be some degree of noise and possibly bias in the lesion segmentations.

Third, in addition to the structural changes, the functional organization of the brain can change over time. That is, a brain region that is not structurally affected by a lesion (either acutely or due to longer-term changes) can take on a new role in a function that was previously supported by a lesioned brain region. Evidence regarding the specific pattern of functional reorganization is highly variable, as might be expected considering the heterogeneity of deficit profiles, treatment strategies, and predominance of small samples [19–21]. The most common pattern of treatment-induced functional reorganization appears to be increased bilateral recruitment of spared neural tissue, though some studies show unilateral upregulation in either the left or right hemisphere and some specifically emphasize increased recruitment of spared perilesional tissue [23]. Other studies have found that successful treatment outcome was associated with reduced neural activation, possibly because of improved processing efficiency [24–26]. Neural structure-function relationships are not fixed, but neither do they change arbitrarily. Any prospective prediction based only on structural lesions will need to make assumptions about functional reorganization (such as implicitly making the simplifying assumption that there is no reorganization); a better strategy would be to include longitudinal functional (and perhaps perfusion) neuroimaging to measure the functional reorganization.

Although these structural and functional changes present challenges for prospective lesion-based symptom prediction, they also provide opportunities for addressing important questions in neuroscience and neurology. Studying structural changes after stroke may offer new insights into the neurobiology of brain damage, recovery, and neuroplasticity. Similarly, we are just starting to be able to say which regions will be recruited during functional reorganization, how effective that reorganization will be, and how to promote functional recovery. This is to say that rather than seeing these challenges as barriers to prospective lesion-based symptom prediction, we can treat them as additional research questions to be addressed. Tackling the problem of prospective lesion-based symptom prediction provides a clinically relevant opportunity to address these fundamental questions about the neuroscience of brain damage, recovery, plasticity, and (re)organization. The next section reviews several recent reports on prospective lesion-based symptom prediction.

3.2 Current State of Art

The critical goal of prospective prediction is to identify the key demographic, behavioral, or lesion-based factors, that, at or near injury onset, can be used to provide a reliable long-term prognosis for patients. Inherent to this challenge is consideration of the clinical utility of both the predictions being made and the data used to make them. For instance, is the variance accounted for by a predictor or set of predictors clinically detectable, and is the improvement in prediction accuracy significant enough to justify the collection of additional behavioral or neuroimaging data? Despite the challenges in conducting prospective studies, there is an emerging consensus regarding the types of data that improve long-term prediction. A 2015 review of lesion-based prediction studies found that lesion location and lesion size were the most robust predictors of recovery. Other important factors were initial severity, treatment received, and aphasia type, whereas demographic factors such as age, gender, level of education, and handedness were not reliable predictors of outcomes [27]. Although lesion-based information appears to improve prediction, there is still considerable variability in the predictive value of lesion variables in recent prospective studies.

A major factor that impacts prediction, both in the prospective and inferential sense, is the outcome or deficit under evaluation. If the outcome is overall severity, as is common in prospective studies, lesion size may be the best lesion-based predictor because larger lesions tend to impact a greater number of brain regions, resulting in increased severity, both at onset and over time. Larger lesions also destroy more of the neural resources that might be recruited for functional reorganization, leaving less opportunity for recovery. Inversely, smaller lesions are associated with better long-term prognosis [28]. In other words, with a general outcome such as overall severity, any variance explained by other lesion-based factors may be subsumed by the variance accounted for by lesion size alone. Despite this, previous reports have questioned the value of lesion size in predicting recovery from stroke on the basis that it accounts for very little variance in models with other traditional predictors such as age and initial severity [29]. This is due, in part, to the fact that patients with initial moderate or severe impairment tend to display substantial variability in cognitive outcomes, making prediction particularly challenging. Initial behavioral deficit severity has been suggested as a better indicator of long-term outcomes [30], although recent reports have estimated the variance accounted for by initial severity alone to be between 50 and 55% [31, 32]. The converse is also a limiting factor: individuals with relatively mild deficits may perform near ceiling on behavioral measures, which severely restricts the opportunity to detect improvement [33, 34]. Near-ceiling performance does not always mean that patients do not have clinical complaints—in our anecdotal experience, participants who had left hemisphere strokes and score above the clinical cut-off on the *Western Aphasia Battery* (i.e., are clinically categorized as recovered or non-aphasic) nevertheless often complain that their language ability is noticeably reduced.

Initial deficit severity may be particularly useful when predicting dichotomized recovery outcomes (e.g., good versus poor), but may fail to provide an accurate prediction of functional deficits [35]. For instance, the magnitude of initial deficit, beyond lesion size and demographic factors, has been shown to be the strongest predictor of outcomes in the chronic stage, but lesion location information alone explains between 4% and 14% of the variance within specific cognitive domains [36]. Prospective studies may have the most impact in using more fine-grained lesion-based measures, such as percent damage in a gray matter region or white matter tract, to predict specific cognitive outcomes. In fact, lesion site information, compared to lesion size alone, appears to improve prediction of deficits within functional domains [31, 37–39], likely because functional outcomes may be more readily localized to and predicted by damage to a critical brain region or tract. In one case, lesion size and demographic variables accounted for 35% of the variance in speech production recovery, and atlas-based lesion location predictors accounted for an additional 17% [38]. Similarly, lesion location, compared to lesion size, age, and initial severity, was the strongest independent predictor of 3-month cognitive outcomes in a separate large sample [39]. There are also consistent reports of the limited utility of demographic data, such as age, handedness, and gender, in predicting specific functional outcomes. One large sample study found that demographic factors only accounted for about 1% of the variance in cross-sectional speech production outcomes, though time since stroke was an important longitudinal predictor [38].

Taken together, recent evidence suggests that lesion size and lesion location information may improve outcome prediction for both general recovery and functional domain outcomes. However, the relationship between these lesion variables and initial severity is complex. Initial severity is a good predictor of dichotomized outcomes in patients with mild initial severity, but lesion size and location information improve prediction for patients with initial moderate or severe impairment, where recovery is much more variable [32, 40]. There is substantial room for improvement on both the outcome and input side of prediction. Some functional deficits may be more predictable than others [e.g., 41], and it is not yet clear what kind of lesion descriptions are optimal, for example, whether finer-grained information (e.g., voxel-wise lesion location, graded lesion values) produces better predictions than simplified lesion descriptions (functional or atlas-based regions of interest, binary lesion values). Any prediction model is necessarily limited by the quality of the input data, so sub-optimal lesion descriptions may lead to underestimating the predictive value of lesion information.

Long-term prediction has been shown to improve with combinations of neuroimaging data that quantify the structural and functional impact of lesions in different ways[3] [35]. One promising avenue is diffusion-weighted imaging (DWI), which provides information about damage to the underlying white matter. Conceptually, knowing that damage occurred to a white matter tract critical to language processing should improve long-term language recovery predictions. In fact, measures of white matter integrity derived from diffusion imaging have been shown to improve predictions of 3-to-6-month post-stroke recovery outcomes [28, 40, 43] and long-term motor outcomes [43], accounting for an additional 6–12% of variance beyond that explained by lesion volume and standard demographic predictors (though lesion location information from traditional MRI/CT methods was not included). The presence and initial severity of leukoaraiosis (i.e., stroke-associated periventricular white matter hyperintensities) in conjunction with initial aphasia severity have also been shown to account for 33% of the variance in longitudinal aphasia severity outcomes [44]. Estimating white matter damage by aligning lesion maps with a white matter atlas (i.e., without diffusion imaging to directly measure the integrity of white matter) has produced mixed results [31, 36, 45], so direct measures of connection integrity should be used for testing whether white matter damage makes unique contributions to deficit prediction.

Comparatively fewer prospective prediction studies have utilized functional MRI data, and the results are mixed. Task-based

[3] This is true in cross-sectional predictive inference studies [13, 42].

fMRI data collected in the acute phase to pinpoint lesion-associated functional motor outcomes improved prediction accuracy compared to standard behavioral and demographic factors. However, data from the same task collected in the subacute phase did not improve prediction of chronic motor outcomes [46]. Longitudinal functional imaging of language network activation has also been shown to improve predictions of 1-year outcomes after stroke with task-based activation explaining an additional 12% of variance beyond initial language performance [47].

Despite these recent encouraging attempts, diffusion data do not always improve prediction [48–50], and, in cases where they do, the amount of variance explained is still relatively small. In addition, the few studies that have used fMRI data in prediction models had very small sample sizes, thus limiting the generalizability of the findings. The evidence suggests that multimodal neuroimaging data improve outcome prediction, but it is not particularly useful to merely reject the null hypothesis that multimodal neuroimaging data provide exactly zero information. Continued research in this area will help determine whether the gain in clinical prediction outweighs both the time and cost associated with collecting these types of data.

There is considerable variability in the period of time across which predictions are made, and, given the nature of recovery, some variables are more suitable for short-term (i.e., acute stage) versus long-term (i.e., chronic stage) prediction. This is most evident in studies utilizing lesion-based information to predict outcomes at different time points. The nature and extent of lesion damage changes over time, and recovery is variable, especially in the case of moderate to severe damage, so predictions made in the acute stage will likely differ from those made in the chronic stage [51]. Standard measures such as initial severity, demographic factors, and lesion size do not appear to predict recovery within the first 1–2 weeks post stroke, although this may be due to limited recovery within this short time window [52] and because ongoing decreases in swelling and hypoperfusion may be having a bigger effect on behavioral symptoms. Conversely, time since stroke, lesion size, and lesion location improve predictions made for long-term outcomes and functional recovery within the chronic stage (i.e., 6 months to several years post-stroke) [38]. A related concern is lesion growth over time because predictions made from lesion topography estimated from MRI data collected weeks or months prior may not capture the full extent of the structural damage. Recent work attempted to address this issue by generating a model of lesion growth using a large longitudinal sample, and these lesion projections improved prediction in an independent data set with variable periods of delay between scan acquisition and language assessment [53].

A clear consensus is hard to establish, however, because these studies are, in theory, examining natural recovery after stroke, but, in practice, predicting changes within a group of patients who have been exposed to diverse treatments. There are few prospective prediction studies examining treatment-related responses. One such study targeted single-word reading skills in patients with an acquired reading disorder secondary to aphasia and found that lesion-based predictors explained 94% of the variance in treatment responses. The out-of-sample prediction decreased to 24%, however, illustrating how in-sample effect sizes can be inflated due to small sample size (in this case, $N = 23$) with high variability [37]. Resting-state fMRI was recently found to be a good predictor of aphasia severity following speech-language therapy, especially for patients with agrammatism or dysgraphia, where the prediction models had $R^2 > 0.8$ [54]. There may be other mechanisms of recovery beyond natural recovery and cognitive rehabilitation, with one study reporting that initial damage within critical white matter tracts and the use of antidepressants (selective serotonin reuptake inhibitors) after stroke was associated with recovery of naming ability [31]. Lesion-based mechanisms appear to play a role in treatment-response, but more prospective studies examining response to treatment are needed.

Most of the reviewed studies utilized a standard regression framework to make predictions, but several studies relied on machine learning methods with cross-validation techniques to train and evaluate predictive models [for a comparison of machine learning techniques in lesion-based prediction, *see* 50]. Support vector machine (SVM) classifiers achieved 82.8% accuracy in discriminating 3-month dichotomized outcomes (i.e., good versus poor) using diffusion maps [28] and 86% accuracy in classifying 4–6 month outcomes using fMRI activation patterns [46]. Gaussian process models accounted for 59% of the variance in a speech production composite score [38] and, in a separate study, yielded strong correlations ($r = 0.68$–0.71) between predicted and actual performance on a spoken language task [53]. Elastic net regression was also very effective at combining behavioral and resting-state fMRI predictors of recovery [54]. Importantly, cross-validation techniques can be used with the standard regression framework, and efforts in this direction illustrate the value of out-of-sample prediction. In one application, prediction models were developed through cross-validation, achieving 70–84% classification accuracy in dichotomized functional and cognitive outcomes, and the model produced comparable prediction accuracy (66–78%) in an independent replication sample [39]. A separate study used automatic linear modeling to isolate the model that resulted in the highest in-sample prediction accuracy given the data and variables ($R^2 = 0.94$) and then used leave-one-out cross-validation to estimate the out-of-sample prediction ($R^2 - 0.23$) [37]. Cross-validation improves generalization to separate samples, overcomes many of the limitations associated with small sample sizes, and provides more realistic effect size estimates.

Prospective investigations provide valuable insight into the recovery mechanisms involved after brain injury. This information can be leveraged to better understand long-term prognosis and aid in the development of individualized treatment plans. Several of the studies reported that lesion-based information, and, in particular, knowledge of where the damage occurred, improved outcome prediction. Nevertheless, a cautionary note is in order: the models that accounted for a relatively large portion of the variance in outcomes also had a large number of predictor variables, often including demographic variables, behavioral scores, and a number of lesion features such as size and atlas-based lesion load estimates. Many prospective studies also have small sample sizes (e.g., $N < 50$) which may result in inflated effect estimates and poor out-of-sample generalization. The value of lesion site information may depend on the properties of the outcome variable: dichotomized recovery outcomes are better predicted by initial deficit severity and lesion size, while long-term deficits within functional domains are better predicted by lesion location information. The location of lesion damage was quantified in different ways across studies, with some studies utilizing structural scans to calculate percent damage within gray and white matter atlas regions, while others used other types of neuroimaging data such as diffusion data, resting-state fMRI data, and even task-based fMRI data.

4 Concluding Remarks and Future Directions

Predictive inference is an important step toward achieving the kinds of inferences that we typically want to make. Full-sample analyses describe the observed data and, consequently, provide overly optimistic estimates of how much variance would be accounted for in a new sample. Predictive inference allows estimating out-of-sample prediction accuracy from an observed sample. The methods used for predictive inference—leave-one-out and k-fold cross-validation—are already standard in machine learning and relatively easy to implement. Although it poses some additional technical and computational challenges, cross-validation lends itself well to parallelization (the folds can be run in parallel and their results combined at the end), so increases in computational power and wide availability of multi-processor high-performance computing clusters mean that use of predictive inference need not be time-prohibitive. Therefore, predictive inference is a powerful analytical technique that should be readily available to most LSM research groups.

Prospective lesion-based symptom prediction is of substantial translational value for both research and clinical practice. If lesion patterns can be used to predict natural recovery and response to treatment, it could also make a substantial contribution to personalized medicine. Structural lesion information is the most widely available because CT and/or MRI scans are part of standard clinical practice for brain injury and neurological disease. Incorporating other kinds of data, such as connectivity data from DWI or functional MRI, is likely to improve prediction accuracy (the role of connectivity in lesion-symptom relationships is discussed in more detail in Chapters 8, 9, and 10).

A major challenge for longitudinal lesion-based symptom prediction is that lesions change over time. However, if they change in systematic ways, then this can be incorporated into the prediction model. Critically, determining the systematic ways that lesions change would make an important contribution to basic neuroscience, so tackling this challenge provides an opportunity to address both basic science and clinical research questions. Natural and treatment-related reorganization of function also need to be considered in predictive models. Ultimately, it would be valuable to be able to predict what kind of functional reorganization will be most effective for a given individual in order to tailor treatment strategies. Surviving a stroke is a life-altering event, sometimes radically so; providing a glimpse into what the future holds would be a precious service to patients and their loved ones.

References

1. Yarkoni T, Westfall J (2017) Choosing prediction over explanation in psychology: lessons from machine learning. Perspect Psychol Sci 12:1100–1122. https://doi.org/10.1177/1745691617693393
2. Mirman D, Strauss TJ, Brecher A, Walker GM, Sobel P, Dell GS, Schwartz MF (2010) A large, searchable, web-based database of aphasic performance on picture naming and other tests of cognitive function. Cogn Neuropsychol 27:495–504. https://doi.org/10.1080/02643294.2011.574112
3. Mirman D, Zhang Y, Wang Z, Coslett HB, Schwartz MF (2015) The ins and outs of meaning: behavioral and neuroanatomical dissociation of semantically-driven word retrieval and multimodal semantic recognition in aphasia. Neuropsychologia 76:208–219. https://doi.org/10.1016/j.neuropsychologia.2015.02.014
4. Mirman D, Chen Q, Zhang Y, Wang Z, Faseyitan OK, Coslett HB, Schwartz MF (2015) Neural organization of spoken language revealed by lesion–symptom mapping. Nat Commun 6:6762. https://doi.org/10.1038/ncomms7762
5. Thye M, Mirman D (2018) Relative contributions of lesion location and lesion size to predictions of varied language deficits in post-stroke aphasia. Neuroimage Clin 20:1129–1138. https://doi.org/10.1016/J.NICL.2018.10.017
6. Mirman D, Thye M (2018) Uncovering the neuroanatomy of core language systems using lesion-symptom mapping. Curr Dir Psychol Sci. https://doi.org/10.1177/0963721418787486
7. Kohavi R (1995) A study of cross-validation and bootstrap for accuracy estimation and model selection. IJCAI 2:1137–1143
8. Zhang Y, Yang Y (2015) Cross-validation for selecting a model selection procedure. J Econ 187:95–112. https://doi.org/10.1016/j.jeconom.2015.02.006
9. Poldrack RA, Huckins G, Varoquaux G (2019) Establishment of best practices for evidence for

prediction: a review. JAMA Psychiatry 77(5): 534–540

10. Zhang Y, Kimberg DY, Coslett HB, Schwartz MF, Wang Z (2014) Multivariate lesion-symptom mapping using support vector regression. Hum Brain Mapp 35:5861–5876. https://doi.org/10.1002/hbm.22590
11. DeMarco AT, Turkeltaub PE (2018) A multivariate lesion symptom mapping toolbox and examination of lesion-volume biases and correction methods in lesion-symptom mapping. Hum Brain Mapp. https://doi.org/10.1002/hbm.24289
12. Pustina D, Avants B, Faseyitan OK, Medaglia JD, Coslett HB (2017) Improved accuracy of lesion to symptom mapping with multivariate sparse canonical correlations. Neuropsychologia. https://doi.org/10.1016/J.NEUROPSYCHOLOGIA.2017.08.027
13. Pustina D, Coslett HB, Ungar L, Faseyitan OK, Medaglia JD, Avants B, Schwartz MF (2017) Enhanced estimations of post-stroke aphasia severity using stacked multimodal predictions. Hum Brain Mapp 38:5603–5615. https://doi.org/10.1002/hbm.23752
14. Seghier ML, Ramsden S, Lim L, Leff AP, Price CJ (2014) Gradual lesion expansion and brain shrinkage years after stroke. Stroke 45:877–879. https://doi.org/10.1161/STROKEAHA.113.003587
15. Schlaug G, Marchina S, Norton A (2009) Evidence for plasticity in white-matter tracts of patients with chronic Broca's aphasia undergoing intense intonation-based speech therapy. In: Annals of the New York Academy of Sciences. Blackwell Publishing Inc., pp 385–394
16. Wan CY, Zheng X, Marchina S, Norton A, Schlaug G (2014) Intensive therapy induces contralateral white matter changes in chronic stroke patients with Broca's aphasia. Brain Lang 136:1–7. https://doi.org/10.1016/j.bandl.2014.03.011
17. Gauthier LV, Taub E, Perkins C, Ortmann M, Mark VW, Uswatte G (2008) Remodeling the brain: plastic structural brain changes produced by different motor therapies after stroke. Stroke 39:1520–1525. https://doi.org/10.1161/STROKEAHA.107.502229
18. Särkämö T, Ripollés P, Vepsäläinen H, Autti T, Silvennoinen HM, Salli E, Laitinen S, Forsblom A, Soinila S, Rodríguez-Fornells A (2014) Structural changes induced by daily music listening in the recovering brain after middle cerebral artery stroke: a voxel-based morphometry study. Front Hum Neurosci 8. https://doi.org/10.3389/fnhum.2014.00245
19. Kiran S, Thompson CK (2019) Neuroplasticity of language networks in aphasia: advances, updates, and future challenges. Front Neurol 10:295
20. Crinion JT, Leff AP (2015) Using functional imaging to understand therapeutic effects in poststroke aphasia. Curr Opin Neurol 28: 330–337
21. Berthier ML, Pulvermüller F (2011) Neuroscience insights improve neurorehabilitation of poststroke aphasia. Nat Rev Neurol 7:86–97. https://doi.org/10.1038/nrneurol.2010.201
22. Crinion JT, Holland AL, Copland DA, Thompson CK, Hillis AE (2013) Neuroimaging in aphasia treatment research: quantifying brain lesions after stroke. NeuroImage 73: 208–214
23. Fridriksson J, Richardson JD, Fillmore P, Cai B (2012) Left hemisphere plasticity and aphasia recovery. NeuroImage 60:854–863. https://doi.org/10.1016/j.neuroimage.2011.12.057
24. Gotts SJ, Chow CC, Martin A (2012) Repetition priming and repetition suppression: a case for enhanced efficiency through neural synchronization. Cogn Neurosci 3:227–237. https://doi.org/10.1080/17588928.2012.670617
25. Gotts SJ (2016) Incremental learning of perceptual and conceptual representations and the puzzle of neural repetition suppression. Psychon Bull Rev 23:1055–1071. https://doi.org/10.3758/s13423-015-0855-y
26. Naito E, Hirose S (2014) Efficient foot motor control by Neymar's brain. Front Hum Neurosci 8. https://doi.org/10.3389/fnhum.2014.00594
27. Watila MM, Balarabe B (2015) Factors predicting post-stroke aphasia recovery. J Neurol Sci 352:12–18. https://doi.org/10.1016/j.jns.2015.03.020
28. Moulton E, Valabregue R, Lehéricy S, Samson Y, Rosso C (2019) Multivariate prediction of functional outcome using lesion topography characterized by acute diffusion tensor imaging. Neuroimage Clin 23. https://doi.org/10.1016/j.nicl.2019.101821
29. Lazar RM, Speizer AE, Festa JR, Krakauer JW, Marshall RS (2008) Variability in language recovery after first-time stroke. J Neurol Neurosurg Psychiatry 79:530–534. https://doi.org/10.1136/jnnp.2007.122457
30. Lazar RM, Minzer B, Antoniello D, Festa JR, Krakauer JW, Marshall RS (2010) Improvement in aphasia scores after stroke is well predicted by initial severity. Stroke 41:1485–1488. https://doi.org/10.1161/STROKEAHA.109.577338

31. Hillis AE, Beh YY, Sebastian R, Breining B, Tippett DC, Wright A, Saxena S, Rorden C, Bonilha L, Basilakos A, Yourganov G, Fridriksson J (2018) Predicting recovery in acute post-stroke aphasia. Ann Neurol 83:612–622. https://doi.org/10.1002/ana.25184
32. Benghanem S, Rosso C, Arbizu C, Moulton E, Dormont D, Leger A, Pires C, Samson Y (2019) Aphasia outcome: the interactions between initial severity, lesion size and location. J Neurol 266:1303–1309. https://doi.org/10.1007/s00415-019-09259-3
33. Bonkhoff AK, Hope T, Bzdok D, Guggisberg AG, Hawe RL, Dukelow SP, Rehme AK, Fink GR, Grefkes C, Bowman H (2019) Bringing proportional recovery into proportion: Bayesian hierarchical modelling of post-stroke motor performance. medRxiv:19009159. https://doi.org/10.1101/19009159
34. Hope TMH, Friston K, Price CJ, Leff AP, Rotshtein P, Bowman H (2019) Recovery after stroke: not so proportional after all? Brain 142:15–22. https://doi.org/10.1093/brain/awy302
35. Stinear CM, Ward NS (2013) How useful is imaging in predicting outcomes in stroke rehabilitation? Int J Stroke 8:33–37
36. Ramsey LE, Siegel JS, Lang CE, Strube M, Shulman GL, Corbetta M (2017) Behavioural clusters and predictors of performance during recovery from stroke. Nat Hum Behav 1. https://doi.org/10.1038/s41562-016-0038
37. Aguilar OM, Kerry SJ, Ong Y-H, Callaghan MF, Crinion JT, Woodhead ZVJ, Price CJ, Leff AP, Hope TMH (2018) Lesion-site-dependent responses to therapy after aphasic stroke. J Neurol Neurosurg Psychiatry 89: 1352–1354. https://doi.org/10.1136/jnnp-2017-317446
38. Hope TMH, Seghier ML, Leff AP, Price CJ (2013) Predicting outcome and recovery after stroke with lesions extracted from MRI images. Neuroimage Clin 2:424–433. https://doi.org/10.1016/j.nicl.2013.03.005
39. Munsch F, Sagnier S, Asselineau J, Bigourdan A, Guttmann CR, Debruxelles S, Poli M, Renou P, Perez P, Dousset V, Sibon I, Tourdias T (2016) Stroke location is an independent predictor of cognitive outcome. Stroke 47:66–73. https://doi.org/10.1161/STROKEAHA.115.011242
40. Wu O, Cloonan L, Mocking SJT, Bouts MJRJ, Copen WA, Cougo-Pinto PT, Fitzpatrick K, Kanakis A, Schaefer PW, Rosand J, Furie KL, Rost NS (2015) Role of acute lesion topography in initial ischemic stroke severity and long-term functional outcomes. Stroke 46:2438–2444. https://doi.org/10.1161/STROKEAHA.115.009643
41. Landrigan J-F, Zhang F, Mirman D (2021) A data-driven approach to post-stroke aphasia classification and lesion-based prediction. Brain. https://doi.org/10.1093/brain/awab010
42. Kristinsson S, Zhang W, Rorden C, Newman-Norlund R, Basilakos A, Bonilha L, Yourganov G, Xiao F, Hillis A, Fridriksson J (2021) Machine learning-based multimodal prediction of language outcomes in chronic aphasia. Hum Brain Mapp 42:1682–1698. https://doi.org/10.1002/hbm.25321
43. Moulton E, Magno S, Valabregue R, Amor-Sahli M, Pires C, Lehéricy S, Leger A, Samson Y, Rosso C (2019) Acute diffusivity biomarkers for prediction of motor and language outcome in mild-to-severe stroke patients. Stroke 50:2050–2056. https://doi.org/10.1161/strokeaha.119.024946
44. Basilakos A, Stark BC, Johnson L, Rorden C, Yourganov G, Bonilha L, Fridriksson J (2019) Leukoaraiosis is associated with a decline in language abilities in chronic aphasia. Neurorehabil Neural Repair 33:718–729. https://doi.org/10.1177/1545968319862561
45. Geller J, Thye M, Mirman D (2019) Estimating effects of graded white matter damage and binary tract disconnection on post-stroke language impairment. NeuroImage 189:248–257. https://doi.org/10.1016/J.NEUROIMAGE.2019.01.020
46. Rehme AK, Volz LJ, Feis DL, Eickhoff SB, Fink GR, Grefkes C (2015) Individual prediction of chronic motor outcome in the acute post-stroke stage: behavioral parameters versus functional imaging. Hum Brain Mapp 36: 4553–4565. https://doi.org/10.1002/hbm.22936
47. van Oers CAMM, van der Worp HB, Kappelle LJ, Raemaekers MAH, Otte WM, Dijkhuizen RM (2018) Etiology of language network changes during recovery of aphasia after stroke. Sci Rep 8. https://doi.org/10.1038/s41598-018-19302-4
48. Doughty C, Wang J, Feng W, Hackney D, Pani E, Schlaug G (2016) Detection and predictive value of fractional anisotropy changes of the corticospinal tract in the acute phase of a stroke. Stroke 47:1520–1526. https://doi.org/10.1161/STROKEAHA.115.012088
49. Forkel SJ, Thiebaut de Schotten M, Dell'Acqua F, Kalra L, Murphy DGM, Williams SCR, Catani M (2014) Anatomical predictors of aphasia recovery: a tractography study of bilateral perisylvian language

networks. Brain 137:2027–2039. https://doi.org/10.1093/brain/awu113

50. Halai AD, Woollams AM, Lambon Ralph MA (2020) Investigating the effect of changing parameters when building prediction models for post-stroke aphasia. Nat Hum Behav 4: 725–735. https://doi.org/10.1038/s41562-020-0854-5

51. Zavanone C, Samson Y, Arbizu C, Dupont S, Dormont D, Rosso C (2018) Critical brain regions related to post-stroke aphasia severity identified by early diffusion imaging are not the same when predicting short- and long-term outcome. Brain Lang 186:1–7. https://doi.org/10.1016/j.bandl.2018.08.005

52. Dunn LE, Schweber AB, Manson DK, Lendaris A, Herber C, Marshall RS, Lazar RM (2016) Variability in motor and language recovery during the acute stroke period. Cerebrovasc Dis Extra 6:12–21. https://doi.org/10.1159/000444149

53. Loughnan R, Lorca-Puls DL, Gajardo-Vidal A, Espejo-Videla V, Gillebert CR, Mantini D, Price CJ, Hope TMH (2019) Generalizing post-stroke prognoses from research data to clinical data. Neuroimage Clin 24. https://doi.org/10.1016/j.nicl.2019.102005

54. Iorga M, Higgins J, Caplan D, Zinbarg R, Kiran S, Thompson CK, Rapp B, Parrish TB (2021) Predicting language recovery in post-stroke aphasia using behavior and functional MRI. Sci Rep 11:8419. https://doi.org/10.1038/s41598-021-88022-z

Chapter 13

Selecting and Handling Behavioral Measures for Lesion-Symptom Mapping

Peter E. Turkeltaub, Andrew T. DeMarco, and Maryam Ghaleh

Abstract

The success of a lesion-symptom mapping (LSM) study in answering a specific research question critically depends on how behavioral variables are selected and handled in the analysis. In this chapter, we discuss the theoretical and practical considerations regarding the use of behavioral measures in LSM studies. We begin by addressing theoretical considerations that impact both the choice of behavioral measures and the design of LSM studies. This includes determining whether the main aim of the study is to address a basic science or a clinical research question, considering what type of brain-behavior relationship is hypothesized, and adopting a testing theory framework for the study. We then address practical aspects of selecting behavioral measures, including considering the psychometric properties of measures, deciding whether to use published tests or to develop new measures, deciding whether to categorize participants or to treat scores as continuous measures, examining the distribution of performance on the tests, and considering how to develop a battery of tests to address the research questions. We then discuss analysis decisions that impact the behavioral and anatomical specificity of LSM results, including methods to address lesion volume confounds and methods to compare multiple behaviors. We end by discussing extensions of LSM that examine the relationship between lesions and non-behavioral variables. We hope this chapter provides guidance to investigators on major issues that should be considered when selecting behavioral measures and developing LSM analysis approaches. Attention to these issues will improve the validity and reliability of LSM results.

Key words Neuropsychology, Behavioral neurology, Lesion studies, Psychometrics, Neuroimaging methods

1 Introduction

Since its inception [1], methodological advances have iteratively improved and expanded the tools available to investigators for lesion-symptom mapping (LSM). These advances have mainly focused on how lesions are measured [2, 3] and the statistical approaches used to relate lesion damage to behavioral outcomes [4–7]. Relatively little attention, however, has been given to the nature of the behavioral variables being investigated. Careful

Dorian Pustina and Daniel Mirman (eds.), *Lesion-to-Symptom Mapping: Principles and Tools*, Neuromethods, vol. 180, https://doi.org/10.1007/978-1-0716-2225-4_13, © Springer Science+Business Media, LLC, part of Springer Nature 2022

consideration of the characteristics of these variables is vital to optimize LSM studies so that results are valid and are interpreted correctly.

In this chapter, we will outline various factors to consider when selecting behavioral variables for examination in LSM studies. There are a limitless number of specific behaviors that could be examined with LSM. Therefore, it is beyond the scope of this chapter to suggest specific assessment instruments for specific behaviors. Instead, we will focus on principles that are important to consider when selecting behavioral variables, irrespective of the specific behavior being examined. First, we discuss the nature of the two main types of questions asked using LSM and how the type of question being addressed impacts selection of behavioral variables. We then discuss major hypotheses regarding the nature of lesion-behavior associations, which ideally determine the selection of both behavioral variables and LSM techniques. Next, we discuss practical considerations regarding variable selection, including both the general psychometric properties of tests and key issues specific to LSM analyses. Then we discuss LSM implementation decisions that can impact the behavioral specificity of the results achieved. Finally, we consider non-behavioral variables that have been used in place of behaviors in LSM-style analyses. We hope that highlighting these considerations will help investigators to make principled choices when developing new batteries of tests for use in LSM analyses, selecting variables for examination using LSM, and deciding how to implement the analyses.

2 Theoretical Considerations That Impact Choice of Behavioral Variables and LSM Designs

The first step in any scientific investigation is to define the question and/or hypothesis to be addressed. In LSM, this step impacts both the nature of the behavioral variable to be investigated and the design of the analysis. Aligning these decisions with the specific question or hypothesis of interest will help investigators to ensure that their analyses will address their research aims as precisely as possible.

2.1 Main Types of Questions Addressed Using LSM

When selecting behavioral variables to examine using LSM, it is first important to consider the overarching goal of the LSM analysis. There are two main types of questions addressed using LSM: basic and clinical. First, LSM is often used as an updated version of the classic lesion method to identify relationships between behaviors and the brain structures that underlie them. This type of study is best considered “basic science.” Although the participants have a clinical disorder, the goal of the analysis is not to learn about the disorder itself. Rather, the effect of the disorder on behavior is used to make inferences about brain-behavior relationships that are expected to generalize beyond the disorder. For instance, an investigator may wish to understand the brain structures underpinning auditory comprehension and perform a voxel-based lesion-symptom mapping (VLSM) analysis on stroke survivors to address this question [e.g., 8]. In this case, the question is not about stroke; stroke only serves as a model of brain lesions and is used to understand which brain structures are critical for auditory comprehension. For studies of this type, behaviors needn't have any clinical relevance, and many common clinical assessments may actually have unattractive features if they offer fairly gross measures of behavior that are not specific at the level of individual cognitive processes (e.g., NIH Stroke Scale [NIHSS], *Western Aphasia Battery* [WAB] [9]). Sometimes because of practical limitations on the data available, these types of clinical measures may be used to address basic science questions, but in an ideal prospectively designed study, investigators may wish to choose assessments specifically designed to address the basic science question at hand.

The nature of the basic science questions addressed using LSM also impacts decisions about how to implement the analysis. For basic science questions, the investigator typically wishes to achieve some level of specificity both in terms of behavior and brain anatomy. For example, in the hypothetical study on auditory comprehension, the investigators may wish to demonstrate that there are brain structures important for auditory, but not written, comprehension. Thus, addressing this type of basic science question often requires comparisons between multiple behavioral variables. When the behavioral score of interest correlates with lesion volume, correction for lesion volume is also often critical to achieving anatomical and behavioral specificity of results [6]. We will discuss various methods for accomplishing behavioral specificity, including handling of multiple behaviors and lesion volume correction, in the section, "Methods to achieve behavioral specificity in LSM results."

The second main type of question addressed with LSM is clinical. In this case, the purpose of the analysis is to understand a brain-behavior relationship that has clinical relevance for the population under investigation. For example, the investigator may wish to know if lesions to particular parts of the brain are associated with deficits in auditory comprehension. The purpose is to help clinicians know when to expect auditory comprehension deficits based on the lesion distribution or perhaps to know where to expect a lesion to be found when an auditory comprehension deficit is observed. More formally, an investigator may wish to predict the auditory comprehension outcome based on the lesion distribution or vice versa (for more discussion of lesion-symptom prediction see Chap. 12). We use this example to point out how subtle the difference is between the basic science and clinical questions addressed by LSM. Such subtle difference though has important implications for the design of the study. Different from the hypothetical basic science study discussed above, the behavioral variable used to address this clinical question should have relevance to clinically important outcomes and needn't be specific or even theoretically motivated in terms of cognitive processes. Depending on the nature of the question, one may or may not wish to control for other behavioral variables in a clinical study; doing so would be less typical of a clinical study than a basic one. For studies aimed at addressing clinical questions like those discussed above, the lesion size confound is not relevant. In fact, if one wishes to know which lesion distributions produce a particular behavioral deficit, the size of the lesion is potentially irrelevant, and perhaps should not be controlled. To address the clinical type of question, the ideal control group might include individuals without lesions so as to be sensitive to the relationship between very small lesions and minor deficits. As such, for this type of question, we believe it is sometimes appropriate to include participants without lesions in the LSM analysis.

As described above, many design decisions for an LSM study—the choice of behavioral measures, the inclusion criteria for participants, the use of behavioral control variables, and the handling of lesion size—depend on whether the aim of the LSM analysis is to address basic questions about brain-behavior relationships or clinical questions about the behavioral outcomes of brain lesions. Thus, the first and perhaps most critical decision investigators must make is what type of question they intend to address with LSM. A question that is posed precisely will lead to clear decisions regarding the selection of behavioral measures and the design of the experiment as a whole.

2.2 Types of Theorized Brain-Behavior Relationships

As discussed above, LSM often serves as a modern version of the classical lesion method to address basic science questions regarding the brain structures that underlie behavior. There are several theories regarding the types of neural architectures that relate to behaviors. When conducting an LSM analysis on a given behavior, the specific hypothesis for that behavior should be considered, as it may impact decisions about how the analysis is conducted. We will discuss a few prominent hypotheses below as examples.

Perhaps the most widely recognized hypothesis regarding the neural underpinnings of behavior is the localized processing module. This hypothesis states that specific brain structures perform specific behaviorally relevant processes, i.e., region X performs process x. A region that performs a specific function is sometimes termed a localized processing module. Well-known examples include the category-selective regions of the ventral visual stream [10, 11]. If one conducts an LSM study examining visual processing of faces or tools, a reasonable hypothesis might be that the associated category-specific region of ventral visual cortex is critical for processing the specific type of stimulus in question. This hypothesis suggests a standard LSM analysis of anatomical lesions, use of a population that often has lesions in the ventral visual pathway, and perhaps examining a behavioral test of face identification as compared to tool identification or vice versa. In general, the localized processing module hypothesis tends to apply to fairly circumscribed behaviors and, as such, benefits from precise behavioral measures with controls for all non-specific sources of potential deficits in performance (e.g., sensory input, motor output). This approach illustrates how the nature of the hypothesis suggests the type of behavioral variables and the analysis approach to be employed.

Another example is the disconnection hypothesis. Some types of deficits are thought to occur not because of damage to a specific processing module but because of a lesion that severs connections between two or more processing modules [12]. Pure alexia is a behavioral syndrome classically hypothesized to occur due to disconnection of a word form lexicon from visual input [13]. If an investigator wishes to examine the brain basis of a behavior thought to be performed via communication between localized modules or wishes to address a clinical question about a deficit thought to arise through disconnection, then perhaps a connectome lesion-symptom mapping analysis on diffusion-weighted imaging data would be ideal [2; see Chaps. 9 and 10]. This provides an example of how the choice of a behavioral variable and the theory of how deficits in the behavior arise impact the choice of lesion measures and LSM methods.

Many behaviors are thought to arise from distributed processing, the coordination of activity among many, often dispersed, brain structures [14, 15]. Here, the hypothesis is that the architecture of the network as a whole, which may be disrupted by lesions, is critical to the behavior of interest or that the placement of a given processor within a network determines its importance to the behavior [16]. There are ongoing debates regarding the degree to which cognitive processes are localized versus distributed, but in general, broadly construed processes like attention, executive control, and memory, which modulate a variety of other specific processes, are widely thought to arise from distributed processing [17]. If an investigator aims to test such a hypothesis, it would not be ideal to use univariate LSM methods that are intended to relate behavioral deficits to individual structures. Multivariate LSM methods that are capable of identifying multiple brain structures that simultaneously contribute to behavioral scores would be preferred (see Chap. 11). Another approach that addresses the distributed processing hypothesis more directly is the use of graph theoretical measures of connectomes to assess the effect of lesions on network architecture [e.g., 18; see also Chap. 8]. This provides another example of how the specific hypothesis regarding the brain basis of a given behavior can guide the type of LSM approach used.

2.3 Theory of Testing

A broad consideration for choosing a behavioral test for any study, including LSM, is the theoretical model of testing that it adopts. The field of test psychometrics offers two primary frameworks for testing: classical test theory and item response theory (IRT) [19–21]. Classical test theory is the more common framework used in LSM studies to date. Classical test theory conceptualizes the problem of testing by assuming that there is a hypothetical true score to be measured but that the true score is unavoidably measured with error. It focuses on understanding and minimizing the error at the level of the whole test to get closer to the true score. One limitation of the classical test theory framework is that an individual's test score depends on the specific test administered. For example, an individual will achieve a lower score on a difficult test than on an easy test. Although the individual's ability has not changed, their measured ability is dependent on the test used. Consequently, the properties of the individual being tested and the properties of the testing instrument cannot be dissociated.

Item response theory conceptualizes that each individual has a latent ability score corresponding to the probability of responding correctly to an item of a specific difficulty level that is independent of the test in which it was administered. Item difficulty and individual ability fall on the same scale. This framework shifts the unit of analysis from the whole test to the item, where individual item performance can be examined so that items that don't perform well can be detected and removed. IRT can also be used for computerized adaptive tests and is the framework used by many common educational assessments. Its popularity is growing in the assessment of individuals with lesions, for instance, in measurement of aphasia [22–25].

IRT offers a number of potential benefits [21]. First, scores can be compared across tests. If applied correctly, IRT may allow the duration of testing to be shorter, since the precision of ability estimation does not depend on the number of items administered. IRT also gives an idea of the precision of score estimates. However, IRT requires a more rigorous norming process that may not be feasible for many investigators. In summary, the choice of a classical or IRT framework depends on the purpose of testing and the available resources. While classical test theory is much more common, there may be cases where an IRT approach is advantageous, for example, when tests designed based on classical test theory have large ceiling or floor effects in a population of interest.

3 Practical Considerations in the Selection of Behavioral Variables

There are a number of practical considerations that impact the choice of behavioral variables for LSM analyses. Some of these are general considerations that apply to any behavioral study of lesion participants, and some are considerations that are more specific to LSM studies.

3.1 Psychometric Properties

Some tests are better than others, and the properties that make them better fall into two broad categories: validity and reliability [26]. The validity of a test is the degree to which it measures what it is intended to measure. Reliability is the ability of the instrument to measure something multiple times and get the same answer. There are various subtypes of both validity and reliability, and a full description is beyond the scope of this chapter. The main point is that basic properties of tests that apply to all clinical or basic research studies also apply to LSM studies. For example, using a score with poor reliability will yield scores with substantial measurement error. LSM analyses of these scores will have suboptimal sensitivity and may also be prone to false-positive errors when sample sizes are small. Using a more reliable measure or taking multiple measurements from each individual may be beneficial in these cases.

3.2 Standardized vs. Investigator-Designed Tests

A very common decision made by investigators is whether to use a published standardized test or to develop their own instrument. As noted above, this decision depends to some degree on the goals of the study. In general, addressing clinical questions favors using widely available measures because clinicians not active in research are often familiar with these measures allowing them to apply results directly to their clinical practice. Further, these measures typically have clear administration and scoring rules that allow for comparisons across research groups. Standard clinical measures also often have known psychometric properties, allowing investigators to make informed decisions about their use. In contrast, instruments designed by individual investigators are often not thoroughly vetted for their psychometric properties, and doing so may not be practical for every investigator. Published measures often (although not always) have standardized scores, which can be helpful especially when the norming accounts for important demographic variables such as age and education.

With all of the advantages of published standardized tests, why would an investigator choose to create their own test? Often this is done because available measures do not adequately assess the specific question being asked by the investigators or because the test is not appropriate to the population under investigation. For example, the norming sample used to standardize a clinical instrument and to determine its psychometric properties may differ in important demographic features from an individual study's population, diminishing the validity of this standardization. This is particularly important for assessments of language, where accents, dialects, and vocabulary can differ even between two cultural groups living in the same region, not to mention obvious issues with norming tests across languages. Some published clinical instruments are designed to measure clinically important deficits (e.g., NIHSS) or to diagnose patients (e.g., WAB [9]), but suffer from ceiling effects in individuals with minor deficits. These measures also often rely on many sensory, cognitive, and motor processes for performance, making them disadvantageous for basic science studies addressing questions about individual mental processes. Sometimes, although a test measures the general process of interest, the individual items may not be carefully controlled for properties of particular relevance to a given study (e.g., transitive vs. intransitive gestures in an apraxia test, word frequency, or spelling regularity in a reading test). For all of these reasons, investigators often choose to design their own measures, especially when addressing basic science questions about brain-behavior relationships. Because LSM research needs to be conducted on a relatively large scale (at least compared to cognitive neuropsychology more generally, which relies heavily on case studies and small group studies), there is a lot at stake in measure selection. This paragraph provided good reasons for investigators to design their own measures, but investigators need to consider the validity, reliability, and score distribution of these measures just as they would for published standardized measures.

3.3 Classification vs. Quantification

There are two general approaches to behavioral measurement used by lesion studies. The first aims to classify lesion patients into diagnostic groups. Examples of this include the classical aphasia syndromes, types of hemispatial neglect, or any other disorder that can be classified as present vs. absent in individuals. Such diagnoses are sometimes made based on single test scores, but more often a pattern of performance across multiple tests is used to categorize patients. For example, impaired irregular word reading relative to regular word reading and the commission of regularization errors leads to a diagnosis of surface alexia [27]. The classification approach is advantageous in some cases because using multiple scores together can often isolate narrowly construed cognitive processes hypothesized by cognitive models of behavior, which are likely to rely on relatively specific anatomical structures [e.g., 28]. From a theoretical standpoint, this approach assumes that individuals with the same diagnosis will have similar lesions and that variations in the severity of deficits are related to factors that are not of interest to the LSM analysis, such as demographic factors, premorbid behavioral differences, recovery via brain plasticity, or measurement error. This assumption is aligned with a modular hypothesis of localization, in that a lesion to a specific processing module is expected to cause the loss of that process in an all-or-nothing manner. There may be alternative pathways or other means of compensation in the network, but under the framework of the classification approach, these are not of interest to the lesion-behavior association question. One limitation of the classification approach is that an arbitrary criterion must be used to determine whether a test score represents impairment. These criteria are typically based on comparisons with control subjects, but still the choice of the cutoff is arbitrary (e.g., how many standard deviations from the normative distribution, $Z = -2$ vs. -3).

The alternative approach, introduced with the VLSM method [1], is to treat behavioral scores as continuous variables. Rather than classifying or categorizing patients, the deficit is measured quantitatively using a test score, and the LSM analysis determines whether particular lesions are associated with variations in this continuous score. An implicit assumption of this approach is that gradations in the severity of deficits in a given ability may be related to lesion distribution. In contrast to the classification approach above, this quantitative approach to behavioral measurement is better aligned with a network architecture framework in which processing is at least somewhat distributed and multiple lesion locations may lead to deficits in the same behavior, some more severe than others. However, the quantification of deficits using continuous scores is not incompatible with the hypothesis of all-or-nothing effects of lesions on a given process. One could consider continuous scores as measuring the probability of a binary deficit being present. Thus, one may hypothesize an all-or-nothing effect of lesions on a given process, but assume that the worse the score, the more confidently one can diagnose a deficit in the process of interest. The continuous quantification of deficits is advantageous because it may be more sensitive than the classification approach to mild deficits. Using continuous measures often allows for inclusion of a wider range of patients and could thus increase the power compared to categorical diagnoses (Baldo, Wilson, & Dronkers, 2012). Moreover, from a statistical perspective, it is more difficult to model dichotomous outcomes than continuous ones, such that, in general, use of continuous variables will provide more power to detect associations of lesions with behavior than use of diagnostic classifications. Behavioral specificity when using continuous behavioral scores can be achieved by comparing the score of interest to other related scores using various methods (see "Methods to Achieve Anatomical and Behavioral Specificity in LSM Results"). This range of analysis options provides another advantage for quantification over classification of deficits.

3.4 Variability and Distribution of Performance

When choosing a behavioral score for LSM analyses, one major consideration is the sensitivity of the measure to lesion-related deficits. An ideal test for detecting lesion-behavior relationship would be a test in which controls perform with very low variability. Since one cannot know the premorbid abilities of the lesion participants with certainty, any variability in performance among healthy controls suggests that performance of the lesion participants would have been similarly variable even in the absence of lesions. Thus, variability in performance among controls provides a measure of the degree of noise inherent to the task, i.e., the variability due to either measurement error or individual differences in abilities that are unrelated to lesions. The greater the variability in performance among controls, the more difficult it will be to detect the variability in performance that is related to lesions. One simple way to achieve minimal variability in healthy controls is to use a test on which unimpaired participants perform at ceiling. However, if all controls perform at ceiling on a measure, the measure will likely also be insensitive to mild deficits in individuals with lesions. The ideal measure for LSM then is one for which all controls are at ceiling, but even minor lesions will reduce performance. Further, an ideal measure captures the entire range of possible performance in patients, including not only mild deficits but also severe deficits. In practice, however, many tests have both ceiling and floor effects in populations with brain lesions. If not ceiling and floor effects, many scores have non-normal distributions that may affect LSM results depending on the analysis approach. For example, a test with a non-normal distribution would fail one of the crucial assumptions of a Student's t-test, so this would not be an appropriate statistical method for LSM. Balancing all of the above considerations is difficult, and many behavioral variables do not achieve the ideal distributional properties for LSM. As in any behavioral analysis, it is thus important for investigators to examine the individual scores of the participants being studied and to report the distributional properties when appropriate. When ceiling effects, floor effects, and skewed non-normal distributions are observed, the analysis can sometimes be adjusted to account for these factors, for instance, by transforming skewed scores or using analysis approaches that do not assume a normal distribution. The results of the LSM should accordingly be interpreted with regard to the distribution of the measured scores. For instance, when results are driven by a bimodal distribution of participants who perform at the ceiling and floor on a task, significant LSM results should not be interpreted as lesion locations that relate to *degree* of impairment.

3.5 Batteries of Tests

Typically, investigators ask participants in LSM studies to perform a battery of tests, not just a single measure of interest. These batteries are often used to characterize the broad range of deficits in the participants. Beyond this, a battery may be designed to collect multiple measures of the behavior of interest to ensure that measurement is not biased by the characteristics of any individual test. Any individual test relies on multiple abilities, including the ability of interest and often other prerequisites for performance that are not of interest (e.g., vision, hearing, hand dexterity). Every individual test also introduces some biases based on its particular attributes (e.g., the specific items tested). Composite scores of multiple tests of the same ability that have different prerequisites and different biases will provide a more robust measure of the ability of interest than any individual test. Composite scores may be developed based on theoretical assumptions regarding the processes required for performance of tests or can be developed using data-driven approaches like principal components analysis to discover latent variables measured by a battery [e.g., 29–31].

Batteries of tests are also often used to achieve behavioral specificity in LSM analyses by including tests that overlap in some processing requirements and not others. As noted above, every test relies on multiple abilities, so measuring a single ability requires comparison between two or more scores that overlap in prerequisites but differ in the process of interest. Many tests include control tasks within the assessment. For example, in the Trail Making Task [32], Part B asks individuals to connect dots on a page alternating between letters and numbers. Although Trails B is often used to measure executive function related to switching, the task also requires visual search, recognition of letters and numbers, and motor output. To control for these task requirements, participants also complete Trails A, in which the examinee connects dots based only on letters or numbers, without switching between them. A comparison between the time needed to complete Trails B and Trails A provides a measure of cognitive cost of alternating between letters and numbers, controlling for basic prerequisites of task performance [33]. This approach is sometimes called "cognitive subtraction." Care must be taken to develop control tasks that are well matched to the task of interest. If the control task requires processes that the task of interest does not, this can induce differences in performance that are unrelated to the process of interest. In the analysis, different approaches may be used to compare performance on tasks of interest and control tasks, as discussed in the next section.

4 Methods to Achieve Anatomical and Behavioral Specificity in LSM Results

4.1 Controlling for Lesion Size

In people with brain damage, behavioral scores are often strongly related to lesion size [34]. Naturally, damage to the brain is generally deleterious to behavioral performance, with a larger area of damage resulting in a greater impact on behavior, but there are at least two more specific reasons that behavioral scores are often related to lesion size. First, performance on many tests requires coordination between multiple brain structures that form a large-scale brain network [14]. The larger the lesion, the greater the potential disruption of information flow through the network responsible to process the function. As discussed above, some behaviors may rely only on a single localized processing module in the brain rather than a large-scale network, but even these behaviors often relate to lesion volume. This is because the likelihood that the processing module is damaged is directly related to the size of the lesion. With a small lesion, chances are low that the structure underlying the behavior is damaged; with a large lesion, chances are high that the structure is damaged. Consequently, many scores, whether performed by large-scale networks or by localized processors, are strongly related to lesion volume [6].

A prototypical LSM analysis asks at each location in the brain: "Do people with lesions involving this structure perform worse on the behavior of interest than people with lesions elsewhere in the brain?" The LSM analysis is thus fundamentally a comparison between subgroups of the lesioned participants in the analysis. The mass-univariate VLSM analysis is akin to a series of prospective lesion studies comparing two groups, one with a lesion at a location of interest in the brain and another with lesions elsewhere in the brain. If one imagines a prospective study of this type, it is clearly a major confound if the group of interest has larger lesions than the control group. Unfortunately, throughout the brain, the "lesion here" subgroup at each location in an LSM analysis tends to have larger lesions than the "lesion elsewhere" group. Thus, when the behavior of interest correlates with total lesion size, LSM analyses tend to produce spatially non-specific maps unless lesion size is controlled [6]. Because many scores are related to lesion volume, they are also related to each other. Thus, when conducting LSM analyses, failing to control for lesion volume yields similar maps for many behaviors. The significance of the maps tend to be inflated, and localization is biased toward voxels in which lesion status is strongly related to lesion volume [Fig. 1; 6]. For this reason, when a behavioral score is related to lesion size, this relationship diminishes the specificity of behavioral localization in LSM analyses (see also Chaps. 6 and 11 for more insights on lesion volume confounds). For some clinical applications of LSM, this issue may not adversely affect the ability to address the research question. For most applications of LSM, however, this confound is critical and must be addressed. Researchers may be frustrated that effects disappear when lesion volume is controlled. Some argue that controlling for lesion volume reduces sensitivity to effects in regions in which lesion status is closely related to lesion volume. This is true, but it is a limitation of the dataset, not a reason to leave lesion volume uncontrolled. The loss of sensitivity demonstrates that the collinearity between lesion status in these brain regions and lesion volume precludes making claims regarding the localization of the behavior in question to these regions. The solution to this problem is not leaving lesion volume uncontrolled, but rather recruiting a participant sample in which lesion status in the region in question is less strongly related to lesion size.

It is important to note that lesion volume control is not necessary when the behavior being mapped is unrelated to lesion volume. When behavioral covariates are included in an LSM model, lesion

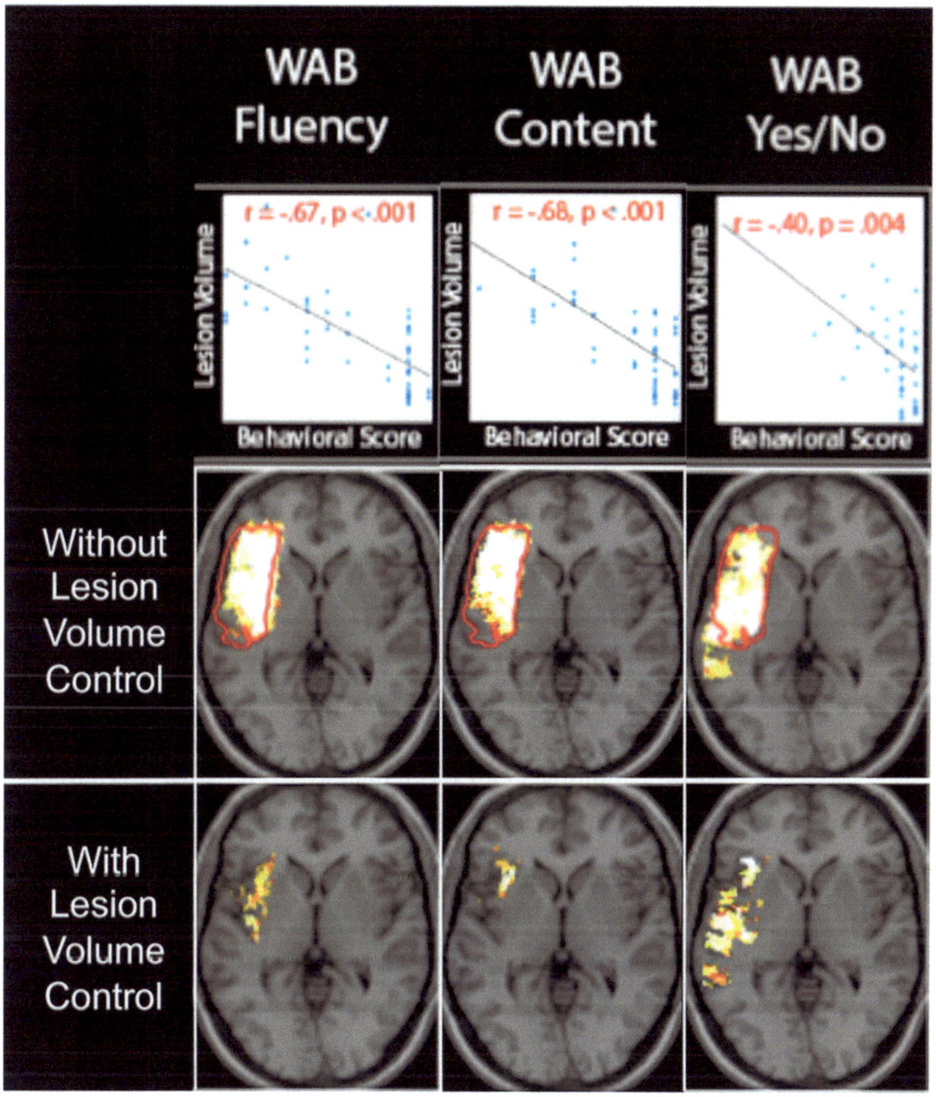

Fig. 1 Effects of lesion volume confounds on LSM results. Results of VLSM analyses on three subtests of the *Western Aphasia Battery* are shown ($P < 0.005$). Scores on all three measures correlate with lesion volume (scatter plots in top row). When VSLM is performed without controlling for lesion volume (middle row), results for all three tests appear similar, identifying voxels in which lesion status is related to lesion volume (red outline). When VLSM is performed using an appropriate method to control for lesion volume (bottom row), more specific and different localization is identified for the three measures. Figure adapted from DeMarco and Turkeltaub (2018), Supplemental Fig. 2. Three subtests from the *Western Aphasia Battery* (WAB) are shown. Fluency is a rating of the flow of narrative speech, "Content" refers to the Information Content subtest, which is a rating of the information conveyed in response to questions and during description of a picture. "Yes/No" refers to a test of the ability to answer yes/no questions

volume control may not be necessary if both the behavior of interest and the covariate are strongly related to lesion volume. In this case, the question is whether or not the residual behavioral score after regressing out the covariate relates to lesion volume. Some investigators may choose to examine the relationship between the relevant behavioral score and lesion volume before deciding whether or not to control for lesion volume in the LSM analysis. However, this strategy is not straightforward in practice because it requires a judgment regarding how much of a relationship between the behavioral score and lesion volume to tolerate. For example, should a correlation of $r = 0.26$ with a significance of $p = 0.11$ trigger a decision to control for lesion volume? While the relationship is not statistically significant, it is still non-zero and will likely induce a mild bias in LSM results. We therefore prefer a cautious approach in which lesion volume is controlled in all cases so as to preclude any concerns about bias.

Depending on the statistical approach being used for LSM, lesion volume can be controlled in several ways. Lesion volume can be controlled by including it as a covariate in the analysis or using a nuisance model that regresses it out of the behavioral score, the lesion maps, or both (see Chap. 6). Regressing lesion volume out of both the behavior and the lesion maps is akin to a partial correlation, in which the contribution of a confounding third variable (in this case, lesion volume) is held constant while assessing the relationship between two variables of interest (in this case, lesion status and the behavioral score). This is also consistent with how a multiple regression analysis would treat a covariate of no interest. In practice, this approach is implicit when lesion volume is included as a covariate in a mass-univariate lesion-symptom mapping analysis that employs multiple linear regression (e.g., VLSM2). We have recently found that for VLSM and support vector regression lesion-symptom mapping (SVR-LSM) analyses of continuous behavioral scores, regressing lesion volume out of both the behavioral score and the lesion maps provides better sensitivity than regressing it out of only the behavioral score or the lesion maps [6]. An alternative method, "direct total lesion volume control," which weights each lesion map by the inverse of lesion size [4], does not alone provide adequate control for the lesion volume confound [6].

4.2 Comparing Multiple Behaviors

As noted above, every behavioral test relies on multiple sensory, cognitive, and/or motor processes. So, to isolate a particular process of interest, comparisons between two or more behavioral tests are needed. This behavioral specificity can be achieved in several ways. First, as noted above, participants can be categorized based on scores on multiple different tests. In this case, the diagnoses are often guided by cognitive models (e.g., phonological vs. surface alexia, apperceptive vs. associative visual agnosia), and so making a specific diagnosis serves to isolate deficits in a particular process. When continuous scores are used to quantify deficits, quantitative comparisons between scores on two or more tests may serve to improve behavioral specificity of results. Using the Trails example described above, an LSM analysis intending to identify lesions associated with deficits in mental flexibility may examine differences in performance between Trails B and Trails A. This can be accomplished in at least two ways. First, a simple difference score or a ratio may be calculated. These approaches have the advantage of being simple, and so the interpretation is fairly straightforward, although differences and ratios have been known to provide different relationships with brain injury [e.g., see 35] for evidence in Trail Making Test]. A disadvantage is that only one control score can be subtracted or used in a ratio. In this context, it is important when quantitatively comparing two scores that there be a strong hypothesis regarding the differences in processes required to complete each task. Ideally, the control task is identical to the task of interest in all respects but does not require the process of interest. If the control task requires additional processes that the task of interest does not, then a difference score or ratio may result in inverse lesion relationship, that is, lesions that produce worse performance on the control task than the task of interest. This may disrupt the ability to identify relationships between lesions and the process of interest. In these cases, it may be advisable to use two-tailed tests to examine bidirectional effects on the difference score.

A second, more flexible, way to achieve behavioral specificity using control behaviors is to add these control task scores as covariates in the LSM analysis [e.g., 36, 37]. Depending on the statistical approach used for LSM, this can be accomplished either by adding them directly as covariates in the statistical model or by using a nuisance model in which they are regressed out of the behavior of interest, the lesion maps, or both, prior to conducting the LSM analysis. This approach has the advantage of allowing for the inclusion of multiple control variables. Again, it is important to carefully consider the differences between the score being examined using LSM and the control variables to ensure that unintended lesion-behavior relationships are not introduced by including the control variables. If there are known associations between demographic factors and test scores (e.g., spelling accuracy with years of education), then demographic covariates can also be included. Their inclusion should reduce the noise related to non-lesion-related individual differences in performance and hence improve the specificity of the analysis to lesion-behavior relationships.

5 Using the LSM Approach with Non-behavioral Variables

LSM has traditionally been used to examine the correlation between performance on a behavioral task and lesion location. However, the LSM approach needn't just examine lesion effects on behavior; in principle, any variable can replace the behavioral score in an LSM-style analysis. Combining LSM with other neuroscience methods can provide information about both remote and local effects of brain damage as well as recovery following injury. For example, LSM has been integrated with EEG [38] to examine how lesions contribute to modulation of event-related potentials (ERPs) after stroke. This approach has the potential to identify the anatomical source of ERPs and to provide mechanistic information to explain how lesions disrupt brain processes electrophysiologically. LSM has also been combined with transcranial direct current stimulation [39] to investigate how lesion location affects response to treatment. Lesion-symptom mapping can also be integrated with functional magnetic resonance imaging (fMRI) to study how lesions in different locations affect brain activity in non-lesioned areas [40]. For example, a study combining fMRI with LSM found that right superior temporal lobe fMRI activity was greater in people with post stroke aphasia than controls and was associated with a good naming outcome. LSM using right superior temporal lobe activity as the "behavior" demonstrated that lesions of the left inferior frontal gyrus were associated with increased right temporal lobe activity, whereas left superior temporal lobe lesions were associated with a failure to recruit the right superior temporal lobe [41]. These types of analyses are able to reveal not only lesion effects on behavior but also lesion effects on the activity of residual brain tissue, potentially addressing important questions about diaschisis, reorganization, and compensation after brain injury.

Although we are not aware of examples of this application of LSM to date, the LSM approach could also be used to determine whether clinical variables such as stroke risk factors (e.g., diabetes, hypertension, atrial fibrillation), etiology (large vessel atherosclerosis vs. cardioembolus; oligodendroglioma vs. astrocytoma), or exposure to certain treatments (e.g., endovascular clot retrieval) are associated with particular lesion distributions. This approach, coupled with the understanding of brain-behavior relationships gained through traditional LSM analyses, could help clinicians monitor for symptoms most likely to occur given the particulars of the case at hand.

6 Conclusions

The LSM approach has proven to be a valuable part of both the clinical and basic neuroscientist's toolbox, and applications of the method are continually expanding. The success of the approach depends on how behavioral variables are selected and handled in analyses. It is critical to consider the specific research question at hand and align the choice of behavioral variables and the analysis approach with the question. The psychometric properties of test instruments and distributional properties of measured variables should be considered whenever possible. When behavioral specificity is needed to address the research question, comparisons between the behavior of interest and control variables, as well as correction for lesion volume confounds, should be considered. Attention to these issues will improve the validity and reliability of LSM analyses.

Acknowledgments

The work was supported by the NIH/NIDCD (R01DC014960 to P.E.T.) and NIH/NCATS via GHUCCTS (TL1TR001431 to A.T. D.).

References

1. Bates E, Wilson SM, Saygin AP, Dick F, Sereno MI, Knight RT, Dronkers NF (2003) Voxel-based lesion-symptom mapping. Nat Neurosci 6(5):448–450. https://doi.org/10.1038/nn1050
2. Yourganov G, Fridriksson J, Rorden C, Gleichgerrcht E, Bonilha L (2016) Multivariate connectome-based symptom mapping in post-stroke patients: networks supporting language and speech. J Neurosci 36(25): 6668–6679. https://doi.org/10.1523/JNEUROSCI.4396-15.2016
3. Fox MD (2018) Mapping symptoms to brain networks with the human connectome. N Engl J Med 379(23):2237–2245. https://doi.org/10.1056/NEJMra1706158
4. Zhang Y, Kimberg DY, Coslett HB, Schwartz MF, Wang Z (2014) Multivariate lesion-symptom mapping using support vector regression. Hum Brain Mapp 35(12): 5861–5876. https://doi.org/10.1002/hbm.22590
5. Pustina D, Avants B, Faseyitan OK, Medaglia JD, Coslett HB (2018) Improved accuracy of lesion to symptom mapping with multivariate sparse canonical correlations. Neuropsychologia 115:154–166. https://doi.org/10.1016/j.neuropsychologia.2017.08.027
6. DeMarco AT, Turkeltaub PE (2018) A multivariate lesion symptom mapping toolbox and examination of lesion-volume biases and correction methods in lesion-symptom mapping. Hum Brain Mapp 39(11):4169–4182. https://doi.org/10.1002/hbm.24289
7. Mirman D, Landrigan JF, Kokolis S, Verillo S, Ferrara C, Pustina D (2018) Corrections for multiple comparisons in voxel-based lesion-symptom mapping. Neuropsychologia 115: 112–123. https://doi.org/10.1016/j.neuropsychologia.2017.08.025
8. Dronkers NF, Wilkins DP, Van Valin RD, Redfern BB, Jaeger JJ (2004) Lesion analysis of the brain areas involved in language comprehension. Cognition 92(1-2):145–177. https://doi.org/10.1016/j.cognition.2003.11.002
9. Kertesz A (1982) Western aphasia battery test manual. Grune & Stratton, New York
10. Martin A, Wiggs CL, Ungerleider L, Haxby J (1996) Neural correlates of category-specific knowledge. Nature 379:649–652
11. Kanwisher N, McDermott J, Chun MM (1997) The fusiform face area: a module in human extrastriate cortex specialized for face perception. J Neurosci 17(11):4301–4311
12. Geschwind N (1965) Disconnexion syndromes in animals and man. I. Brain Journal of Neurology 88(2):237–294
13. Dejerine J (1892) Contribution a l'etude anatomo-pathologique et clinique des differentes varietes de cecite verbale. Memoires Societe Biologique 4:61–90
14. Bressler SL, Menon V (2010) Large-scale brain networks in cognition: emerging methods and principles. Trends Cogn Sci 14(6):277–290. https://doi.org/10.1016/j.tics.2010.04.004
15. Sporns O, Chialvo DR, Kaiser M, Hilgetag CC (2004) Organization, development and function of complex brain networks. Trends Cogn Sci 8(9):418–425. https://doi.org/10.1016/j.tics.2004.07.008
16. Warren DE, Power JD, Bruss J, Denburg NL, Waldron EJ, Sun H, Petersen SE, Tranel D (2014) Network measures predict neuropsychological outcome after brain injury. Proc Natl Acad Sci U S A 111(39):14247–14252. https://doi.org/10.1073/pnas.1322173111

17. Mesulam M-M (1990) Large-scale neurocognitive networks and distributed processing for attention, language, and memory. Ann Neurol 28(5):597–613. https://doi.org/10.1002/ana.410280502
18. Marebwa BK, Fridriksson J, Yourganov G, Feenaughty L, Rorden C, Bonilha L (2017) Chronic post-stroke aphasia severity is determined by fragmentation of residual white matter networks. Sci Rep 7(1):8188. https://doi.org/10.1038/s41598-017-07607-9
19. Embretson SE, Reise SP (2013) Item response theory. Psychology Press, Hove
20. Crocker L, Algina J (1986) Introduction to classical and modern test theory. ERIC, New York
21. Hambleton RK, Jones RW (1993) An NCME instructional module on. Educ Meas Issues Pract 12(3):38–47. https://doi.org/10.1111/j.1745-3992.1993.tb00543.x
22. Hula WD, Fergadiotis G, Martin N (2012) Model choice and sample size in item response theory analysis of aphasia tests. Am J Speech Lang Pathol 21(2):S38–S50. https://doi.org/10.1044/1058-0360(2011/11-0090)
23. Hula WD, Doyle PJ, Stone CA, Hula SNA, Kellough S, Wambaugh JL, Ross KB, Schumacher JG, Jacque AS (2015) The aphasia communication outcome measure (ACOM): dimensionality, item Bank calibration, and initial validation. J Speech Lang Hear Res 58(3): 906–919. https://doi.org/10.1044/2015_JSLHR-L-14-0235
24. Fergadiotis G, Hula WD, Swiderski AM, Lei C-M, Kellough S (2019) Enhancing the efficiency of confrontation naming assessment for aphasia using computer adaptive testing. J Speech Lang Hear Res 62(6):1724–1738. https://doi.org/10.1044/2018_JSLHR-L-18-0344
25. Walker GM, Hickok G, Fridriksson J (2018) A cognitive psychometric model for assessment of picture naming abilities in aphasia. Psychol Assess 30(6):809–826. https://doi.org/10.1037/pas0000529
26. Carmines EG, Zeller RA (1979) Reliability and validity assessment, vol 17. Sage, Newbury Park
27. Friedman RB, Glosser G (1997) Aphasia, alexia, and agraphia. In: Encyclopedia of mental health. Academic, Oxford
28. Rapp B, Purcell J, Hillis AE, Capasso R, Miceli G (2016) Neural bases of orthographic long-term memory and working memory in dysgraphia. Brain 139(Pt 2):588–604. https://doi.org/10.1093/brain/awv348
29. Mirman D, Chen Q, Zhang Y, Wang Z, Faseyitan OK, Coslett HB, Schwartz MF (2015) Neural organization of spoken language revealed by lesion-symptom mapping. Nat Commun 6:6762. https://doi.org/10.1038/ncomms7762
30. Halai AD, Woollams AM, Lambon Ralph MA (2017) Using principal component analysis to capture individual differences within a unified neuropsychological model of chronic post-stroke aphasia: revealing the unique neural correlates of speech fluency, phonology and semantics. Cortex 86:275–289. https://doi.org/10.1016/j.cortex.2016.04.016
31. Lacey EH, Skipper-Kallal LM, Xing S, Fama ME, Turkeltaub PE (2017) Mapping common aphasia assessments to underlying cognitive processes and their neural substrates. Neurorehab Neur Repair:1–9. https://doi.org/10.1177/1545968316688797
32. Reitan RM (1958) Validity of the trail making test as an indicator of organic brain damage. Percept Mot Skills 8(3):271–276
33. Arbuthnott K, Frank J (2000) Trail making test, part B as a measure of executive control: validation using a set-switching paradigm. J Clin Exp Neuropsychol 22(4):518–528
34. Thye M, Mirman D (2018) Relative contributions of lesion location and lesion size to predictions of varied language deficits in post-stroke aphasia. NeuroImage Clin 20: 1129–1138. https://doi.org/10.1016/j.nicl.2018.10.017
35. Corrigan JD, Hinkeldey NS (1987) Relationships between parts a and B of the trail making test. J Clin Psychol 43(4):402–409. https://doi.org/10.1002/1097-4679(198707)43:4<402::Aid-jclp2270430411>3.0.Co;2-e
36. Dickens JV, Fama ME, DeMarco AT, Lacey EH, Friedman RB, Turkeltaub PE (2019) Localization of phonological and semantic contributions to Reading. J Neurosci 39(27): 5361–5368. https://doi.org/10.1523/JNEUROSCI.2707-18.2019
37. Leff AP, Schofield TM, Crinion JT, Seghier ML, Grogan A, Green DW, Price CJ (2009) The left superior temporal gyrus is a shared substrate for auditory short-term memory and speech comprehension: evidence from 210 patients with stroke. Brain J Neurol 132 (Pt 12):3401–3410. https://doi.org/10.1093/brain/awp273
38. Obrig H, Mentzel J, Rossi S (2016) Universal and language-specific sublexical cues in speech perception: a novel electroencephalography-lesion approach. Brain 139(6):1800–1816. https://doi.org/10.1093/brain/aww077

39. Campana S, Caltagirone C, Marangolo P (2015) Combining voxel-based lesion-symptom mapping (VLSM) with A-tDCS language treatment: predicting outcome of recovery in nonfluent chronic aphasia. Brain Stimul. https://doi.org/10.1016/j.brs.2015.01.413
40. Fridriksson J, Bonilha L, Baker JM, Moser D, Rorden C (2010) Activity in preserved left hemisphere regions predicts anomia severity in aphasia. Cereb Cortex 20(5):1013–1019
41. Skipper-Kallal LM, Lacey EH, Xing S, Turkeltaub PE (2017) Functional activation independently contributes to naming ability and relates to lesion site in post-stroke aphasia. Hum Brain Mapp. https://doi.org/10.1002/hbm.23504

Chapter 14

Lesion-Behavior Awake Mapping with Direct Cortical and Subcortical Stimulation

Stephanie K. Ries, Kesshi Jordan, Robert T. Knight, and Mitchel Berger

Abstract

Electrical stimulation during awake brain surgery is a classic method yielding direct information about brain-behavior relationships. This technique has provided key leaps in our understanding of the organization of the brain, including the discovery of the somatotopic organization of the primary sensory and motor cortices by Wilder Penfield and his colleagues in the 1950s or the previously unsuspected degree of inter-individual variability in the cortical representations of speech and language functions by George Ojemann and colleagues. The electrical stimulation technique, combined with an array of advanced neuroimaging methods, has expanded allowing us to test not only cortical functions but also the role of underlying white matter tracts in behavior. In this chapter, we will describe the method, its evolution, and the novel types of discoveries enabled by direct cortical and subcortical electrical stimulation of the human brain.

Key words Direct electrical stimulation, Awake brain surgery, Inter-individual variability, Neuroplasticity, Cortical mapping, Subcortical mapping

1 Description of the Method

Direct electrical stimulation (DES) during intraoperative brain mapping is typically performed in patients undergoing brain surgery for tumor resection or in patients suffering from drug-resistant epilepsy undergoing surgical ablation of epileptogenic tissue. The primary aim of DES is to preserve the parts of the brain that are critical for supporting key functions such as motor, sensory, or speech and language functions. These parts of the brain are often referred to as the "eloquent cortex." Some functions, such as simple motor responses, can be tested while the patient is asleep either by observing twitches or with electromyography. However, the mapping of functions such as language requires active participation of the patient. Therefore, these cases are performed with a special protocol that allows the patient to remain awake during DES mapping and, in some cases, during the resection itself. This chapter will focus on the more complex protocol for

Dorian Pustina and Daniel Mirman (eds.), *Lesion-to-Symptom Mapping: Principles and Tools*, Neuromethods, vol. 180, https://doi.org/10.1007/978-1-0716-2225-4_14, © Springer Science+Business Media, LLC, part of Springer Nature 2022

awake surgery, but an asleep version of this surgery is a widely used technique in resections requiring motor mapping.

The procedure is as follows, with slight variations between institutions. We here describe the procedure currently followed at the University of California San Francisco (UCSF).

1.1 Preoperative Evaluations and Scans

The principle of intraoperative DES mapping is to test the effect of a lesion transiently using electrical stimulation prior to the resection of tissue. Every brain is unique, and, while broad organizational principles usually dominate, the precise effect of a resection is difficult to predict at the individual level. There are large differences, for example concerning which hemisphere dominates in language processing, as well as effects of brain reorganization in reaction to the tumor growth. Preoperative mapping helps the surgical team plan an intervention that maximally benefits the patient, determining what functions are important to test during the surgery. Here we outline the series of evaluations and brain scans that the patients are subjected to preoperatively and refer to published studies using these methods.

Tasks

Patients are prepared for the tasks used intraoperatively including counting aloud, repetition, single word reading, and picture naming. For the word reading task, 20–30 words of varying length (usually between one and four syllable-long) are used as stimuli. These words are of varying lexical frequency, can be regular (e.g., cat) or irregular (e.g., yacht), and are presented in a random order. The reason why the lexical frequency, length, and regularity of the words are manipulated is to test various processes associated with word reading. In particular, lexical frequency and regularity are manipulated to insure word reading can take place through both the lexical/semantic pathway and the sublexical/phonological pathway (using letter to sound mapping) [1, 2]. For the picture naming task, around 100 line-drawings from normed picture databases (such as [3]) are used as stimuli. These pictures represent common objects or animals and are selected to have high name agreements (above 75%), to avoid confusion linked to multiple names being associated with a same picture (e.g., sofa and couch). Following this, only the pictures or words produced without hesitation are selected for intraoperative monitoring, and problematic items are excluded to insure that the results seen during awake language mapping are due to the stimulation and not to preoperative difficulties with certain tasks or items.

Brain Scans

Patients undergo magnetic resonance imaging (MRI) and magnetoencephalography (MEG) within the week preceding surgery. The

T1, T2, and FLAIR sequences are used to delineate the contours of the lesion(s) (from the upcoming surgical resection or the existing tumor). Diffusion imaging (high angular resolution diffusion imaging, HARDI) sequences are used with tractography algorithms [4] to model the contours of the major white matter pathways of interest (arcuate fasciculus, AF; inferior longitudinal fasciculus, ILF; inferior frontal-occipital fasciculus, IFOF; superior longitudinal fasciculus, SLF; aslant tract, AT; etc.) using a virtual dissection technique described in [5, 6]. Note that HARDI is preferred over the more traditional diffusion sequences that are analyzed with diffusion tensor imaging (DTI) because HARDI can be fit with more complex models that have been shown to have better concordance with intraoperative stimulation [7], and it provides better results than DTI in reconstructing white matter pathways that are still functional but have been invaded by tumors as well as in crossing regions [5]. Indeed, DTI-derived tractography has been shown to sometimes fail to model crucial pathways present within the tumor [8]. It is important to stress that, while these methods are constantly improving, all tractography methods have unique behavior that cannot accurately model the full spectrum of anatomy, especially when pathology is introduced. These are models that cannot provide an accurate representation of neuron-level anatomy, whose interpretation is highly uncertain [9]. The clinical decision is always based on stimulation, but despite their inherent limitations [10], our experience with these methods over many years has been that consistent application of these algorithms helps guide the mapping process [5].

MEG data is used to determine language laterality, evaluate functional connectivity, and perform source localization using magnetic source imaging (MSI) combined with the MRI scans. MSI is used to find peaks of activity associated with language perception and production using a verb generation task (as in [11, 12]) or picture naming if the patient is unable to perform the verb generation task. Source localization is also performed on data from finger tapping and equivalent sensory tasks to support sensorimotor mapping. The model of the white matter tracts and peaks of activity derived through MSI are overlaid on the anatomical T1 scans that are then used intraoperatively to help orient and guide the surgical resection.

1.2 Intraoperative Testing in Awake Surgery

For the awake surgery, the patient is sedated intravenously with either propofol or dexmedetomidine at the start of the procedure. The patient then undergoes a craniotomy using monitored anesthesia care, involving local anesthetic infiltration applied to the scalp. Surgical exposure is always tailored to the target lesion. After the craniotomy, the patient is awakened by decreasing the intravenous anesthesia. Orientation questions are then asked to

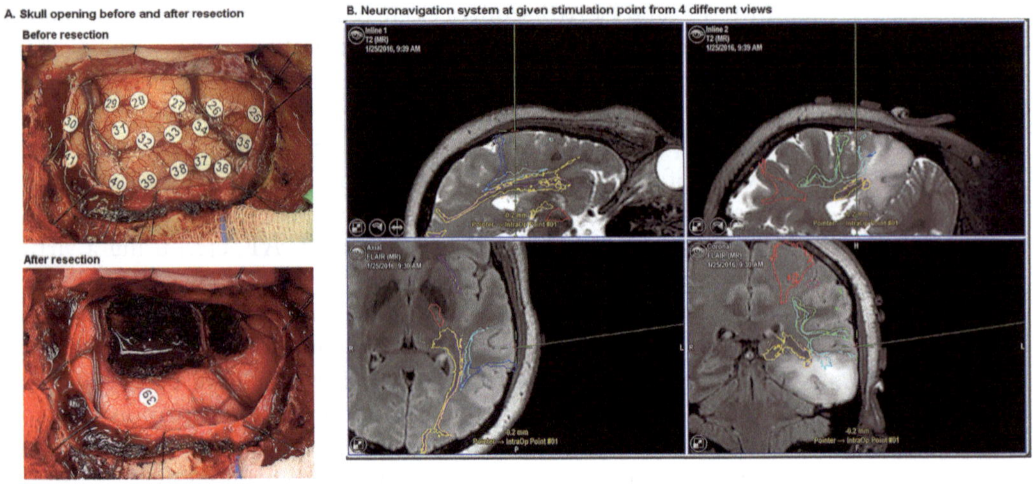

Fig. 1 (**a**) Tested sites are marked using small white numbered tickets. In this patient, no tested site was associated with reliable (i.e., repeated) disruption of function during cortical testing. The after-resection picture shows the resected area. Ticket #39 was kept on for comparison with the pre-resection picture. (**b**) Neuronavigation system showing the modeled white matter tracks in the vicinity of the lesion in the left inferior temporal lobe (with extension in the middle temporal gyrus, occipito-temporal, and parahippocampal gyri): motor tract (red), optic tract (gold), arcuate fasciculus (aqua), inferior fronto-occipital fasciculus (purple), superior longitudinal fasciculus inferior parietal (green), superior longitudinal fasciculus temporoparietal (blue)

verify the patients' responsiveness (e.g., What day is it today? What is your name?). Intraoperative mapping then commences and is performed over the entire exposed surface. Motor and sensory mapping is performed first. Bipolar electrical stimulation (60 Hz) is applied for 2-s intervals to the exposed cortical surface using a manual handheld probe ramping up to a tailored intensity (typically ~3–4 mA adjusted on an individual basis). The patients are asked whether they feel any sensations in any part of their body and are monitored to observe any induced movement. Then, receptive and expressive speech and language testing is performed using the tasks tested preoperatively (i.e., counting, repetition, single word reading, and picture naming). Tested sites are marked using small tickets (Fig. 1a), and positive sites (i.e., sites where stimulation causes a disruption of the tested function) are recorded by the attending neuropsychologist. This part of the method has remained largely unchanged since the seminal works by Penfield (*see* Sect. 2).

Intraoperative electrocorticography (ECoG) is also used to monitor for stimulation-induced after-discharges. This monitoring limits the occurrence of seizures induced by the electrical stimulation and allows the neurosurgeon to adapt the stimulation intensity on a by-patient basis. Cold Ringer's lactate solution is kept on-hand to control seizure activity. This technique has been shown to be a simple and effective way to terminate partial seizures induced by electrical stimulation during intraoperative brain mapping, thereby limiting the need for intravenously administering antiepileptic barbiturates [13]. Intraoperative ECoG can be performed using electrodes placed individually and held within a metal ring clamped to the skull temporarily or using grids of electrodes placed directly on the cortical surface. High-density grids are more often used in epilepsy cases rather than brain tumor cases, as ECoG grids can provide more accurate localization of the epileptogenic zone as compared to electrodes placed individually on the cortical surface. Essential sites are defined as those resulting in a loss of function in at least two of three stimulations. Following this cortical mapping, every effort is made to avoid or limit the resection of these essential sensory, motor, and language sites while maximizing tumor or epileptogenic tissue removal.

Resection is performed with an ultrasonic aspirator guided by intraoperative neuronavigation (Fig. 1b). Mapping of subcortical areas can also take place if the resection includes tissue near critical white matter pathways, such as the arcuate fasciculus, and if the patient's condition permits continued stimulation during resection. In this case, receptive and expressive language tasks, such as reading and picture naming, are performed again while stimulating the subcortical pathways using the neuronavigation system, allowing the neurosurgeon to orient themselves with respect to the preoperative MRI scans and white matter tracts models. After mapping, patients are re-sedated with either propofol or dexmedetomidine for the remainder of the surgical procedure.

2 Evolution: From Penfield to Today

The first reports by Wilder Penfield and colleagues of systematic awake mapping of motor, sensory, and behavioral functions during brain surgery date back to the 1950s [14]. However, electrical stimulation was being performed earlier, although typically reported in isolated cases (*see* review by [15]). For example, in 1874, Roberts Bartholow reported electrical stimulation being performed on a patient whose dura had been exposed by a cancerous ulcer (causing an approximately 2-inch-diameter (51 mm) hole in her skull) in experiments widely considered to be a breach of medical ethics [16]. Muscular contractions and paresthesias were observed in the patient's contralateral arm and leg [17]. With the development of stimulation equipment and techniques, Penfield and colleagues' investigations and thorough documentation provided systematic findings in larger cohorts of patients propelling the field forward on several fronts. Penfield was trained partly by Otfrid Foerster who was an expert in the pathogenesis and surgical management of epilepsy and who was using cortical stimulation in the excision of brain scars caused by gunshot wounds [18, 19]. Penfield then started applying this technique to patients with epilepsy initially to try to reproduce seizures intraoperatively and to outline sensory and motor areas. In particular, he used Galvanic current to outline motor and sensory areas and a faradic coil to try to reproduce features of a seizure. He used a thyratron stimulator with unipolar or bipolar electrodes which were made of a platinum wire in a glass holder [20, 21]. This technique led to the description of cortical maps for sensory and motor responses, now depicted as the sensory and motor homunculi [22]. In addition, Penfield along with Rasmussen and Roberts described detailed maps of cortical regions associated with different types of speech disruptions which remain highly relevant for our understanding of speech and language cortical organization to this day [14, 23] (Fig. 2a). Following Penfield's groundbreaking work, the use of DES spread worldwide.

George Ojemann, followed by Mitchel Berger and others, further developed the technique and the understanding of the cortical organization of speech and language functions by performing large-scale studies involving over 100 patients each [24–26]. These large-scale studies largely confirmed the seminal findings by Penfield and colleagues but also allowed the development of maps of inter-individual variability in the cortical organization of speech and language (Fig. 2b). These studies revealed that even brain areas thought to be critical for speech and language, such as Broca's and Wernicke's areas, were not always associated with speech or language disruptions when stimulated. In fact, it is this inter-individual variability that makes DES critical in neurosurgical resections, as the location of essential language sites cannot be simply inferred based on textbook knowledge of speech and language organization in the brain. Factors influencing this inter-individual variability include inter-individual variability in language organization in healthy adults and the effects of tumor grade on cerebral reorganization ([27]; *see* Sect. 3). These factors continue to be studied to this day in the hopes of better understanding the human language system and improving treatment of these complex cases (*see* Sect. 3).

Due to the high degree of inter-individual variability in the cortical representation of speech and language, the original approach pioneered by Penfield of finding positive sites to avoid eloquent areas required large craniotomies and extensive mapping over a large section of tissue. Negative mapping (which corresponds to the approach described in sect. 1) was introduced to accomplish the same purpose of preserving eloquent cortex, but this approach repeats stimulation to establish confidence that tissue is not involved in the tested function prior to resection [26]. This approach can be performed with tailored craniotomies that do not necessarily expose positive sites (as in the example displayed in Fig. 1) and constitutes a paradigm shift in the language-mapping techniques. Adoption of the negative mapping technique allows for

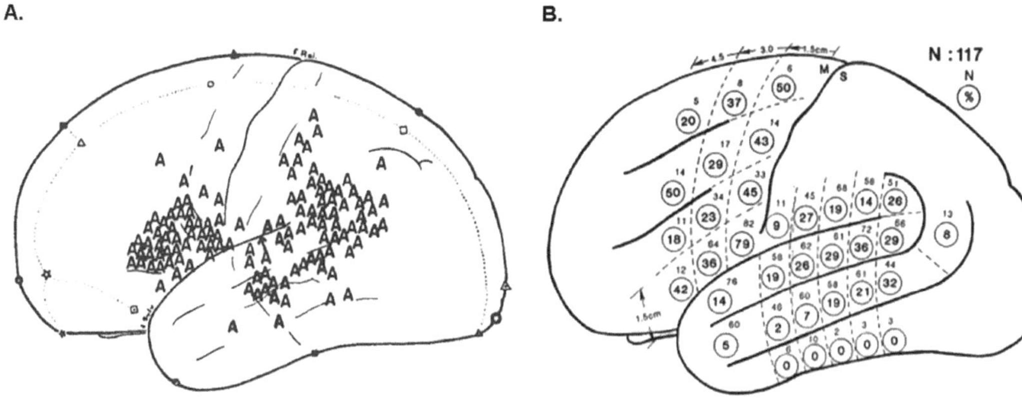

Fig. 2 Cortical maps associated with speech or language disruption by Penfield and Roberts, 1959 (**a**), and Ojemann et al., 1989 (**b**). (**a**) "A"'s mark the locations associated with speech arrest on a schematic left lateral view of the brain. This map largely replicates the initial Wernicke-Geschwind model of speech and language by showing that speech arrest can occur after stimulation of the left inferior frontal cortex, left posterior superior temporal cortex, and angular gyrus. (**b**) In each quadrant, the small number indicates the number of patients tested in each area, and the bigger circled number indicates the percentage of patients for whom the site was positive (i.e., associated with a disruption of function in at least two out of three stimulations). A total of 117 patients were tested. This map shows the high degree of inter-individual variability in the cortical representation of speech and language function in the human brain as no area was associated with 100% disruption in all patients

smaller craniotomies, less extensive mapping, and shorter surgeries. This approach has been shown to be successful as only 1.6% of over 200 patients tested over a 10-year period at UCSF were found to have a persistent language deficit 6 months after surgery using the negative mapping approach [26].

More recently, awake speech and language mapping has expanded to subcortical mapping. Hugues Duffau has been a leader in this novel approach and has published numerous studies highlighting the central causal role of white matter pathways in language and cognition more generally [28–32]. These investigations have led to the proposal of a new hodotopical (i.e., delocalized) model of language and speech in the brain emphasizing the role of large-scale parallel cortico-subcortical subnetworks underlying different aspects of language processing, where white matter pathways play a central role in the neurobiology of language [33]. The importance of preserving white matter pathways in awake language mapping has been echoed by other groups. In particular, [5] have developed and tested a virtual dissection method using residual bootstrap q-ball fiber tracking [4] for the segmentation of eight white matter tracts associated with language that has been applied to most awake language cases performed at UCSF since 2012. Their pilot study showed that preserving connectivity through the AF and the temporoparietal component of the SLF (tp-SLF) is associated with preserving language functions postoperatively with damage to these two pathways predictive of long-term language deficits [34].

As a result of combining these techniques, a growing number of case studies have furthered the understanding of human brain function and neuroplasticity and helped guide neurosurgeries for optimal preservation of complex functions [35, 36]. For example, Piai and colleagues combined DES, diffusion imaging, and electrocorticography data in a violinist who was being monitored to preserve musical abilities during resection of a medial frontal tumor involving the left supplementary motor area [36], a region previously associated with musical ability [37]. Garcea and colleagues furthered the understanding of the causal role of the right posterior superior temporal gyrus in music processing by combining preoperative behavioral testing and functional magnetic resonance imaging (fMRI) using a series of functional localizers with intraoperative DES in a saxophone player [35]. In both of these studies, the patients played their musical instrument in the operating room to verify that their musical abilities were preserved.

Finally, technological improvements are continuing to bring in new possibilities for awake brain mapping. For example, intraoperative MRI (iMRI) can now be performed in some centers, and iMRI has been used to verify the extent of pathological tissue remaining after initial resection [38]. In this study, the authors found that further resection was needed after the iMRI scan in 40.5% of cases, allowing surgeries to be more efficient. New paradigms are also being developed to test more complex functions or different functions during awake brain mapping. Those include musical abilities, like in the cases abovementioned, but also other functions which we discuss in the next section.

3 Potential of Direct Cortical and Subcortical Electrical Stimulation for the Understanding of the Human Brain

The potential contribution of DES for advancing the understanding of the human brain is unique in several respects. For instance, DES is one of the only techniques able to directly test the causal role of brain regions and white matter pathways in supporting human brain function by transiently disrupting neuronal activity at the cortical surface and subcortically. As outlined above, the use of DES has enabled outlining maps of the primary somatosensory and motor cortices, speech, and language regions but also highlights the large degree of inter-individual variability between patients in the cortical representations of speech and language. DES has also been used to further the understanding of other brain functions and neuroplasticity in general. The combined use of advanced neuroimaging techniques such as ECoG, MEG, fMRI, and diffusion imaging is adding to this understanding and to improvements in surgical outcome.

3.1 Mapping More Specific Human Functions

Some of the functions recently investigated using DES include music, vision, but also specific language processes such as syntactic processing, word retrieval with competing semantic neighbors, or action naming [33, 35, 36, 39–43].

Although speech and language functions were among the first human brain functions to be mapped using DES, recent studies have extended the depth and breadth of these investigations by targeting specific linguistic processes and by moving from single words to sentences [39, 42, 43] or from object to action naming [40]. For example, Riès and colleagues asked participants to perform two tasks: picture naming with distractor words on top of the pictures and generating sentences based on action pictures [42]. The picture naming task was a picture-interference paradigm (PWI), which manipulated meaning dependencies between the picture and the distractor word [44, 45]. This task was chosen as it had been used extensively in the psycholinguistics literature to study specific linguistic processes, such as word retrieval [46–49]. In particular, the semantic-relatedness between the picture and the distractor word was manipulated such that the picture and the word were sometimes semantically related (e.g., picture of a pear with the distractor word "lemon" on top, Fig. 3a) and sometimes not (e.g., picture of a pear with the distractor word "horse" on top). Participants are slower to name the picture when the distractor word is semantically related, an effect referred to as the semantic interference effect which has been tied to the difficulty of word retrieval. In this study, the use of this task was contrasted to

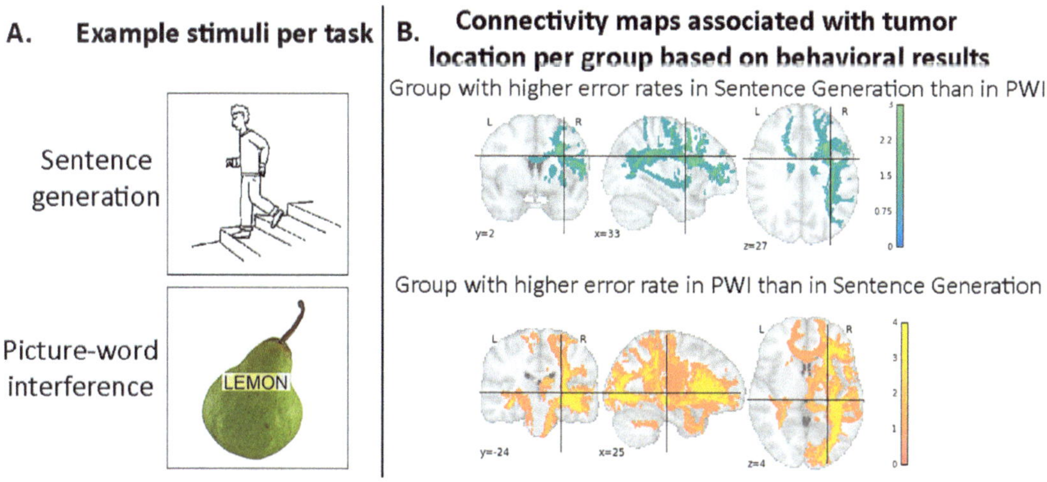

Fig. 3 (**a**) Example stimuli used per task in [42]. (**b**) Connectivity maps in MNI space associated with tumor locations for the patients with higher error rates in sentence generation than in PWI during subcortical testing (top) and for the patients with higher error rates in PWI than sentence generation during subcortical testing (bottom). The tumor reconstruction volume (i.e., mask) was targeted from a whole-brain streamline dataset; voxels connected to the tumor volume were mapped here as described in the original publication. Lighter colors indicate higher overlap between patients

a second task in which patients were asked to generate simple sentences in response to action pictures (e.g., the man is walking down the stairs, Fig. 3a). This task also involved word retrieval, although it did not manipulate semantic interference. Instead, this task involved the need to order words in a sequence using syntax. The aim was to target different essential linguistic processes by contrasting these tasks. Patients were tested on these tasks preoperatively and intraoperatively using DES at the cortical surface and, in some patients, subcortically using the neuronavigation system in relation to pre-identified white matter tracts around the lesion (as in Fig. 1). In addition, tractography was performed using the diffusion imaging data and tumor masks drawn on each patient's anatomical MRI scans in all patients in whom subcortical testing was performed. The results showed that patients showing more disruptions in the sentence generation task than in the PWI task had a white matter connectivity profile, as determined by tractography and seeded in the tumor site (hereafter referred to as tumor connectivity profile), that involved the dorsal stream of language processing. By contrast, patients showing more disruptions in the PWI than in the sentence generation task had a tumor connectivity profile which involved the ventral stream more than the dorsal stream of language processing [50–52] (Fig. 3b). In addition to enhancing our understanding of the neuroanatomy of different language functions, the results also showed that disruptions were more frequent in the sentence generation than in the picture naming task, emphasizing the importance of testing patients at the sentence-level in addition to the single word level [39]. Finally, disruptions in both tasks were more prominent during subcortical than during cortical testing, underlying again the importance of performing subcortical DES to preserve white matter pathways during awake language mapping.

Other functions, such as visual functions, are typically not mapped extensively during awake brain stimulation, although the ability to perceive visual stimuli is always checked prior to speech and language testing using visual stimuli pre- and intraoperatively. Using combined subcortical DES and diffusion imaging data, Gras-Combe and colleagues conducted a study mapping optic radiations during visual perception of pictures presented in all four visual quadrants in patients with tumors infiltrating the optic radiations [41]. The challenge was to successfully resect the tumors while limiting the occurrence of long-term visual impairment in these patients who did not have any major preoperative visual deficits. The optic radiations were mapped functionally during the surgery by using DES subcortically to define the edges of the area that was safe to resect. If visual disturbances were caused during DES in a specific subcortical area, the neurosurgeon stopped tumor resection in that region. Even though 12 out of the 14 total cases would have been expected to show postoperative quadranopsia based on their tumor locations, the use of this intraoperative monitoring technique led to only one patient experiencing permanent hemianopia post-surgery. These studies along with an increasing number of recent studies are aiming to provide a more detailed understanding of human brain functions through the unique lens of DES, which will in turn enable a more tailored intraoperative approach to awake brain mapping. In addition, these studies further confirm the importance of conducting awake DES to take into account inter-individual differences in brain organization and preserve critical brain regions for better surgery outcomes.

3.2 Understanding and Making Use of Neuroplasticity

Cortical and subcortical DES have provided insight on the impressive degree of neuroplasticity in the adult human brain [53]. Indeed, brain regions which have been put at the core of linguistic ability, such as Broca's and Wernicke's areas for speech and language, are not always associated with language disruption when stimulated [25]. According to Duffau, neuroplasticity is enabled by the presence of parallelism and redundancy in brain networks supporting brain function [27]. According to this hodotopic principle, the functions of the brain are supported by both the cortical epicenters (i.e., topos or sites) and the connections between those centers (i.e., hodos or pathways). The same cortical center can be involved in more than one function depending on which other activated centers and connected pathways are co-engaged at any given time.

Robles and colleagues introduced an innovative development in the field of awake brain mapping directly making use of this high degree of neuroplasticity [54]. These investigations used a multistage surgical approach in two patients with World Health Organization Grade II gliomas who had tumors involving eloquent areas including the dominant left premotor cortex. These tumors are classically not removed entirely because of the risk of causing permanent deficits postoperatively. Robles and colleagues used functional MRI to identify brain regions associated with premotor functions preoperatively and then again postoperatively after the first resection, which preserved the critical areas. In the two patients, they identified a shift in the brain areas associated with these functions on the postoperative MRIs several years later, which allowed them to perform a second resection to completely remove the tumor.

In the view of Ghinda and Duffau, the brain network or connectomics perspective constitutes a paradigm shift for understanding brain function in neuro-oncology approaches [55]. These types of cases suggest that a possible future development would be the implementation of platforms allowing neurosurgeons to link intraoperative cortical stimulation results with neuronal network models during surgery. Connectivity-based modeling could be used to predict functional changes after surgery. We note that related approaches are being developed in the field of epilepsy using for example cortico-cortical evoked potentials to map directional connectivity of large-scale brain networks [56, 57].

In conclusion, DES was one of the first method used to study brain function and has undergone a fast and promising recent evolution. Today, the combined use of other human brain imaging techniques with DES during awake brain mapping, along with novel perspectives from the fields of neuroscience and psychology, is contributing to our understanding of human brain function, neuroplasticity, and improved surgery outcomes.

References

1. Coltheart M, Rastle K (1994) Serial processing in reading aloud: evidence for dual-route models of reading. J Exp Psychol Hum Percept Perform 20(6):1197–1211. https://doi.org/10.1037/0096-1523.20.6.1197
2. Plaut DC, McClelland JL, Seidenberg MS, Patterson K (1996) Understanding normal and impaired word reading: computational principles in quasi-regular domains. Psychol Rev 103(1):56–115. https://doi.org/10.1037/0033-295x.103.1.56
3. Snodgrass JG, Vanderwart M (1980) A standardized set of 260 pictures: norms for name agreement, image agreement, familiarity, and visual complexity. J Exp Psychol Hum Learn 6(2):174–215. https://doi.org/10.1037/0278-7393.6.2.174
4. Berman J, Chung S, Mukherjee P, Hess C, Han E, Henry R (2008) Probabilistic streamline q-ball tractography using the residual bootstrap. NeuroImage 39(1):215–222. https://doi.org/10.1016/j.neuroimage.2007.08.021
5. Caverzasi E et al (2016) Identifying preoperative language tracts and predicting postoperative functional recovery using HARDI q-ball fiber tractography in patients with gliomas. J Neurosurg 125(1):33–45. https://doi.org/10.3171/2015.6.JNS142203
6. Jordan KM, Amirbekian B, Keshavan A, Henry RG (2018) Cluster confidence index: a streamline-wise pathway reproducibility metric for diffusion-weighted MRI tractography. J Neuroimaging 28(1):64–69. https://doi.org/10.1111/jon.12467
7. Bucci M et al (2013) Quantifying diffusion MRI tractography of the corticospinal tract in brain tumors with deterministic and probabilistic methods. NeuroImage Clin 3:361–368. https://doi.org/10.1016/j.nicl.2013.08.008
8. Leclercq D et al (2010) Comparison of diffusion tensor imaging tractography of language tracts and intraoperative subcortical stimulations. J Neurosurg 112(3):503–511. https://doi.org/10.3171/2009.8.JNS09558
9. Jones DK, Knösche TR, Turner R (2013) White matter integrity, fiber count, and other fallacies: the do's and don'ts of diffusion MRI. NeuroImage 73:239–254. https://doi.org/10.1016/j.neuroimage.2012.06.081
10. Thomas C et al (2014) Anatomical accuracy of brain connections derived from diffusion MRI tractography is inherently limited. Proc Natl Acad Sci U S A 111(46):16574–16579. https://doi.org/10.1073/pnas.1405672111
11. Englot DJ et al (2015) Epileptogenic zone localization using magnetoencephalography predicts seizure freedom in epilepsy surgery. Epilepsia 56(6):949–958. https://doi.org/10.1111/epi.13002
12. Traut T et al (2019) MEG imaging of recurrent gliomas reveals functional plasticity of hemispheric language specialization. Hum Brain Mapp 40(4):1082–1092. https://doi.org/10.1002/hbm.24430
13. Sartorius CJ, Berger MS (1998) Rapid termination of intraoperative stimulation-evoked seizures with application of cold Ringer's lactate to the cortex. Technical note. J Neurosurg 88(2):349–351. https://doi.org/10.3171/jns.1998.88.2.0349
14. Penfield W, Roberts L (1959) Speech and brain-mechanisms. Princeton University Press, Princeton
15. Bulsara KR, Johnson J, Villavicencio AT (2005) Improvements in brain tumor surgery: the modern history of awake craniotomies. Neurosurg Focus 18(4):e5
16. Harris LJ, Almerigi JB (2009) Probing the human brain with stimulating electrodes: the

story of Roberts Bartholow's (1874) experiment on Mary Rafferty. Brain Cogn 70(1): 92–115. https://doi.org/10.1016/j.bandc.2009.01.008

17. Morgan JP (1982) The first reported case of electrical stimulation of the human brain. J Hist Med Allied Sci 37(1):51–64. https://doi.org/10.1093/jhmas/xxxvii.1.51
18. Eccles J, Feindel W (1978) Wilder Graves Penfield, 26 January 1891–5 April 1976. Biogr Mem Fellows R Soc 24:473–513
19. Feindel W (1977) Wilder Penfield: his legacy to neurology. To praise an absent friend. Can Med Assoc J 116(12):1365–1367
20. Penfield W (1936) Epilepsy and surgical therapy. Arch Neurol Psychiatr 36(3):449–484. https://doi.org/10.1001/archneurpsyc.1936.02260090002001
21. Penfield W (1941) Epilepsy and cerebral localization: a study of the mechanism, treatment and prevention of epileptic seizures, 1st edn. Charles C. Thomas, Springfield
22. Penfield W, Boldrey E (1937) Somatic motor and sensory representation in the cerebral cortex of man as studied by electrical stimulation. Brain J Neurol 60:389–443. https://doi.org/10.1093/brain/60.4.389
23. Penfield W, Rasmussen T (1950) The cerebral cortex of man. Macmillan, New York
24. Ojemann GA (1979) Individual variability in cortical localization of language. J Neurosurg 50(2):164–169. https://doi.org/10.3171/jns.1979.50.2.0164
25. Ojemann G, Ojemann J, Lettich E, Berger M (1989) Cortical language localization in left, dominant hemisphere. An electrical stimulation mapping investigation in 117 patients. J Neurosurg 71(3):316–326. https://doi.org/10.3171/jns.1989.71.3.0316
26. Sanai N, Mirzadeh Z, Berger MS (2008) Functional outcome after language mapping for glioma resection. N Engl J Med 358(1):18–27. https://doi.org/10.1056/NEJMoa067819
27. Duffau H (2014) Diffuse low-grade gliomas and neuroplasticity. Diagn Interv Imaging 95(10):945–955. https://doi.org/10.1016/j.diii.2014.08.001
28. Duffau H (2015) Stimulation mapping of white matter tracts to study brain functional connectivity. Nat Rev Neurol 11(5):255–265. https://doi.org/10.1038/nrneurol.2015.51
29. Duffau H (2017) Mapping the connectome in awake surgery for gliomas: an update. J Neurosurg Sci 61(6):612–630. https://doi.org/10.23736/S0390-5616.17.04017-6
30. Duffau H et al (2003) Usefulness of intraoperative electrical subcortical mapping during surgery for low-grade gliomas located within eloquent brain regions: functional results in a consecutive series of 103 patients. J Neurosurg 98(4):764–778. https://doi.org/10.3171/jns.2003.98.4.0764
31. Duffau H et al (2002) Intraoperative mapping of the subcortical language pathways using direct stimulations. An anatomo-functional study. Brain J Neurol 125(Pt 1):199–214
32. Duffau H, Peggy Gatignol ST, Mandonnet E, Capelle L, Taillandier L (2008) Intraoperative subcortical stimulation mapping of language pathways in a consecutive series of 115 patients with Grade II glioma in the left dominant hemisphere. J Neurosurg 109(3):461–471. https://doi.org/10.3171/JNS/2008/109/9/0461
33. Duffau H, Moritz-Gasser S, Mandonnet E (2014) A re-examination of neural basis of language processing: proposal of a dynamic hodotopical model from data provided by brain stimulation mapping during picture naming. Brain Lang 131:1–10. https://doi.org/10.1016/j.bandl.2013.05.011
34. Catani M, Mesulam M (2008) The arcuate fasciculus and the disconnection theme in language and aphasia: history and current state. Cortex J Devoted Study Nerv Syst Behav 44(8):953–961. https://doi.org/10.1016/j.cortex.2008.04.002
35. Garcea FE et al (2017) Direct electrical stimulation in the human brain disrupts melody processing. Curr Biol 27(17):2684.e7–2691.e7. https://doi.org/10.1016/j.cub.2017.07.051
36. Piai V, Vos SH, Idelberger R, Gans P, Doorduin J, ter Laan M (2019) Awake surgery for a violin player: monitoring motor and music performance, a case report. Arch Clin Neuropsychol 34(1):132–137. https://doi.org/10.1093/arclin/acy009
37. Zatorre RJ, Chen JL, Penhune VB (2007) When the brain plays music: auditory-motor interactions in music perception and production. Nat Rev Neurosci 8(7):547–558. https://doi.org/10.1038/nrn2152
38. Maldaun MVC et al (2014) Awake craniotomy for gliomas in a high-field intraoperative magnetic resonance imaging suite: analysis of 42 cases: clinical article. J Neurosurg 121(4): 810–817. https://doi.org/10.3171/2014.6.JNS132285
39. Chang EF, Kurteff G, Wilson SM (2017) Selective interference with syntactic encoding during sentence production by direct electrocortical stimulation of the inferior frontal gyrus. J Cogn Neurosci 30(3):411–420. https://doi.org/10.1162/jocn_a_01215

40. Corina DP, Gibson EK, Martin R, Poliakov A, Brinkley J, Ojemann GA (2005) Dissociation of action and object naming: evidence from cortical stimulation mapping. Hum Brain Mapp 24(1):1–10. https://doi.org/10.1002/hbm.20063
41. Gras-Combe G, Moritz-Gasser S, Herbet G, Duffau H (2012) Intraoperative subcortical electrical mapping of optic radiations in awake surgery for glioma involving visual pathways: clinical article. J Neurosurg 117(3):466–473. https://doi.org/10.3171/2012.6.JNS111981
42. Ries SK et al (2019) Roles of ventral versus dorsal pathways in language production: an awake language mapping study. Brain Lang 191:17–27. https://doi.org/10.1016/j.bandl.2019.01.001
43. Rofes A, Miceli G (2014) Language mapping with verbs and sentences in awake surgery: a review. Neuropsychol Rev 24(2):185–199. https://doi.org/10.1007/s11065-014-9258-5
44. Glaser WR, Düngelhoff FJ (1984) The time course of picture-word interference. J Exp Psychol Hum Percept Perform 10(5):640–654
45. Lupker SJ (1979) The semantic nature of response competition in the picture-word interference task. Mem Cogn 7(6):485–495. https://doi.org/10.3758/BF03198265
46. Bürki A (2017) Electrophysiological characterization of facilitation and interference in the picture-word interference paradigm. Psychophysiology 54(9):1370–1392. https://doi.org/10.1111/psyp.12885
47. Costa A, Alario FX, Caramazza A (2005) On the categorical nature of the semantic interference effect in the picture-word interference paradigm. Psychon Bull Rev 12(1):125–131
48. Piai V, Roelofs A, Acheson DJ, Takashima A (2013) Attention for speaking: domain-general control from the anterior cingulate cortex in spoken word production. Front Hum Neurosci 7:832. https://doi.org/10.3389/fnhum.2013.00832
49. Piai V, Roelofs A, Jensen O, Schoffelen J-M, Bonnefond M (2014) Distinct patterns of brain activity characterise lexical activation and competition in spoken word production. PLoS One 9(2):e88674. https://doi.org/10.1371/journal.pone.0088674
50. Hickok G, Poeppel D (2004) Dorsal and ventral streams: a framework for understanding aspects of the functional anatomy of language. Cognition 92(1–2):67–99. https://doi.org/10.1016/j.cognition.2003.10.011
51. Rauschecker JP, Tian B (2000) Mechanisms and streams for processing of 'what' and 'where' in auditory cortex. Proc Natl Acad Sci U S A 97(22):11800–11806. https://doi.org/10.1073/pnas.97.22.11800
52. Saur D et al (2008) Ventral and dorsal pathways for language. Proc Natl Acad Sci U S A 105(46):18035–18040. https://doi.org/10.1073/pnas.0805234105
53. Tracy JI, Hampstead BM, Sathian K (2014) Cognitive plasticity in neurologic disorders. Oxford University Press, Oxford
54. Robles SG, Gatignol P, Lehéricy S, Duffau H (2008) Long-term brain plasticity allowing a multistage surgical approach to World Health Organization Grade II gliomas in eloquent areas. J Neurosurg 109(4):615–624. https://doi.org/10.3171/JNS/2008/109/10/0615
55. Ghinda CD, Duffau H (2017) Network plasticity and intraoperative mapping for personalized multimodal management of diffuse low-grade gliomas. Front Surg 4. https://doi.org/10.3389/fsurg.2017.00003
56. Keller CJ, Honey CJ, Mégevand P, Entz L, Ulbert I, Mehta AD (2014) Mapping human brain networks with cortico-cortical evoked potentials. Philos Trans R Soc B Biol Sci 369(1653). https://doi.org/10.1098/rstb.2013.0528
57. Kunieda T, Yamao Y, Kikuchi T, Matsumoto R (2015) New approach for exploring cerebral functional connectivity: review of cortico-cortical evoked potential. Neurol Med Chir (Tokyo) 55(5):374–382. https://doi.org/10.2176/nmc.ra.2014-0388

Chapter 15

Transcranial Magnetic Stimulation Mapping for Perceptual and Cognitive Functions

Samantha Strong and Edward H. Silson

Abstract

Transcranial magnetic stimulation (TMS) is a powerful neuroscience technique that provides a mechanism to study causal brain-behavior relationships in both healthy and damaged/diseased brains with a high-degree of both spatial and temporal precision. In this chapter, we take you through the mechanisms of TMS and discuss practical aspects of using TMS for mapping perceptual and cognitive functions in both basic science and clinical contexts.

Key words Perception, Cognition, Lesion symptom mapping, TMS, Causal mechanism

1 Introduction

This chapter is intended to provide the reader with an overview of the use of transcranial magnetic stimulation (TMS) for mapping perceptual and cognitive functions in both basic science and clinical contexts. The empirical evidence and methodological examples covered in this chapter have been sampled largely from the visual perception literature, meant as a selection and are by no means exhaustive. In order to appreciate the experimental and clinical significance of TMS, it is first vital to gain a handle on the basic components of TMS. A good starting point then is the question: What *is* TMS?

2 What Is TMS?

TMS operates on the principle that a rapidly changing electrical current within a conducting coil can induce a magnetic field, with flux lines (magnetic lines of force/direction) orientated perpendicular to the plane of the coil (Fig. 1a). When the stimulating coil is placed onto the participant's head (i.e., the coil is in contact with

Dorian Pustina and Daniel Mirman (eds.), *Lesion-to-Symptom Mapping: Principles and Tools*, Neuromethods, vol. 180,
https://doi.org/10.1007/978-1-0716-2226-1_15,

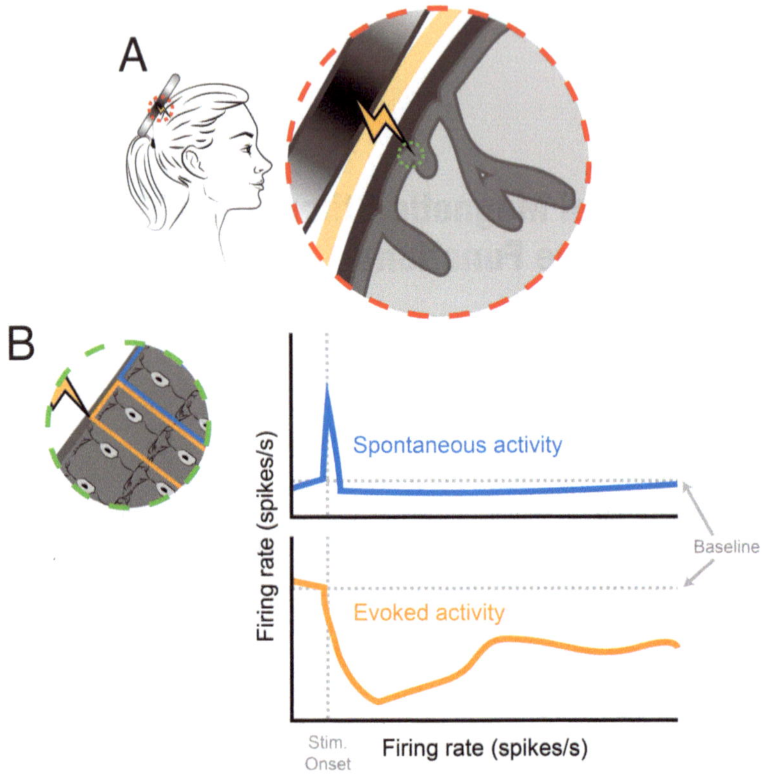

Fig. 1 Schematic of TMS effect. (**a**) During TMS the stimulating coil is placed flush with the participants scalp. The magnetic flux lines extend perpendicular to the orientation of the coil and penetrate unimpeded through the scalp, inducing electrical current in the underlying brain tissue. (**b**) Recorded neuronal effects of TMS of visual cortex. Brief stimulation of visual cortex induced both a sharp but transient increase in the spontaneous firing of non-encoding neurons (blue-line, top) and also a decrease in the firing rate of encoding neurons (orange-line, bottom) that persisted for a longer duration [7]

the scalp), the resulting magnetic field penetrates unimpeded through the skull and into the brain, inducing electrical currents inside the brain that interfere with normal neuronal functioning.

This interference can result in either inhibitory or excitatory effects on neuronal function. Inhibitory effects can be thought of as "transient" functional lesions as they reduce the performance (relative to normal) of the targeted area. However, these effects retain some differences when compared to a "true" cortical lesion because inhibitory TMS reduces sensitivity in the area, whereas cerebral damage would remove the neural capability completely.

Whether this interference results in inhibitory (most commonly associated effect of TMS) or excitatory outcomes appears to depend on a number of factors including the temporal structure of the TMS pulses, the precise timing of these pulses relative to neuronal latency, and the intensity of stimulation. We will discuss these and other experimental factors in the following sections.

3 Types of TMS

There are a variety of ways to administer TMS. The particular paradigm one chooses depends upon (1) the desired impact on neuronal functioning (excitatory/inhibitory) and (2) the type of measurement one is interested in recording (e.g., reaction times, behavioral performance, neuronal spike rates). The following section is meant to provide just a brief description of some of the most commonly employed stimulation paradigms.

3.1 Single-Pulse TMS

The classical design of TMS experiments is to deliver single pulses at specific intervals, which are administered "live" during the performance of a task (e.g., visually discriminating the orientation of a stimulus). Such single-pulse TMS paradigms are largely considered to elicit excitatory effects but may interact with other factors to produce inhibition. Walsh and Cowey [1] demonstrated the excitatory potential of single-pulse TMS by administering TMS pulses to the occipital cortex and measuring the phosphenes (transient, hallucinatory spots of light) elicited by the resultant excitation of visual parts of the brain.

3.2 1 Hz Paradigm/ "Distal" TMS

In contrast, the 1 Hz paradigm, whereby repetitive single pulses (delivered at a frequency of 1 Hz for a fixed duration, e.g., 10 min) are delivered whilst the participant sits passively, is generally considered to *inhibit* neuronal processing. The length of inhibition is difficult to estimate accurately [2], but it is generally considered to be a window approximating somewhere between half and twice the length of the stimulation (e.g., 10 mins of 1 Hz = 5–20 min inhibition window).

3.3 Repetitive TMS (rTMS)

Repetitive TMS (rTMS), at higher frequencies, such as 5 and 10 Hz have been shown to produce both excitatory and inhibitory effects, although inhibitory effects are more commonly reported [3]. During rTMS, pulses are administered in pulse "trains" (e.g., 5 pulses, 1 every 100 ms for 500 ms), separated by inter-train-intervals (of, e.g., > 1 s), and have the advantage that it can be used concurrently with the performance of a task.

3.4 Theta Burst TMS (cTBS)

A fourth commonly used TMS paradigm is continuous theta burst (cTBS). During cTBS, three pulses are typically delivered at 50 Hz in 200 ms, which repeats for up to a minute (900 pulses total). Intermittent TBS (ITBS) involves the same pulse structure (three 50 Hz pulses every 200 ms), but stimulation switches on and off (e.g., 2 s of stimulation, 8 s of rest). Both types of TBS are considered inhibitory and have reported effects lasting well beyond the length of stimulation [4].

4 How Does TMS Work?

How TMS interacts with ongoing neuronal processes in the brain is a fundamental question at the heart of TMS studies. Several TMS experiments have attempted to investigate the impact of TMS on cortical "noise" in humans, with mixed results. Some studies report that application of TMS decreases the strength of neuronal signals [5], whilst others report that it increases the degree of neural noise in the stimulated system [6].

Perhaps the clearest example of the neuronal mechanism of TMS stimulation comes from direct measurements of visual responses in the cat visual cortex [7]. Allen and colleagues applied short TMS pulses to the visual cortex of anesthetized cats ($n = 8$) whilst simultaneously recording neuronal activity and tissue oxygenation. During these experiments, TMS pulses were interleaved with the presentation of a visual stimulus (an orientated grating) and blank periods to account for baseline activity. TMS not only elicited substantial increases in the spontaneous firing or background noise of neurons (~200% increase), which remained elevated for close to 1 min post stimulation (blue-line, Fig. 1b), but also caused a substantial decrease (~50%) in the firing rate of the encoding neurons, that is, those responsive to the visual stimulus (orange-line, Fig. 1b). Such inhibition had a longer temporal envelope than the excitation (~5 min versus 1 min), but nevertheless the net effect of TMS was found to be a significant change in the signal-to-noise ratio (SNR), which is indicative of a functional lesion.

This direct demonstration of the effect of TMS on neuronal firing is important in the context of cognitive explanations of the TMS mechanism, which have generally posited that TMS interferes with ongoing cognitive processes by reducing SNR either via a reduction in the strength of the encoding signal (e.g., firing rate of activated neurons) or via the increase in strength of the noise signal (e.g., increase in firing-rate of non-encoding neurons). Allen and colleagues demonstrated, at least in the cat visual cortex, that the neuronal impact of TMS is both possibilities simultaneously, with the precise effect in a given TMS experiment ultimately depending on the ratio between excitation and inhibition.

5 Spatial Specificity of TMS

The spatial specificity of TMS is often at the forefront of researchers' minds when considering starting a TMS experiment. Indeed, it is very difficult to provide concrete answers regarding the exact anatomical specificity of TMS, as this will vary with coil type, size, stimulation duration, and stimulation intensity (to name but a few factors). It is also the case that TMS can have both local and distal consequences. We will consider each in turn in the following sections.

5.1 Coil Type and Size

When designing a TMS experiment, it is important to consider the necessary requirements, particularly in terms of spatial specificity. For example, if the experiment requires a great deal of functional specificity, then it will be necessary to ensure that the coil produces a narrow enough "beam" (aperture of stimulation) in order to stimulate the function of the targeted area without overlapping with any adjacent cortical areas. One easy way to control this is to choose a TMS coil that is most appropriate. For example, H-coils produce a very wide breadth of stimulation, whereas figure-of-eight coils produce increasingly narrow "beams" as the distance between the windings decreases in physical size [8–10].

5.2 Local Effects of TMS

One hallmark of TMS is that it can elicit spatially specific and transient changes to normal neuronal functioning and thus provides a method for producing localized and transient *virtual lesions* by stimulating small areas of cortex, providing valuable information about normal cognitive and perceptual functions. But how spatially specific is it?

It is perhaps more useful to consider *functional specificity* when starting a TMS protocol. By *functional specificity*, we refer to the idea that each individual brain region (target site for the TMS) has a specific function. TMS is able to interfere with these regions providing the distance between target sites is large enough that one can reliably measure differential effects when applying TMS, thereby targeting one *specific* function of the cortex. An early example of such functional specificity is [11] whereby stimulation of the hand representation of the motor cortex induced movement of the contralateral hand, but not corresponding movement of the lower arm or face, which are adjacent to the hand representation in motor cortex. Whether or not these regions were stimulated at a subthreshold level remains an open question, but for how the effect of TMS was operationalized in that study (i.e., eliciting a movement of the contralateral hand), such stimulation was *functionally specific*. Similarly, in the visual cortex, Walsh and Cowey [1] demonstrated that TMS of targets separated by 1 cm on the scalp surface elicited phosphenes at increasingly eccentric visual field positions, coinciding with the retinotopic organization of visual cortex. Again, this example serves as an excellent source of TMS' functional specificity (~1 cm).

The advent of functional magnetic resonance imaging (fMRI) has enabled researchers to push the boundaries of the spatial specificity of TMS in order to demonstrate that stimulation of high proximity targets (~10 mm in Euclidean space, possibly closer in cortex) can differentially alter performance on complex visual perception tasks. One study [12] stimulated three close proximity targets on the lateral surface of the occipital lobe that were defined as key locations of (a) object-selectivity (lateral occipital cortex), (b) face-selectivity (occipital face area), and (c) body-selectivity (extrastriate body area) in each individual participant. These target sites were then stimulated with TMS during object, face, and body discrimination experiments. A series of double dissociations were identified following TMS of these specific targets and the behavioral discrimination of their preferred categories (e.g., TMS of object-selective cortex impaired object but not face discrimination). This work demonstrated the feasibility of stimulating very close proximity targets within an individual and underscored the importance, for certain types of questions, for individually analyzing fMRI data for target localization.

A second study [13] employing a similar paradigm selectively stimulated adjacent retinotopic maps, defined on the basis of prior fMRI retinotopic mapping experiments. In contrast to the peak selective voxel used previously [12], the centers of mass in each retinotopic map were selected as targets, and these were separated on average by ~10 mm. Like Pitcher and colleagues' prior work, this study identified a double dissociation. Stimulation of the two retinotopic maps elicited specific deficits in orientation (lateral occipital area 1, LO1) and shape discrimination (lateral occipital area 2, LO2), respectively. These studies, and many more [14–17], extended the work of Walsh and Cowey [1] to demonstrate the *functional specificity* of TMS to elicit differential behavioral effects for different perceptual tasks.

5.3 Distal Effects of TMS

Until here, we have discussed the effects of TMS on neuronal firing [7] and a handful of studies that demonstrate the local spatial specificity of TMS [1, 12, 13] on perceptual performances. However, TMS can also produce distal cortical effects, and these need to be considered carefully. Stimulation of a given brain area will likely not only affect the neuronal firing of regions directly under and immediately adjacent to the coil, but it also has the potential to affect regions that are most heavily connected to the stimulated region. This means the inter- and intra-hemispheric connections of each area are important considerations when designing a TMS experiment.

The potential distal effects of TMS are of considerable theoretical importance to cognitive/perceptual experiments. Whilst many TMS studies have demonstrated a remarkable degree of spatial specificity in terms of primary stimulation, it is worth noting that TMS can also produce distal effects in anatomically connected regions that, in cortical terms, fall far from the stimulated region. For example, Paus and colleagues [18] showed that changes in local cerebral blood flow following TMS of the frontal eye fields not only produce a change underneath the TMS coil (local effect) but also in posterior parietal regions of the ipsilateral hemisphere (TMS'd hemisphere) known to be strongly anatomically connected with the frontal eye fields (distal effect). Such distal effects are also not necessarily restricted to the TMS'd hemisphere. Illmoniemi et al. [19] stimulated either the visual or motor cortices of one hemisphere whilst simultaneously recording responses using electroencephalography (EEG). Following TMS, EEG responses from the stimulated area were measured within a few milliseconds, but responses in the homogenous region of the contralateral hemisphere (non TMS'd hemisphere) were observable ~20–30 ms later.

Recently, very distal effects of TMS have been reported [20]. Here, researchers first used resting-state functional connectivity in fMRI to identify the posterior parietal region showing the strongest functional connectivity with the hippocampus. cTBS was then delivered to this posterior parietal site over the course of several days and were reported to increase the connectivity between the stimulated area and the hippocampus.

6 How Intense Should the Stimulation Be?

How does one decide the level of stimulation appropriate, necessary, and more importantly safe, to investigate the impact of TMS-induced virtual lesions on perceptual and cognitive functioning? Several methodological papers exist that outline what stimulation parameters are considered safe, but assuming once selected TMS protocol falls within these guidelines, there are three main options available to researchers: (a) stimulate all participants at a fixed intensity relative to some internal measure such as motor threshold; (b) stimulate all participants at a fixed machine output (absolute intensity), which will be above some participants' motor threshold and below others; or (c) stimulate at an intensity which is determined by calculating the distance between the coil and the cortical target [21]. From this, one could argue that (c) is the most principled, but most labour-intensive method, whereas (b) might be the quickest and most simple to implement.

The most common method for selecting stimulus intensity is arguably (a)—percentage of motor threshold (MT), but this is likely a historical factor. Indeed, there is little empirical evidence to suggest that MT is relevant anywhere outside of the motor cortex. For instance, Stewart and colleagues [22] used single-pulse TMS to establish and compare both MT and phosphene threshold in 15 participants, with a re-test in seven participants a week later. These data had several important implications: (1) phosphene threshold was more variable (phosphene threshold range 50%, min = 35%, max = 85%) than MT (MT threshold range 23%, min = 22%, max = 45%) across participants, (2) thresholds were stable across time within participants (Pearson's correlation coefficient of $r = 0.7$ & $r = 0.8$, for phosphene and MT, respectively), and (3) there appeared no clear positive relationship across measurements ($r = 0.3$). Thus, MT, although more stable, and slightly more reliable than phosphene threshold, may be inappropriate for determining the simulation intensity for regions outside of motor cortex.

7 Experimental Control

Perhaps the greatest benefit of TMS is a reversal of the lesion-function mapping, which dominates neuropsychological studies. With TMS, one can select the lesion-site based on the research question without having to select the questions based on the location of the lesion. Every aspect of the TMS experiment is under experimental control, from the method of localization (fMRI-guided to distances on the scalp) to the orientation of the coil, including the area of stimulation as well as the intensity, structure, and duration of the pulse trains. This degree of experimental control is attractive to researchers and offers an alternative approach to inferring causal relationships from the spatial overlap of lesions across patients.

A welcome advancement in TMS experimenter control is the availability of neuronavigation software (e.g., BRAINSIGHT , Rogue Research). Such software allows one to track the position of both the subject and stimulating coil in real time with respect to the subjects' anatomy. The real-time tracking of both coil and subjects allow for a number of metrics to be computed, and these can be analyzed to provide additional information regarding the precision of TMS delivery. Some examples include the coil-target distance in Euclidean space and the coil-target angle.

Given that the target location is defined as a fixed location in the brain (and thus cannot move), the coil-target distance provides an index of the precision of the stimulation and also the quality of the calibration (which determines the hotspot of the coil from which the distance metrics are calculated). In studies where the same target may be stimulated in different sessions, this distance should not change, and this measurement provides an account of that. Similarly, the coil-target angle provides a summary statistic for precision. As mentioned at the beginning of the chapter, the flux lines of the magnetic field produced by the stimulating coil run perpendicular to the coil's plane. Ideally then, the coil should be placed flush with the scalp ensuring that the coil-target angle is 90° or as close as possible. Again, this is potentially most important when stimulating either the same location in different sessions or different target sites. In an ideal situation, such angle measurements would be equivalent in all conditions.

These measurements provide researchers with a high-level of experimental control and can be used to facilitate the interpretation of TMS-induced effects on performance. For example, suppose a study was conducted in which the experimenter targets two regions (A & B), whilst subjects are performing two tasks (X & Y)—a total of four sessions. It is possible that site A was stimulated precisely during task X, but not Y and site B were stimulated precisely during Task Y, but not X. The result might very well look like a classic double dissociation but could just as easily represent precise stimulation during two out of four sessions. One study [13] analyzed these summary statistics to rule out the above possibility as an explanation for the observed double dissociation.

8 The Control Site

Of equal importance is the selection of the control site for stimulation. The interpretability of TMS results are due, in large part, to the appropriateness of the control site. A study comparing visual cortex stimulation during a visual task against stimulation of the auditory cortex tells us little more about the perceptual processes involved beyond that visual cortex is needed for visual tasks.

The vertex is often selected as a control site, and this serves as a reasonable control condition for the sensation of stimulation and the noise that the TMS coil makes, but it is a recognizable control paradigm in participants who have an awareness of cortical anatomy and does not directly stimulate any cortex. A better, more convincing control is an active control site, and ideally, such a control site would be within the same "network" as the target area (e.g., both active and control sites are within visual cortex, if studying a visual task).

Again, we can look at the vision literature for excellent examples of active control sites that have strengthened considerably the inferences made. Pitcher et al. [12] demonstrated, through a series of double dissociations, that stimulation of three independent sites induced selective impairments for the stimulated regions' preferred category. In each case, a different category-selective area was the active control site. The beauty of this paradigm was that both target and control sites were within the same network (e.g., visual cortex) and within close proximity, which largely controlled for scalp twitches and noise of the TMS coil. The fact that selective impairments were induced despite the active control sites being in close proximity bolstered the evidence in favor of specialized processing within those regions.

Similarly, Silson et al. [13] selected an active control site by moving the TMS coil closer to the primary visual cortex (V1) by the same distance that separated the two sites of interest (LO1 and LO2, ~10 mm). In this case, the location of the active control site was crucial to the interpretation of the data. Based on prior fMRI work [23], it was hypothesized that LO1 and LO2 would be differentially sensitive to stimulus orientation and shape, respectively. But anatomically LO1 falls closer to orientation-selective V1 than LO2, and so in order to conclude that TMS of LO1 selectively impaired orientation discrimination, it was necessary to demonstrate that such an effect was not simply due to LO1's closer proximity to V1. The lack of an effect at the control site on either task significantly bolstered the evidence in favor of independent computations of orientation and shape in LO1 and LO2, respectively.

9 Can the Brain Be TMS'd Everywhere?

Unfortunately, not all cortical areas are amenable to TMS intervention. Estimates of the depth of penetration from TMS pulses vary [24], but in general are thought to directly stimulate only a few cm into the cortex. Of course, distal effects of TMS are possible, with the stimulated area altering the functioning of downstream regions, but the precise nature of these distal effects is still not clear.

Given the relatively shallow depth of direct local stimulation, much of the cerebral cortex is thus essentially out of range for TMS. Regions along the medial wall, for example, cannot be stimulated directly with TMS. Likewise, much of the ventral surface of the brain is beyond the reach of TMS (although *see* [1] and [25] for examples of posterior ventral regions being stimulated). One of the considerations here is how much cortex one is comfortable stimulating in order to get at the area of interest. Of the easily stimulated regions of cortex, those directly underneath the skull, stimulation of some may be experienced as more uncomfortable than others, particularly if such stimulation inadvertently activates superficial nerves on the scalp. This could lead to unpleasant activations of facial nerves, jaw clenching, or blinking, depending on the location of the coil.

Despite not being able to stimulate medial or deep structures directly, relatively recent studies have employed TMS in an attempt to influence the activity of deep structures, such as the hippocampus [20]. Here, the paradigm was first to use fMRI resting-state functional connectivity in order to identify the superficial node (i.e., voxel on the cortical surface) that exhibited the strongest connectivity with the hippocampus. This voxel was then targeted for TBS across 5 consecutive days. After which, changes in the hippocampus and more broadly in the memory network were reported. It is possible that similar paradigms might be explored for other networks, for example, stimulation of pSTS or OFA to measure response in the amygdala [26].

10 Where to Stimulate?

The question of where to stimulate is a central issue in TMS experiments, but one that allows a good deal of experimental control. Indeed, as mentioned before, perhaps the greatest benefit of TMS is a reverse of the function-lesion mapping. With TMS, the location to stimulate can be decided upon based on the question being asked. These focal, controlled TMS effects provide a powerful complement to the larger lesions defined by disease etiology.

So, location matters, of course, but exactly how much? For example, it is not necessarily the case that all TMS experiments require individual target sites defined on the basis of prior fMRI data. Indeed, for some research questions, gross anatomical landmarks or rough approximations based on distances across the scalp are sufficient. An excellent empirical exemplification of this is provided by Sack et al. [27]. In this study, the researchers investigated the well-known impact of right intraparietal TMS on numerical processing. The crucial manipulation here was in the method of localization which were (a) individually specific fMRI scans, (b) individually specific anatomical scans, (c) group-based Talairach coordinates, or (d) position P4 on a standard 10–20 EEG system. A significant effect of TMS was observed with each localization methods, but the number of required subjects increased markedly from $n = 5$ in the case of individually specific fMRI targets to $n = 47$ for the P4 localization method ($n = 9$ for anatomy alone and $n = 13$ for Talairach coordinates, respectively). All of these localization methods were *correct* in so much as a significant effect of TMS was observed, but the sharp increase in the number of required subjects when moving away from at least individually specific anatomical scans is noteworthy.

11 How Long Do the Effects Last and When Should One Stimulate?

One of the major strengths of TMS is that the impact of stimulation (whether excitatory or inhibitory) is transient and largely reversible. That being said, the different TMS protocols have different temporal durations, and the following section aims to provide an overview and some empirical evidence to highlight this aspect.

The temporal duration of TMS effects depends, in part, on the scale of the outcome of your measurement. As Allen and colleagues [7] demonstrated, direct neuronal recordings may show a longer-lasting effect of TMS pulses (e.g., ~5 min suppression of encoding neurons) than, for example, behavioral measurements (e.g., RT, per cent correct, phosphene detection). In addition to considering the temporal duration of the intended TMS effect, the temporal specificity of TMS delivery is also critical. The temporal window of TMS, that is, the optimal time to stimulate, again depends on a number of factors and can be in the range of milliseconds to minutes, but it is worth noting that how one determines a "TMS effect" will also alter this. The onset, offset, recovery time of neurons, and critical period of the task will all interact to determine the most effective temporal window of TMS.

Early TMS work by Amassian and colleagues [28] demonstrated the relationship between the temporal window of TMS delivery, the task being performed, and the measured response. Subjects were presented with trigrams (three randomly generated letters) at fixation and later asked to recall each letter. TMS was delivered to the back of the head at 11 different stimulus onset asynchronies (SOA) from coincident with the visual stimulus ($t = 0$) to 200 ms post stimulus presentation ($t = 200$ ms) in steps of 20 ms. The resulting performance curves demonstrated a sharp drop in performance when TMS was delivered between 80 and 120 ms post stimulus presentation, with performance relatively unaffected outside of this time window.

However, such a time window may not always be maximally disruptive. For instance, McKeefry and colleagues [14] investigated the impact of TMS of V5/MT on motion discrimination thresholds. In these experiments, participants performed a two alternative forced choice paradigm, in which they had to respond whether a test grating was drifting faster or slower than a reference grating. They found that TMS of V5/MT caused a perceptual slowing of these stimuli. In other words, a faster grating was required for subjects to match it to the reference during TMS of V5/MT. Interestingly, unlike the Amassian study [28], when different SOAs were tested, this effect was found to be maximized when test stimulus and TMS pulses were coincident with one another.

12 Language Mapping in Neurotypicals with TMS

Previous paragraphs focused primarily on evidence gleaned from TMS studies of visual cortex and visual perception, but TMS has also made significant contributions to our understanding of other brain systems and behaviors and none more so than language.

In particular, the *virtual lesion* capability of TMS has contributed novel insights into the functional anatomy of one of the most well-studied cortical regions—Broca's area. Located within the left inferior frontal gyrus (LIFG), Broca's area has long been associated with speech production and syntactic processing. When fMRI studies suggested a possible rostro-caudal division of labor within LIFG for sematic and phonological processing, respectively, TMS was able to provide causal evidence in favor of such a dissociation [29]. For instance, stimulation of rostral LIFG during a simple semantic decision-making task (e.g., deciding whether a visually presented word like "dog" referred to a natural object) significantly increased reaction times relative to no stimulation, but not when focusing on visual properties of the presented words [30, 31]. In contrast, stimulation of caudal LIFG increased error rates during a phonological working memory task but left performance on a visual working memory task intact [32].

These prior single dissociation studies were complemented by a later study specifically designed to test the LIFG division of labor hypotheses directly [33]. Here, participants viewed two letter strings simultaneously and performed a series of tasks deciding whether or not the two conveyed the same meaning, sound, or visual percept. Crucially, stimulation of rostral LIFG increased reaction times when focusing on the meaning of words, but not their sound, whereas stimulation of caudal LIFG increased reaction times when focusing on the sound of words but not their meaning—a double dissociation. The visual task was unaltered by stimulation of either LIFG site. Here, TMS was used to provide precise spatial and causal evidence of the roles played by two close proximity sites within LIFG. It is important to note that although this double dissociation was predicted based on prior fMRI results, it required the use of TMS to confirm it [29].

13 Pre-surgical Mapping

In recent years, TMS has moved beyond the sphere of basic science and has been employed as a technique to aid pre-surgical mapping of patients undergoing surgical resection. The following sections will provide a summary, albeit brief, of some of the results of such attempts. Before covering the literature, it is worth noting that a common thread among these papers is that they stress that pre-surgical mapping with TMS is not meant to, nor should it, replace the gold standard of direct cortical stimulation (DCS) for pre-surgical mapping (*see* Chapter 14) but that it could prove a suitable alternative if DCS is not feasible and at the very least could augment DCS data. The use of TMS for pre-surgical mapping has been applied to both language-eloquent and motor-eloquent brain regions and has also been suggested as a valuable tool for measuring potential cortical reorganization following surgery. We will look at examples of each in the following sections.

14 Language-Eloquent Mapping

Ille and colleagues [34] compared directly the impact of fMRI-guided TMS language mapping with DCS—the current gold standard for functional mapping. Thirty-five patients presenting with left-sided perisylvian lesions (tumors) initially underwent structural and functional MRI, which included an object-naming to task to identify language-eloquent regions. During TMS sessions, each patient's resting-motor-threshold was determined and used as a baseline measurement during subsequent TMS mapping sessions, consisting of an object-naming task of 131 colored everyday objects. Initially, three stimulation protocols were employed in which rTMS pulse trains consisting of either five or seven pulses, were delivered at either 5 or 7 Hz and at 100% RMT. This (a-b-c) paradigm was repeated at a higher threshold (110–120%) if object-naming was not clearly altered by any of the preceding paradigms.

Prior to stimulation, baseline recordings were performed twice. Patients were instructed to name the objects as efficiently and accurately as possible, with the number of baseline errors recorded. Misnamed pictures were discarded, and remaining pictures were presented time locked to a train of rTMS pulses. 80–120 sites were stimulated three times (max 360 pulse trains). Error rates were binned into 12 5% bins from 0 to 50%. The direct cortical stimulation (50 Hz 0–20 mA for 4 s) was delivered using the same pictures, and cortical sites were stimulated three times.

For both types of stimulation, language errors were first categorized blinded to either the stimulation or tumor site and then visualized onto a parcellation of the cerebral cortex. A detailed analysis revealed that TMS led to high predictive value of DCS effects, such that TMS sites that produced naming errors (>15% of the time) also produced errors under DCS. These data suggest that the use of rTMS, particularly when augmented by prior fMRI data, is a valuable pre-surgical technique for functional mapping.

Ille and colleagues [35] extended their previous work to investigate the impact of resection of language-eloquent brain areas based purely on rTMS language mapping in the absence of subsequent DCS during awake surgery. Four patients with left-sided perisylvian lesions who did not qualify for awake surgery underwent rTMS language mapping. Two outcomes were measured: the extent of the resection and the functional outcome of patients.

Forty-six pre-determined stimulation sites were tested three times each with the same procedure as described used previously [34] and described above, but here, both hemispheres were stimulated. Preoperatively as well as 5 days and 3 months post-surgery, functional outcomes were assessed using the authors standard aphasia grading (0—no impairment to 3—severe impairment). The actual rTMS mapping was performed 1–8 days prior to surgery.

Three out of four patients (75%) did not suffer any kind of language impairment postoperatively, with language functions unaffected by the lesion resection. Postoperative assessments of the extent of the resection also showed complete resections in these cases. One patient, who suffered from a mild language impairment preoperatively, presented with a transient further impairment of language function 5 days postoperatively. These two studies provide empirical support for the use of TMS as a feasible method for pre-surgical mapping of language eloquent brain areas when the gold standard DCS is not possible.

15 Motor-Eloquent Mapping

TMS has also been explored as an accurate and safe tool for pre-surgical mapping of motor-eloquent regions. For instance, one study [36] reported that pre-surgical mapping with TMS, whilst recording simultaneously from a dense cortical sampling array, produced a mean difference with the intraoperative hotspot of <15 mm. Such close correspondence between the intracranial hotspot and TMS hotspot provide important demonstrations of TMS's *functional specificity* and underscore the utility of TMS for performing pre-surgical mapping, when the gold standard is not feasible. Further, Tarapore et al. [37] reported that in a large cohort of patients ($n = 733$), of which half had prior seizures, pre-surgical mapping with TMS did not induce any instances of additional seizure and was well tolerated by the majority of patients.

Raffa et al. [38] conducted a systematic review of studies analyzing the impact of TMS-based motor mapping on surgery of patients affected by motor-eloquent brain tumors, compared to a series of patients without TMS mapping. The authors considered four outcome measures: occurrence of postoperative new permanent motor deficits (eight studies), gross total resection rate (seven studies), size of craniotomy (four studies), and length of surgery (three studies). TMS significantly reduced the occurrence of postoperative new motor deficits, increased the gross resection rate, reduced the size of the craniotomy, and also reduced surgery time by ~10 mins, although not significantly.

TMS has also been used to measure potential functional plasticity following surgery [39]. Here, rTMS was used to map cortical motor representations in 22 patients with gliomas affecting the precentral gyrus preoperatively and 2–42 months postoperatively. The authors measured the location changes of the primary motor area (M1), defined as the hotspots and centers of mass. On average, hotspots showed an average shift of ~5 mm medially and 10 mm anteriorly. Similarly, centers of mass shifted by 5 and 8 mm, respectively. Overall 9/16 (56%) patients showed a functional shift of >10 mm. Several important factors should be considered in light of this result, including the accuracy of TMS stimulation, and the angle of the coil, but those caveats notwithstanding, the results offer support for the use of TMS in examining functional reorganization as a result of surgical resection.

16 Summary

Over the past few decades, the frequency and importance of TMS for mapping perceptual and cognitive functions have increased dramatically. Advances in software and methodological approaches now enable researchers to stimulate more precisely and selectively than ever before. As the final section outlined, the influence of TMS now extends far beyond basic science and is an important clinical tool for pre-surgical mapping. It is hoped that researchers will continue to use this technique to answer fundamental questions about the human brain.

References

1. Walsh V, Cowey A (2000) Transcranial magnetic stimulation and cognitive neuroscience. Nat Rev Neurosci 1:73–80. https://doi.org/10.1038/35036239
2. Matsuyoshi D, Hirose N, Mima T et al (2007) Repetitive transcranial magnetic stimulation of human MT+ reduces apparent motion perception. Neurosci Lett 429:131–135. https://doi.org/10.1016/j.neulet.2007.10.002
3. Rossini PM, Rossi S (2007) Transcranial magnetic stimulation: diagnostic, therapeutic, and research potential. Neurology 68:484–488. https://doi.org/10.1212/01.wnl.0000250268.13789.b2
4. Huang Y-Z, Edwards MJ, Rounis E et al (2005) Theta burst stimulation of the human motor cortex. Neuron 45:201–206. https://doi.org/10.1016/j.neuron.2004.12.033
5. Ruzzoli M, Marzi CA, Miniussi C (2010) The neural mechanisms of the effects of transcranial magnetic stimulation on perception. J Neurophysiol 103:2982–2989. https://doi.org/10.1152/jn.01096.2009
6. Harris JA, Clifford CWG, Miniussi C (2008) The functional effect of transcranial magnetic stimulation: signal suppression or neural noise generation? J Cogn Neurosci 20:734–740. https://doi.org/10.1162/jocn.2008.20048
7. Allen EA, Pasley BN, Duong T, Freeman RD (2007) Transcranial magnetic stimulation elicits coupled neural and hemodynamic consequences. Science 317:1918–1921. https://doi.org/10.1126/science.1146426
8. Thielscher A, Kammer T (2004) Electric field properties of two commercial figure-8 coils in TMS: calculation of focality and efficiency. Clin Neurophysiol 115:1697–1708. https://doi.org/10.1016/j.clinph.2004.02.019
9. Zangen A, Roth Y, Voller B, Hallett M (2005) Transcranial magnetic stimulation of deep brain regions: evidence for efficacy of the H-coil. Clin Neurophysiol 116:775–779. https://doi.org/10.1016/j.clinph.2004.11.008
10. Roth Y, Amir A, Levkovitz Y, Zangen A (2007) Three-dimensional distribution of the electric field induced in the brain by transcranial magnetic stimulation using figure-8 and deep H-coils. J Clin Neurophysiol 24:31–38. https://doi.org/10.1097/WNP.0b013e31802fa393
11. Barker AT, Jalinous R, Freeston IL (1985) Non-invasive magnetic stimulation of human motor cortex. Lancet 325(8437):1106–1107
12. Pitcher D, Charles L, Devlin JT et al (2009) Triple dissociation of faces, bodies, and objects in extrastriate cortex. Curr Biol 19:319–324. https://doi.org/10.1016/j.cub.2009.01.007
13. Silson EH, McKeefry DJ, Rodgers J et al (2013) Specialized and independent processing of orientation and shape in visual field maps LO1 and LO2. Nat Neurosci 16: 267–269. https://doi.org/10.1038/nn.3327
14. McKeefry DJ, Burton MP, Vakrou C et al (2008) Induced deficits in speed perception by transcranial magnetic stimulation of human cortical areas V5/MT+ and V3A. J Neurosci 28:6848–6857. https://doi.org/10.1523/JNEUROSCI.1287-08.2008
15. McKeefry DJ, Gouws A, Burton MP, Morland AB (2009) The noninvasive dissection of the human visual cortex: using fMRI and TMS to study the organization of the visual brain. Neuroscientist 15:489–506. https://doi.org/10.1177/1073858409334424
16. Strong SL, Silson EH, Gouws AD et al (2017) A direct demonstration of functional differences between subdivisions of human V5/MT+. Cereb Cortex 27:1–10. https://doi.org/10.1093/cercor/bhw362
17. Strong SL, Silson EH, Gouws AD et al (2019) An enhanced role for right hV5/MT+ in the

analysis of motion in the contra- and ipsi-lateral visual hemi-fields. Behav Brain Res 372: 112060. https://doi.org/10.1016/j.bbr.2019.112060
18. Paus T, Jech R, Thompson CJ et al (1997) Transcranial magnetic stimulation during positron emission tomography: a new method for studying connectivity of the human cerebral cortex. J Neurosci 17:3178–3184. https://doi.org/10.1523/JNEUROSCI.17-09-03178.1997
19. Ilmoniemi RJ, Virtanen J, Ruohonen J, Karhu J, Aronen HJ, Näätänen R, Katila T (1997) Neuronal responses to magnetic stimulation reveal cortical reactivity and connectivity. Neuroreport 8(16):3537–3540
20. Wang JX, Rogers LM, Gross EZ et al (2014) Targeted enhancement of cortical-hippocampal brain networks and associative memory. Science 345:1054–1057. https://doi.org/10.1126/science.1252900
21. Stokes MG, Chambers CD, Gould IC et al (2005) Simple metric for scaling motor threshold based on scalp-cortex distance: application to studies using transcranial magnetic stimulation. J Neurophysiol 94:4520–4527. https://doi.org/10.1152/jn.00067.2005
22. Stewart LM, Walsh V, Rothwell JC (2001) Motor and phosphene thresholds: a transcranial magnetic stimulation correlation study. Neuropsychologia 39:415–419. https://doi.org/10.1016/S0028-3932(00)00130-5
23. Larsson J, Heeger DJ (2006) Two retinotopic visual areas in human lateral occipital cortex. J Neurosci 26:13128–13142. https://doi.org/10.1523/JNEUROSCI.1657-06.2006
24. Hilgetag CC, Théoret H, Pascual-Leone A (2001) Enhanced visual spatial attention ipsilateral to rTMS-induced "virtual lesions" of human parietal cortex. Nat Neurosci 4: 953–957. https://doi.org/10.1038/nn0901-953
25. Cohen D, Goddard E, Mullen KT (2019) Reevaluating hMT+ and hV4 functional specialization for motion and static contrast using fMRI-guided repetitive transcranial magnetic stimulation. J Vis 19:11. https://doi.org/10.1167/19.3.11
26. Pitcher D, Japee S, Rauth L, Ungerleider LG (2017) The superior temporal sulcus is causally connected to the amygdala: a combined TBS-fMRI study. J Neurosci 37:1156–1161. https://doi.org/10.1523/JNEUROSCI.0114-16.2016
27. Sack AT, Cohen Kadosh R, Schuhmann T et al (2009) Optimizing functional accuracy of TMS in cognitive studies: a comparison of methods. J Cogn Neurosci 21:207–221. https://doi.org/10.1162/jocn.2009.21126
28. Amassian VE, Cracco RQ, Maccabee PJ et al (1989) Suppression of visual perception by magnetic coil stimulation of human occipital cortex. Electroencephalogr Clin Neurophysiol Potentials Sect 74:458–462. https://doi.org/10.1016/0168-5597(89)90036-1
29. Devlin JT, Watkins KE (2007) Stimulating language: insights from TMS. Brain 130: 610–622. https://doi.org/10.1093/brain/awl331
30. Devlin JT, Matthews PM, Rushworth MFS (2003) Semantic processing in the left inferior prefrontal cortex: a combined functional magnetic resonance imaging and transcranial magnetic stimulation study. J Cogn Neurosci 15: 71–84. https://doi.org/10.1162/089892903321107837
31. Köhler S, Paus T, Buckner RL, Milner B (2004) Effects of left inferior prefrontal stimulation on episodic memory formation: a two-stage fMRI—rTMS study. J Cogn Neurosci 16:178–188. https://doi.org/10.1162/089892904322984490
32. Nixon P, Lazarova J, Hodinott-Hill I et al (2004) The inferior frontal gyrus and phonological processing: an investigation using rTMS. J Cogn Neurosci 16:289–300. https://doi.org/10.1162/089892904322984571
33. Gough PM (2005) Dissociating linguistic processes in the left inferior frontal cortex with transcranial magnetic stimulation. J Neurosci 25:8010–8016. https://doi.org/10.1523/JNEUROSCI.2307-05.2005
34. Ille S, Sollmann N, Hauck T et al (2015) Combined noninvasive language mapping by navigated transcranial magnetic stimulation and functional MRI and its comparison with direct cortical stimulation. J Neurosurg 123: 212–225. https://doi.org/10.3171/2014.9.JNS14929
35. Ille S, Sollmann N, Butenschoen VM et al (2016) Resection of highly language-eloquent brain lesions based purely on rTMS language mapping without awake surgery. Acta Neurochir 158:2265–2275. https://doi.org/10.1007/s00701-016-2968-0
36. Lefaucheur J-P, Picht T (2016) The value of preoperative functional cortical mapping using navigated TMS. Neurophysiol Clin Neurophysiol 46:125–133. https://doi.org/10.1016/j.neucli.2016.05.001
37. Tarapore PE, Picht T, Bulubas L et al (2016) Safety and tolerability of navigated TMS for preoperative mapping in neurosurgical

patients. Clin Neurophysiol 127:1895–1900. https://doi.org/10.1016/j.clinph.2015.11.042

38. Raffa G, Scibilia A, Conti A et al (2019) The role of navigated transcranial magnetic stimulation for surgery of motor-eloquent brain tumors: a systematic review and meta-analysis. Clin Neurol Neurosurg 180:7–17. https://doi.org/10.1016/j.clineuro.2019.03.003

39. Conway N, Wildschuetz N, Moser T et al (2017) Cortical plasticity of motor-eloquent areas measured by navigated transcranial magnetic stimulation in patients with glioma. J Neurosurg 127:981–991. https://doi.org/10.3171/2016.9.JNS161595

Correction to: Voxel-Based Lesion Symptom Mapping

Juliana V. Baldo, Maria V. Ivanova, Timothy J. Herron, Stephen M. Wilson, and Nina F. Dronkers

Correction to:
Chapter 5 in: Dorian Pustina and Daniel Mirman (eds.),
***Lesion-to-Symptom Mapping: Principles and Tools*, Neuromethods, vol. 180,**
https://doi.org/10.1007/978-1-0716-2225-4_5

This chapter was mistakenly published with mistake in chapter 5. The Figure 4 was incorrectly printed. The correct image is mentioned below:

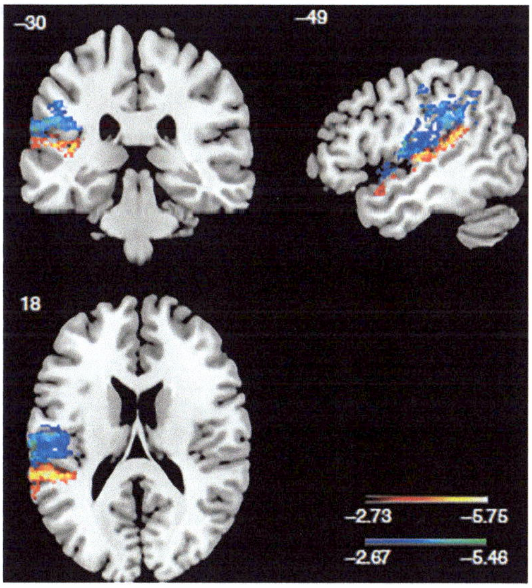

The book has also been updated to reflect this change

The updated original version of this chapter can be found at
https://doi.org/10.1007/978-1-0716-2225-4_5

Dorian Pustina and Daniel Mirman (eds.), *Lesion-to-Symptom Mapping: Principles and Tools*, Neuromethods, vol. 180,
https://doi.org/10.1007/978-1-0716-2225-4_16,

Appendix 1: Lesion-Symptom Mapping—Tools and Examples

Introduction

This appendix will provide six software usage examples for lesion to symptom mapping analyses. All the examples use the same dataset of 131 lesions and one simulated behavioral score (download at https://github.com/dorianps/LESYMAP/tree/master/inst/extdata). The order of the examples is the following (authors listed under each example):

1. NPM software
   Christoph Sperber, University of Tübingen, Tübingen, Germany
2. VOXBO software
   Olufunsho K. Faseyitan, University of Pennsylvania, Philadelphia, PA, USA
3. SVR-LSM GUI software
   Andrew T. DeMarco, Georgetown University Medical Center, Washington, DC, USA
4. NiiStat software
   Stefan Smaczny, University of Tübingen, Tübingen, Germany
   Hannah Rosenzopf, University of Tübingen, Tübingen, Germany
   Lisa Röhrig, University of Tübingen, Tübingen, Germany
5. LESYMAP software
   Melissa Thye, University of Edinburgh, Edinburgh, United Kingdom
6. NeMo software

Dorian Pustina and Daniel Mirman (eds.), *Lesion-to-Symptom Mapping: Principles and Tools*, Neuromethods, vol. 180, https://doi.org/10.1007/978-1-0716-2225-4, © Springer Science+Business Media, LLC, part of Springer Nature 2022

Amy Kuceyeski, Weill Cornell Medical College, New York, NY, USA
Keith Jamison, Weill Cornell Medical College, New York, NY, USA

Lesion-Behavior Mapping Using NPM

NPM is a stand-alone executable for mass-univariate voxel-based lesion-behavior mapping. It is named after non-parametric mapping (NPM; Rorden et al. [1]) and performs both parametric and non-parametric voxelwise analyses. It is provided together with MRIcron (https://www.nitrc.org/projects/mricron). Note that it is not included in the 2019 version of MRIcron, but only in legacy versions as the 2010 version (for Windows, download at https://www.nitrc.org/frs/download.php/2597/mricrond.zip; for other versions, search at https://www.nitrc.org/frs/?group_id=152). NPM is generally able to perform analyses for both binary and continuous imaging data and both binary and continuous behavioral variables. The present tutorial guides through an analysis of binary lesion maps and a continuous behavioral variable.

What We Need

- A sample of binary lesion maps that were normalized to the same brain space; valid are the different NIFTI data formats used by MRIcron (file extensions .nii/.hdr+.img /.voi).
- The behavioral scores; importantly, NPM performs one-sided tests and assumes that higher scores indicate better (=less pathological) performance. In this case, NPM's positive results identify for brain regions where damage leads to more severe pathology, as it is almost always intended.
- Download and unzip MRIcron 2010. Execute mricron/npm.exe:
  - This opens the graphical user interface.

Design Specification

The first step in any analysis of binary images is the design specification:

- Go to VLSM/Design.
- Click on the "Design" panel.
- Set the number of predictors to "1."
- Click on the "Select Images" panel. Use the explorer menu to select all lesions to be included in the analysis.
- The number next to "Ignore voxels damaged in less than N%" excludes rarely damaged voxels from the analysis. It is given in per cent; thus, if we only want to include voxels damaged in at least 10 patients in a sample of 130 patients, we have to calculate

the corresponding percentage ($100/130*10 \approx 7.69$; you can only type integers, hence type 8).

- Click "Ok."
- A table appears with the file names of the lesion maps in the left column. Copypaste or type the behavioral scores into the right column. Caution: the copypaste function might not properly work under all circumstances; copypaste from a .txt file might fail, and a decimal separator might get lost. Copypaste of integer numbers from Excel works.
- Save the design file (.val extension) via File/Save:
  - This file contains the imaging file names and the behavioral values, but not the folder path to the imaging files. Instead, NPM assumes the .val-file to be located in the same folder as the imaging files.
  - Hint: open the .val file with a text editor to sanity check if NPM correctly recognized numerical values.

Performing the Analysis

- Under Options/Permutations we can choose if NPM applies a maximum statistic permutation approach and how many permutations should be used.
  - This is considered by some researchers to be an optimal solution to the multiple comparison problem as it provides an approximately exact family-wise error correction.
  - A permutation analysis will take much longer than a non-permutation analysis; choose the number of permutation as large as your hardware and time allows.
  - A non-permutation analysis by FDR and Bonferroni is also computed by default.
- Under Options/tests we can select what tests should be used in analyses with binary images and continuous behavior—a parametric t-test and/or a non-parametric Brunner-Munzel test.
- Start the analysis under VLSM/Binary images, continuous behavior and select the .val file.
- Select the name and location of the compressed output file.

Output

The output contains multiple NIFTI images and a text file, including:

- An overlap plot of all included lesions ([outputname]Sum[predictorname].nii.gz).
- A Z-transformed topography of voxelwise t-statistics for the t-test ([outputname]ttest[predictorname].nii.gz).
- A Z-transformed topography of voxelwise BM-statistics for the non-parametric Brunner-Munzel test ([outputname]BM[predictorname].nii.gz).

- A text file including information about specifications and statistical thresholds that are necessary to identify significant results.

Results Interpretation

For the visualization and interpretation of results, we need another software tool. MRIcron or MRIcroGL are perfectly suited to do so. To visualize statistical results in MRIcron:

- Open mricron/MRIcroN.exe.
- Select a background for data visualization; statistical maps are shown projected on this image. Often, visually appealing images in the same space as the normalization template are used; however, the normalization template itself might be a less appealing, but honest solution. Open the image via File/Open or choose a default template via File/Open templates. MRIcron's default ch2 T1 is very popular.
- Open the statistical map as an overlay via Overlay/Add. Select the statistical map for the t-test or the BM test.
  - For the present example, we look at the results of the t-test on the ch2 default template

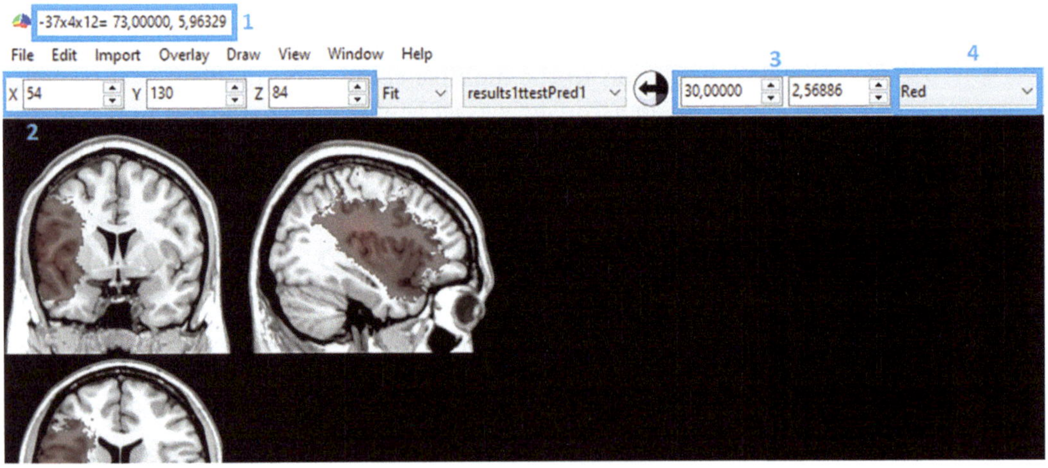

In the graphical user interface of MRIcron, the following information is relevant or helpful:

- 1: If you left-click on a voxel, these numbers provide you with (a) the voxel's XYZ-coordinates in the imaging space, which is $1 \times 1 \times 1\text{mm}^3$ MNI space for the ch2 template, (b) the voxel value of the background image (here: the T1 template shows a greyscale value equivalent to 73), and (c) the voxel's statistic in the overlay map (here: the Z-transformed t-statistic is 5.96).

  - 2: The voxel's XYZ-coordinates measured with an origin in the corner of the image. Type in values here to jump to a specific slice.
  - 3: The color scaling range. The chosen color scale is applied to visualize numerical values in the range between the left and the right value. Values outside this range are not shown, even if they might be present.
  - 4: The color scale. My personal favorites are 5redyell and actc
- in parallel, open the .txt file.

```
Only testing voxels damaged in at least 10 individual[s]
Number of Lesion maps = 131
Permutations = 1000
Analysis began = 2021-Mrz-28 14:59:01
Sum image finished = 2021-Mrz-28 14:59:34
Voxels damaged in at least 10 individuals = 397271
Memory planks = 0,0969367120414972
Generating BM permutation thresholds
Computing plank = 1 of 1
397271 test Std Bonferroni FWE Z 0.050=5,156, 0.025=5,285, 0.01=5,450
214808 test Number of Unique Lesion Patterns Bonferroni FWE Z 0.050=5,040, 0.025=5,171, 0.01=5,340
ttest Range -3,212...12,440 1 2
ttest +FDR Z 0.050=1,83081000, 0.01=2,56886300
ttest -FDR Z 0.050=9,20000000, 0.01=9,20000000
ttest+: permutationFWE , 0.050=4,95379900, 0.025=5,17560600, 0.01=5,37814600
ttest-: permutationFWE , 0.050=-4,66536200, 0.025=-4,81385800, 0.01=-4,99759700
BM Range -3,615...22,283 3
BM +FDR Z 0.050=1,88144400, 0.01=2,59438200
BM -FDR Z 0.050=9,20000000, 0.01=9,20000000
BM+: permutationFWE , 0.050=5,64977600, 0.025=5,93725600, 0.01=6,34091700
BM-: permutationFWE , 0.050=-5,53472900, 0.025=-5,80065000, 0.01=-6,15927000
Analysis finished = 2021-Mrz-28 17:11:30
Processing Time = 7949
```

the relevant information is

  - 1 (blue): the minimum-maximum range of z-transformed test statistics
  - 2 (red): the statistical threshold after FDR-correction for positive lesion-deficit associations (i.e., where lesions induce more severe symptoms), at a one-sided $p < 0.05$
  - 3 (green): the statistical threshold after maximum statistic permutation-correction for positive lesion-deficit associations at a one-sided $p < 0.05$
- These values can now be applied to the statistical map that we loaded as an overlay in MRIcron. For example, if we want to visualize the results of the mapping by a t-test with a permutation-based correction, we have to visualize the range from the threshold to the maximum value, i.e., 4.9537 to 12.440.

- For a non-parametric mapping, load the BM-statistic topography and apply the values from the .txt file on BM thresholds.
- In case of a null result (i.e., no significant voxels remain after correction), NPM mysteriously shows an FDR threshold of 9.2000.

The two lines for each test, e.g., +FDR and –FDR, give thresholds for one-sided tests in both directions. –FDR indicates voxels where a lesion is associated with less severe symptoms.

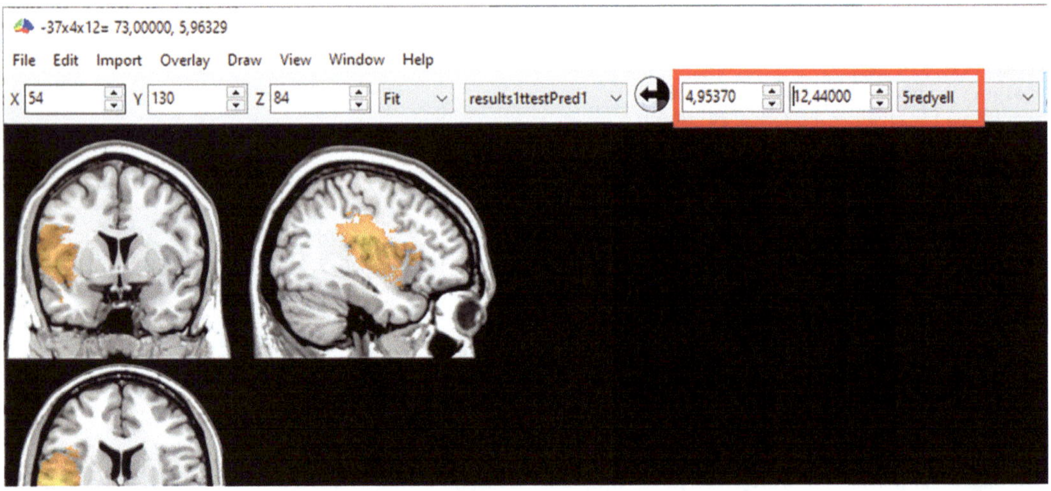

Additional Tips and Information

- The companion software MRIcron and the more recent MRIcroGL (https://www.nitrc.org/projects/mricrogl/) perfectly synergize with NPM. They can be used for a wide range of tasks, such as data visualization, manual lesion drawing, creation of overlap and subtraction plots, NIFTI file conversion, and assessment of descriptive image information.
- NiiStat (https://www.nitrc.org/projects/niistat/) is a more recent MATLAB/SPM12-based lesion mapping toolbox, also created by Christopher Rorden.
- For a binary behavioral variable (e.g., symptom present versus symptom absent), the procedures are almost identical. Again, larger numbers of the variable indicate better performance (1 = normal; 0 = pathological). The analysis is then started under VLSM/Binary images, binary behavior. This computes a Liebermeister test (*see* [1]) with the L-statistic. The output file can be interpreted following the strategy shown above.
- Covariates can be controlled by choosing VLSM/Binary images, multiple regressors. Add ≥2 predictors to the design specification. This results in multiple output files that can be interpreted following the strategy shown above. Choose the output file for

your behavioral variable of interest; this includes results corrected for all other predictors

– Hint: this provides data controlled by FDR and Bonferroni correction, but not by permutation correction, which is challenging in the presence of covariates. A permutation-based solution is the Freedman-Lane procedure, e.g., provided by NiiStat.

- Lesion size can also be controlled by choosing VLSM/Binary images, multiple regressors. Conveniently, you do not have to set up an additional predictor yourself. When starting the analysis, you are asked if you want to add lesion volume as a regressor. Simply click "Yes."
- The overlay plot that is part of NPM's output can be used to visualize the lesion sample in MRIcron; open it as an overlay as you did with the statistical map.

References

1. Rorden, C., Karnath, H. O., & Bonilha, L. (2007). Improving lesion-symptom mapping. *Journal of Cognitive Neuroscience, 19*(7), 1081–1088. https://doi.org/10.1162/jocn.2007.19.7.1081

VLSM with VoxBo

VoxBo is a neuroimaging package with command line tools which can be used to perform functional neuroimaging and VLSM analyses. Using VoxBo requires some familiarity with terminal windows and some knowledge of basic linux commands. To get help type any VoxBo command in the terminal without comment or using the -h flag. Below are some examples of frequently used VLSM commands.

Generating a 4D Nifity Lesion File

```
vbconv lesions/Subject_* -o 131_lesions.nii.gz
```

The above VoxBo command (vbconv) will convert all 131 individual subject files in the lesions directory into a single four-dimensional (X,Y,Z, subject) nifty file called 131_lesion.nii.gz Required inputs include the individual lesion masks and the -o flag to name the output.

Generating Lesion Coverage Map

```
vbim 131_lesions.nii -count -write 131_coveragemap.nii
```

The VoxBo image munge (vbim) command will generate an overlap map showing the extent of lesion coverage in your data and the number of subjects with damage to a given voxel. The required input for this command is the lesion file, followed by the -count flag to generate a count map and last the -write flag to name the output file.

Calculating Total Lesion Volume

```
vbmaskinfo subject_001.nii.gz
```

To calculate the total lesion volume for a subject, use the vbmaskinfo command followed by the a single mask file. VoxBo will display detailed information about the mask including “voxels with included data”; this value multiplied by .001 is your lesion volume in cubic center meters (cc).

Calculating Regional Percent Damage

```
vboverlap -a aal.nii.gz Subject_001.nii.gz
or
vboverlap -a brodmann.nii.gz -m lesions/Subject_* -r 44
```

To calculate the percent damage to a given area, use the vboverlap command displayed above. The first command will provide the percent overlap (i.e., damage) for all regions in the AAL atlas for subject 001. The second command will provide the percent overlap to region 44 in the Brodmann atlas for all 131 subjects in the lesion directory. The input for this command is -a flag followed by the atlas file, the -m flag followed by the individual lesion file(s), and the -r flag for a specific region.

- Quick Linux tip: if you add > output.txt to the end of the command, any screen output generated by the command will be written to a text file called output instead of displayed in the terminal window.

Performing a Chi-Square

```
vbcmp Subject_001.nii.gz Subject_002.nii.gz
```

To conduct a pairwise comparison, use the vbcmp command in VoxBo. The inputs for this command are two masks.

Performing a T-Test

```
vbtmap 131_lesions.nii.gz behavior.txt output_statmap.nii.gz
-n 10 -f -nodup -q 0
```

To run a t-test in VoxBo, use the vbtmap command. As inputs the command requires the lesion file, followed by a text file with behavioral scores, and last the name for the output map. It is important to make sure the order of the lesion files and the behavioral scores are matched. Additional flags are used at the end of the command to refine the analysis output. The -n flag constraints your analysis to a minimum number of subjects with damage to a given voxel to be included in the analysis. The -f flag specifies the relationship between the behavioral data and lesion presence. By including -f, a negative relationship between performance and the presence of a lesion is assumed. An example is accuracy data where lesions produce poor performance scores. Used without the flag, a positive relationship between lesion and behavior is assumed. An example is reaction time where lesion produces longer latencies. The -nodup flag limits identical voxels from counting more than once in FDR threshold calculations, and finally the -q 0 flag will display a range of FDR thresholds for the analysis.

- Note: The vbtmap command is capable of running two tailed tests using the -2 flag and a Welch's t-test for unequal variance. To perform the Welch's test, include the -w flag before the FDR threshold flag.

Performing a T-Test (with Bonferroni)

```
vbtmap 131_lesions.nii.gz behavior.txt output_statmap.nii.gz
-n 10
vbmaskinfo 131_lesions.nii.gz -n 10
vbtcalc
```

To run a t-test with Bonferroni correction for multiple comparisons instead of FDR, use the vbtmap without the -q flag. Use the vbmaskinfo command to compute the total number of voxels that will be analyzed, and then use the vbtcalc command to calculate the corrected threshold. The vbtcalc command will invoke a calculator interface that requires the total number of voxels, voxel size, and effective degrees of freedom to calculate the Bonferroni threshold.

Performing a T-Test (with Permutation)

```
makevlsm -d permidr -p 1000 -l 131_lesions.nii.gz -s behavior.
txt -n 10 -f -o 131_PERMstatmap.nii.gz
```

```
then
 vbperminfo -pm permdir
```

To run a t-test with permutation to correct for multiple comparisons, use the makevlsm command. This is a batching command making it easy to invoke multiple vlsm commands like vbtmap, vbcmp, and vbpermgen in a single line. The command inputs for a t-test with permutation are -d flag, which specifies a directory for permutated images, the -p flag with the number of permutations to perform, -l for the lesion file(s), -s for the behavioral measure, -n which is the minimum number of patients to include in the analysis, -f that specifies the relationship between the lesion and behavior, and -o followed by an output name for the uncorrected statistical map. After makevlsm is done generating the permutated images, use the vbperminfo command to calculate the corrected thresholds. The command syntax is vbperminfo followed by -pm and the directory containing the permuted images. The command will generate a file called resultvalues.ref which contains the corrected thresholds. To read the file, open it in a text editor.

Performing a Regression

```
 vbvolregress -dv behavior.txt -iv 131_lesions.nii.gz -iv
TotalLesionVol_131.txt -int -c 'foo t vec -1 0 0' -m min10-
submask.nii.gz -o regression_statmap.nii.gz.
```

In addition to a t-test, VoxBo allows users to perform regressions using the vbvolregress command. The example above regresses the effect of total lesion volume from the analysis. The command inputs are -dv flag which indicates the dependent variable and -iv flag for the independent variable(s). For VLSM regressions, the behavioral score will always be your dependent variable, and the lesion file will be one of the independent variables. There are no limits to the number of independent variables the command can handle (although a large number of variables may overfit the statistical model). The -int flag is to include an intercept term and -c is for the desired contrast. The contrast syntax used above is foo an arbitrary name for the contrast, t is the desired statistic, vec is always included, and last are weights for each variable with the contrast bounded by single or double quotes. The remaining regression command flags are -m to include a mask (see note below) and -o to name the output.

- Note: The vbvolregress command does not have a flag to easily constrain your analysis to a minimum number of subjects per voxel like vbtmap. Instead use a mask thresholded by the minimum number of subjects. To generate a mask with greater than ten subjects per voxel, use the command below.

```
vbim 131_lesions.nii.gz -count -thresh 10 -quantize 1 -write
min10submask.nii.gz
```

Other Useful VoxBo Commands and Tools

- vbview2—is visualization tool to display results
- vbconv—conversion tool to convert dicoms to nifti or other imaging file types
- dicominfo—can be used to read dicom header files
- niftiinfo—useful too to read nifty header information
- vbim—can be used to perform simple image manipulations like creating masks, flipping an image, thresholding, copying headers, quantizing (changing all non-zero values to a single value), and much more
- vbcmp—will run a chi-square test
- vbmaskinfo—will give basic information about a mask like lesion volume
- vbmaskcompare—will compare the differences between two masks
- makevlsm—for running vlsm analysis with permutation tests instead of FDR
- vbperminfo—used to calculate permutation results

SVR-LSM GUI Software

Get Software Running

Visit the GitHub page https://github.com/atdemarco/svrlsmgui to download the software package. To check requirements, click the Help menu and choose the Requirements options. You will see a list of requirements (some optional) in the list. A check next to the entry indicates that the requirement is detected.

Properly Formatting the Data

Formatting the Imaging Data

Lesions should be in NiFTI (nii) format, and not gzipped. **If you have downloaded the data associated with this tutorial, this means you will need to unzip all of the .nii.gz files so that their file extension is .nii and not .nii.gz.** The location of a lesion within each file is delineated by any value greater than zero. Lesion files should all be of the same image dimensions and in register with each other. Typically, this means they are normalized to a standard template space.

Formatting the Behavioral Design File

Behavioral data should be stored in a CSV file with a row of headers. The file must contain a column with the header row "Registry Code." This column should contain the names of each subject in the analysis and should correspond to lesion filenames (without the NiFTI extension). Other columns should contain relevant behavioral data. Since lesion size is calculated at runtime, it is not necessary to include lesion size in this file. **If you have downloaded the data associated with this tutorial, you will paste the contents of the text file into an Excel spreadsheet and assign the column a title. Then type in the subject names associated with each row, as in the figure below. Then you will save this as a CSV file from within Excel.**

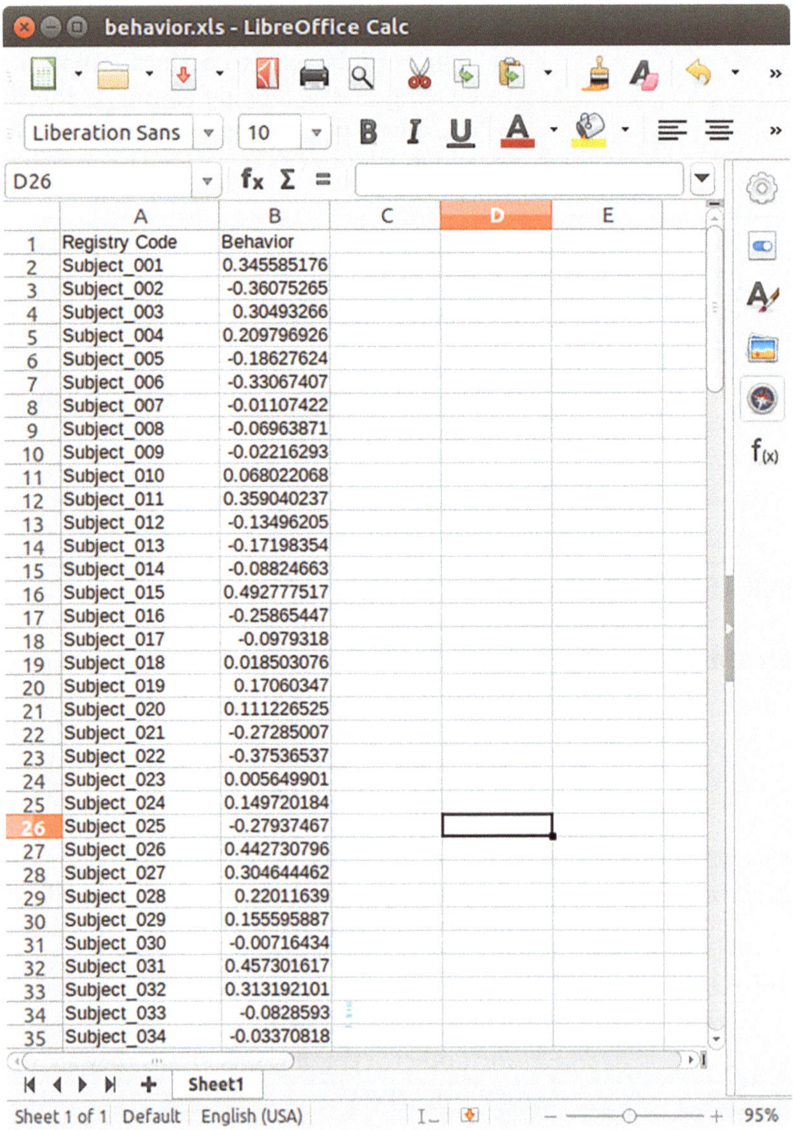

	A	B
1	Registry Code	Behavior
2	Subject_001	0.345585176
3	Subject_002	-0.36075265
4	Subject_003	0.30493266
5	Subject_004	0.209796926
6	Subject_005	-0.18627624
7	Subject_006	-0.33067407
8	Subject_007	-0.01107422
9	Subject_008	-0.06963871
10	Subject_009	-0.02216293
11	Subject_010	0.068022068
12	Subject_011	0.359040237
13	Subject_012	-0.13496205
14	Subject_013	-0.17198354
15	Subject_014	-0.08824663
16	Subject_015	0.492777517
17	Subject_016	-0.25865447
18	Subject_017	-0.0979318
19	Subject_018	0.018503076
20	Subject_019	0.17060347
21	Subject_020	0.111226525
22	Subject_021	-0.27285007
23	Subject_022	-0.37536537
24	Subject_023	0.005649901
25	Subject_024	0.149720184
26	Subject_025	-0.27937467
27	Subject_026	0.442730796
28	Subject_027	0.304644462
29	Subject_028	0.22011639
30	Subject_029	0.155595887
31	Subject_030	-0.00716434
32	Subject_031	0.457301617
33	Subject_032	0.313192101
34	Subject_033	-0.0828593
35	Subject_034	-0.03370818

Configuring Analysis Options

Indicate the Correct Lesion File Source Directory

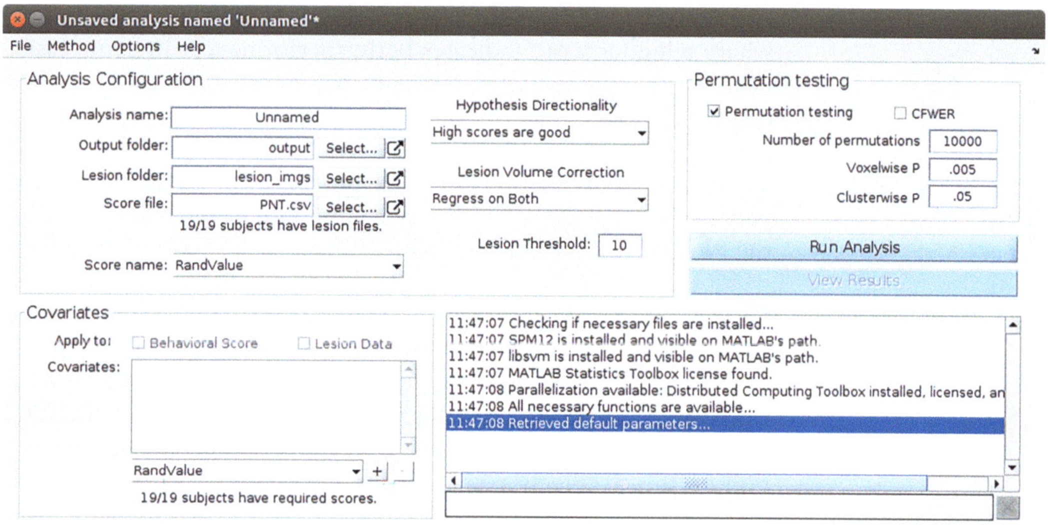

Choose the Correct Behavioral Design File

Choose the desired behavioral design file by clicking the "Select. . ." button beside the "Score file" choice in the "Analysis Configuration" pane (top left pane of the svrlsmgui window). This will be the .csv file you formatted previously.

Choosing Predictor and Covariates

After loading the behavioral design file, select the desired predictor by choosing its name in the "Score name" dropdown list at the bottom of the "Analysis Configuration" pane. Once the lesion folder and behavioral design file are specified, you will see an indication of how many subjects in the behavioral design file were also found to have lesion files in the specified lesion directory. You will want to ensure that this says 131/131, indicating that all files were found. If this does not say 131/131, then you will want to reformat and/or reselect your lesion file directory or behavioral design file.

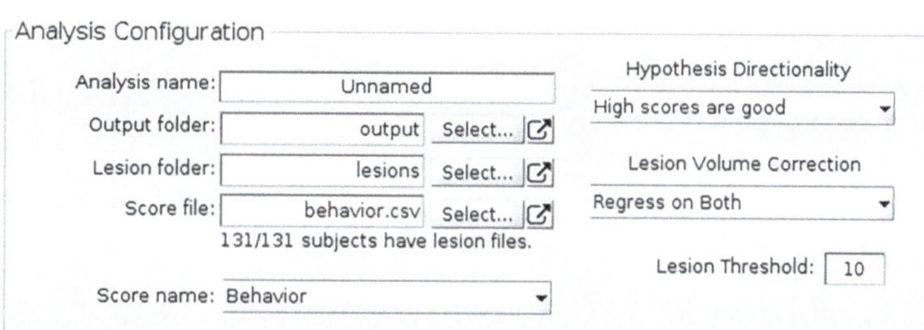

Set the Hypothesis Directionality

Set the hypothesis directionality for your lesion-symptom mapping analysis using the "Hypothesis Directionality" combo box. This has two options: high scores are good, and high scores are bad. High scores are good for some measures like the *WesternAphasiaBattery*, where a higher score indicates better performance. In some cases, however, a high test score indicates bad performance, such as in a count of errors.

Setting Minimum Lesion Overlap

To exclude voxels where lesions have poor overlap across the patient sample, enter the desired minimum overlap threshold in the "Lesion Threshold" editbox.

Lesion Volume Correction

Lesion volume correction is chosen using the "Lesion Volume Correction" combo box in the "Analysis Configuration" pane. This box offers four options, including No correction, Regress on Both, Regress on Lesion, Regression on Behavior, and dTLVC. Please see the relevant chapters in this text for additional information on these correction approaches.

Multiple Comparisons Correction

Correction for multiple comparisons is configured in the "Permutation testing" pane. There are three main configurations. The first is to skip permutation testing, which will skip producing *p*-values for the analysis. To disable permutation testing, uncheck the Permutation testing checkbox. To enable permutation testing, click the checkbox labeled "Permutation testing," and enter the number of permutations desired in the eponymous textbox. The second option is to choose a traditional two-step correction with voxelwise thresholding using a voxelwise P (specified in the editbox) and a clusterwise P (specified in the editbox) that will be applied after voxelwise thresholding. This approach will compare the null distribution of voxel values and cluster sizes to the real data. Finally, click the CFWER checkbox to enable continuous family-wise error correction. With this option selected, you will supply a v value in cubic millimeters and a FWER value in the eponymous textboxes.

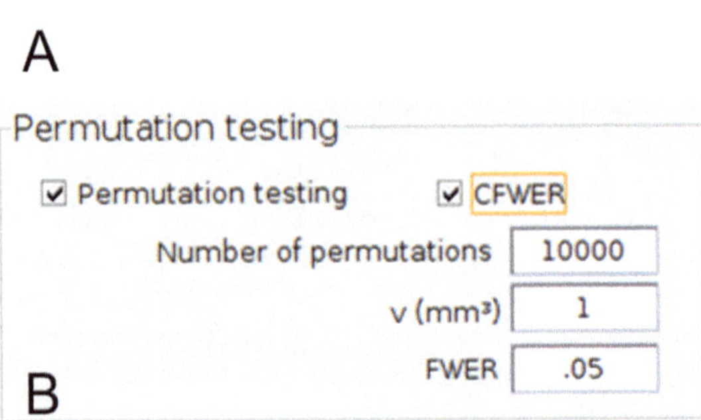

Run an Analysis

Before Starting Your Analysis

The example dataset has a large sample of lesion files at a high resolution. Thus, to hasten the analysis and to ensure we don't run out of memory, we will ask the software to downsample the data prior to the analysis. We can do this by choosing from the toolbar, Options > Image Data > Resample > Manual, and entering 4mm cubes.

Starting Your Analysis

Click the light blue "Run Analysis" button, and then the analysis will run. A progress bar will begin underneath the white progress window. Please be patient, and keep your eye on the messages in the progress window. When the analysis is done, the "View Results" button will become enabled. Click the View Results button to show a narrative summary of the results in HTML format. You can access the output files directory also.

Inspecting the Output

The program makes various output files that can be viewed with the "View Results" button after an analysis has completed. Alternatively, the files can be found in your system file browser. If a Summary Output file was created, the View Results button will launch this HTML file in MATLAB's web browser.

Output File Structure

Output files are created in the output directory specified by the analysis (e.g., "my analyses"). The folder containing the analysis output will match the name of the analysis name specified by the graphic interface (e.g., "my new analysis"). Within this output folder, there will be another subdirectory indicating the date on which the analysis was run (e.g., "25 Oct 2017"), and within the date folder, there will be another subdirectory corresponding to the behavior under investigation and the tails specified in the analysis (e.g., "Naming, one-tailed (negative)"). All together, this path may look like /home/andrew/my analyses/my new analysis/25-Oct-2017/Naming, one-tailed (negative)/.

Output Created with or Without Permutation Testing

Whether or not permutation testing is selected, the analysis will produce a number of files, including:

- Beta map (unthresholded).nii: This file contains SVR-β values before any thresholding. These values should be interpreted tentatively if at all, since SVR-β must be submitted to permutation testing before interpretation.
- All lesion overlap.nii: This file contains a sum of all of the lesions for individuals included in the analysis. The values in this image correspond to how many individuals had lesions at the given point in the image. It will be named to match how many lesions contributed to the image (e.g., "All lesion overlap n=45.nii").

- Minimum lesion mask.nii: This file contains a binary mask indicating where the number of lesion in the patient sample exceeded the minimum lesion cutoff value in the analysis configuration. In other words, the analysis is restricted to voxels that have values of 1 in this image. The file is named to reflect the minimum lesion mask value specified in the analysis (e.g., Minimum lesion mask=9.nii).
- Lesion Volumes.txt: This file contains a list of lesion volumes for the participants included in the analysis. The file contains two columns. The first column contains the image file names, and the second column contains the volumes of each lesion tracing in units of voxels.
- overview.html: This file contains an overview of the analysis specification and results.
- Analysis Parameters.mat: This file contains the parameters chosen for the analysis when it was conducted. This file can be loaded back into SVR-LSM GUI to view the analysis configuration later, shared with colleagues, etc. Note that this file may contain absolute paths to flies, which may need to be reconfigured if files are moved or the parameter file is viewed on a different system from the one on which it was created.
- rtemplate.nii: This file contains an SPM canonical template that is resliced to your sample's lesion space data to be used as an underlay in the summary output file. It simply has the same image dimensions as your lesion data. It can be deleted or ignored and does not play a role in the actual analysis.
- images/ subdirectory: This directory contains images displayed in the overview.html file and plays no role in the actual analysis.

Additional Output Created with Permutation Testing

If permutation testing for voxelwise and clusterwise thresholding is conducted, a subdirectory will be created that is named "Voxelwise p005 based on 1000 permutations," with the *P*-value corresponding to the voxelwise *P*-value and number of permutations specified in the analysis.

Voxelwise Thresholding Output

- Beta value cutoff mask.nii: This file contains SVR-β values that correspond to the voxelwise critical *P*-value specified for voxelwise thresholding in the analysis. Note that these values are different from voxel to voxel, which is different from traditional mass-univariate analyses.
- Thresholded *P*-values: This file contains SVR-β values observed in the analysis translated into *P*-values (based on permutation testing). If "Invert p-map" is selected in the analysis, then the

values are 1-P so that .995 is equivalent to a *P*-value of .005. This can assist with viewing in image viewing software. Otherwise, the values are raw *P*-values, so that .995 corresponds to a *P*-value of .995 (that is, not significant).

- Voxelwise thresholded beta map.nii: This file contains SVR-β values in voxels that meet the voxelwise critical value specified in the analysis. The contents of this file have been thresholded voxelwise using the "Beta value cutoff mask.nii" image, which is derived from the permutation testing that was conducted.
- Voxelwise thresholded beta map_clustidx.nii: This file takes the Voxelwise thresholded beta map.nii and replaces each discrete cluster with a numbered label that can be looked up in the clusterwise thresholding output subdirectory file "Table of clusters.txt." Note that this file does not reflect the application of clusterwise thresholding, so many cluster labels may be present. Most or all of these will not meet the critical cluster threshold criterion.

Clusterwise Thresholding Output

- Largest clusters.mat: This file contains a list of volumes of the single largest clusters that survived voxelwise thresholding at each permutation step of the clusterwise permutation process. This list is sorted to determine the critical cluster size value corresponding to a requested clusterwise *P*-value.
- Table of clusters.txt: This file contains a list of contiguous islands of SVR-β values that survive voxelwise thresholding. Each row corresponds to one cluster, sorted large to small. The first column contains the numbered identity of each voxel, which can be looked at visually in the Voxelwise thresholded beta map_clustidx.nii file. The second column contains *P*-values for the clusters based on the permutation testing performed. A value of .05, for instance, in this column, indicates that a cluster was larger than 95% of those clusters that were generated at random during the permutation process, corresponding to a *P*-value of .05.
- Thresholded by cluster size.nii: This file contains SVR-β values that survive both voxelwise thresholding and clusterwise thresholding.

Overview File

An overview file will be created that can be viewed in an HTML Viewer or web browser. The file contains:

- Analysis summary: Narrative describing the specifications of the analysis and its completion.
- Lesion overlaps: Thumbnails of the lesion overlap of the patient sample and outline of the regions meeting the minimum lesion threshold.

- Unthresholded SVR-β map: Thumbnails showing various slices of the unthresholded SVR-β map. This should be interpreted with caution since an extreme value does not necessarily guarantee that it is reliable (i.e., significant).
- Voxelwise thresholded SVR-β map: If permutation testing was conducted, this section will contain thumbnails showing various slices of the voxelwise thresholded SVR-β map. Voxels in this image meet the specified voxelwise critical value.
- Clusterwise thresholded SVR-β map: If permutation testing was conducted, this section will contain thumbnails showing voxels belonging to clusters that exceed both voxelwise and clusterwise thresholding. The number of significant clusters will be reported and labeled numerically by size. If there are no significant clusters, there will be no results here.
- Behavioral nuisance model diagnostics: If a behavioral nuisance model was specified in the analysis, this section will plot pairwise correlations of the predictors. Specifically, each pane will contain a scatterplot between pairs of variables. A Pearson's correlation coefficient is also displayed in each scatterplot pane. If the correlation is significant, the text appears red. Otherwise the text appears black. Along the diagonal is plotted a histogram of each variable.
- Cluster correction threshold stability: If permutation testing was conducted, this section will show how the largest clusters in the data relate to the cluster size cutoff as it evolved over the course of permutation testing (e.g., every 100 permutations over 10,000 permutations).
- Hyperparameter quality report: This section of the output gives an overview of characteristics of the final SVR model, including whether it converged, the final hyperparameter values used, the number of support vectors used, and the model bias term. Model performance is also reported in terms of three metrics. The first is prediction accuracy, which is a density of correlation coefficients between predicted scores and training scores across N replications of K-fold cross-validation. The default is to perform ten replicates of fivefold cross-validation. The second metric is reproducibility index, which is a distribution of voxelwise correlation coefficients computed pairwise between N replicates of SVR-B maps, each generated using a random percentage P of observations. The default is to perform ten replicates using 80% of observations. This metric was used in Zhang et al. [1]. The third metric is mean absolute difference, which is a density of average difference of the absolute values of training data and

predictions from those data across N replications of K-fold cross-validation. The default is to perform ten replicates of fivefold cross-validation. Parameters N, K, and P for calculating the hyperparameter quality report can be specified in the graphical interface (Method > Multivariate Options > Quality Report).

All of the information in the summary output is available in the raw files that are created by the SVR-LSM analysis.

Image File Output

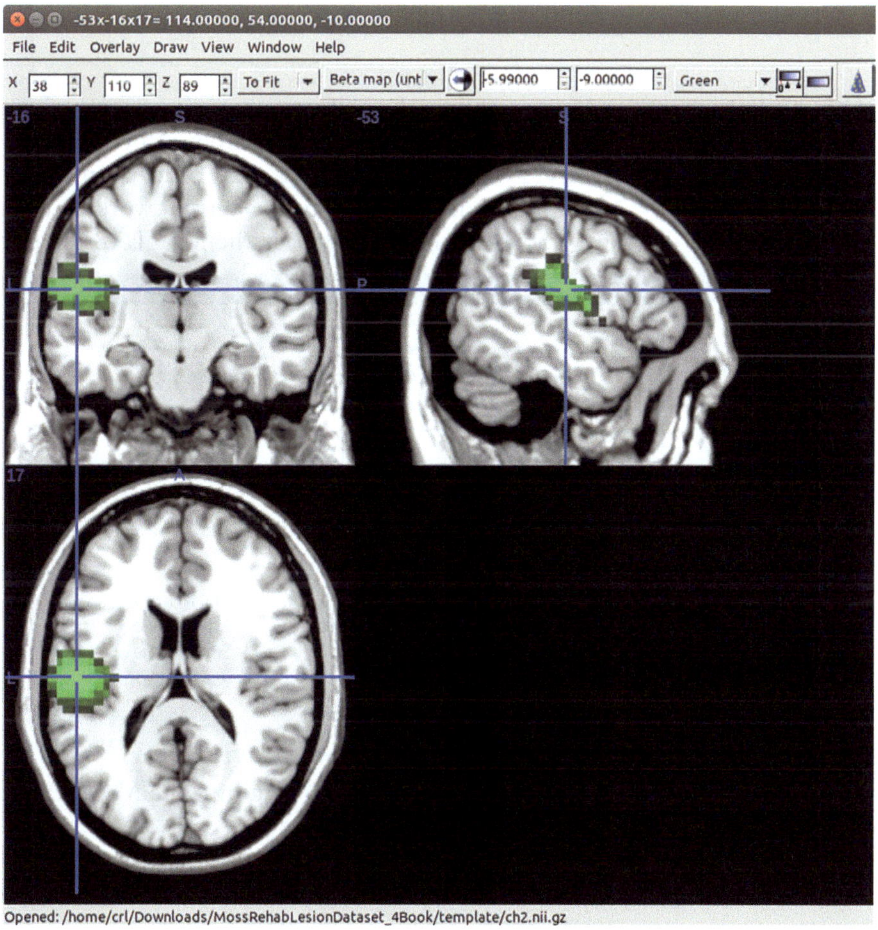

Lesion Analysis with NiiStat Tutorial

First off, you need to download all the NiiStat scripts from the NITRC page, where you can also find further information on NiiStat:

https://www.nitrc.org/projects/niistat

The following example was carried out in MATLAB R2020b with the most recent NiiStat version (Ver 1.1) from the 27.07.2020 GitHub upload. In case you are using another MATLAB version, you may have to adapt the path commands that are used in this tutorial accordingly.

If You Are Unfamiliar with MATLAB

After unzipping the NiiStat-master.zip, we recommend copy-pasting it into your MATLAB-toolbox subdirectory (e.g., "C:\Program Files \MATLAB\R2020b\toolbox"). In order to make the scripts work, you must add their path to the current working directory.

You can either do this via the MATLAB interface:

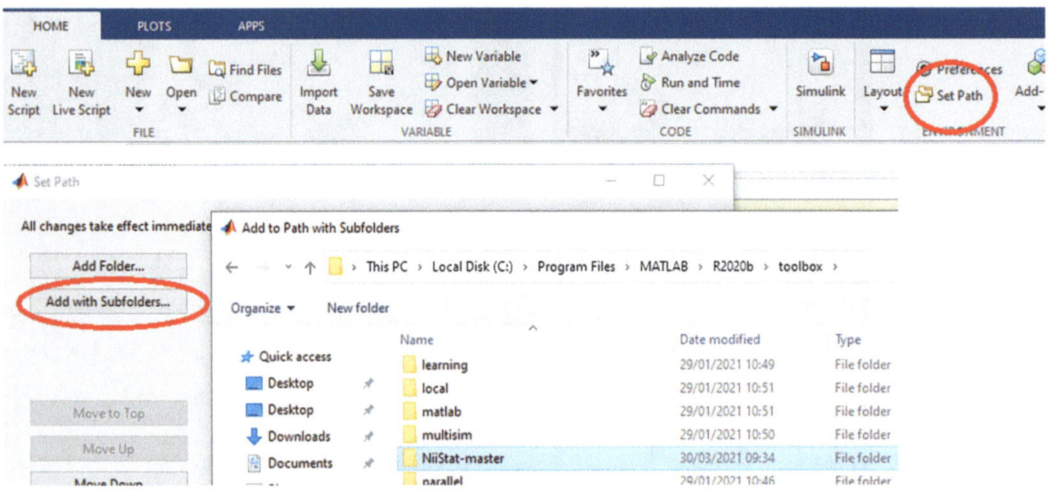

or by typing the following into the MATLAB command line:

```
addpath('C:\Program Files\MATLAB\R2020b\toolbox\NiiStat-mas-
ter')
addpath('C:\Program Files\MATLAB\R2020b\toolbox\NiiStat-mas-
ter\NiiStatGUI')
```

Preparation of Data

You will need to convert all your .nii files into .mat files. This can be done with the **nii_nii2mat** script provided by NiiStat. If you run **nii_nii2mat** in the command line, you will get an extra window (similar to the SPM-toolbox), and then navigate to the folder where you have your lesions, and include all. When you have selected all lesions, click “Done”:

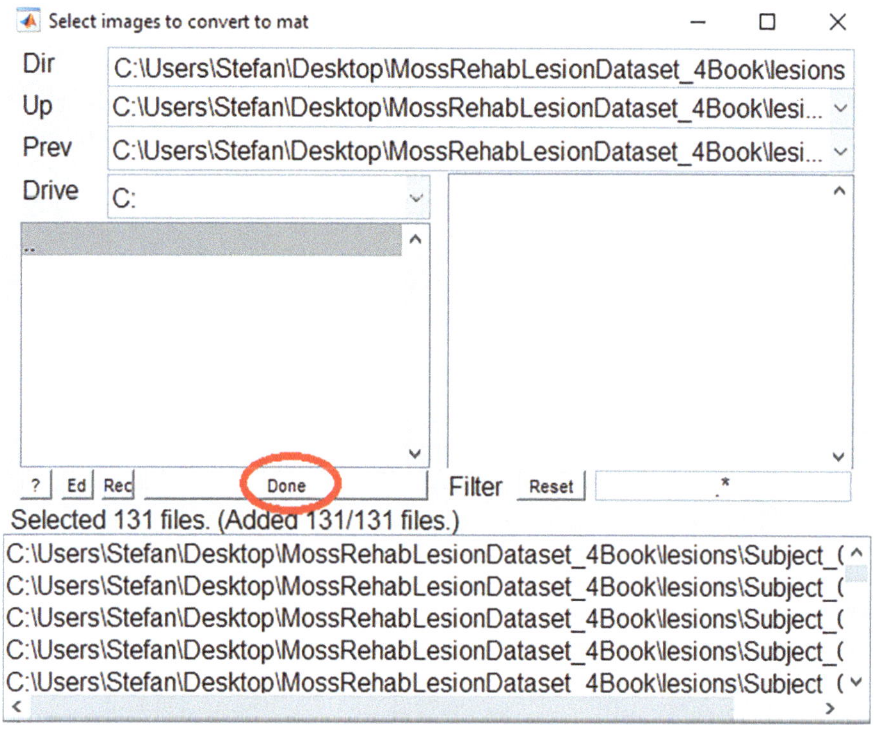

You will then be asked for the modality of the images. In our case, we are using lesions, so we just type “1” and “OK.” This may take a while.

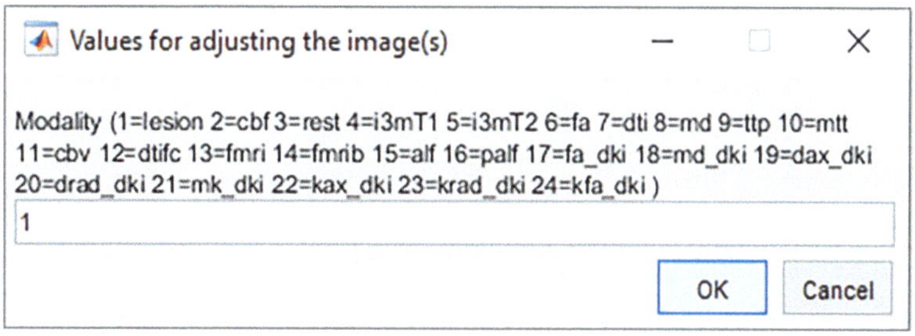

It is necessary to bring the behavioral dataset into the format displayed in the left screenshot below:

- You need the header "participant."
- The following header(s) you can name as you like, except that they may not contain spaces (e.g., "working memory" would not work, it would have to be "WorkingMemory").
- It is important that the participant names are named exactly like the scan files (without the file-type extension ".nii.gz").

Note that NiiStat tests data one-sided (it is assumed that a lesion will only worsen task performance, not improve it). Therefore, you must make sure that better performance is indicated by a higher number, while worse performance is indicated by a lower number.

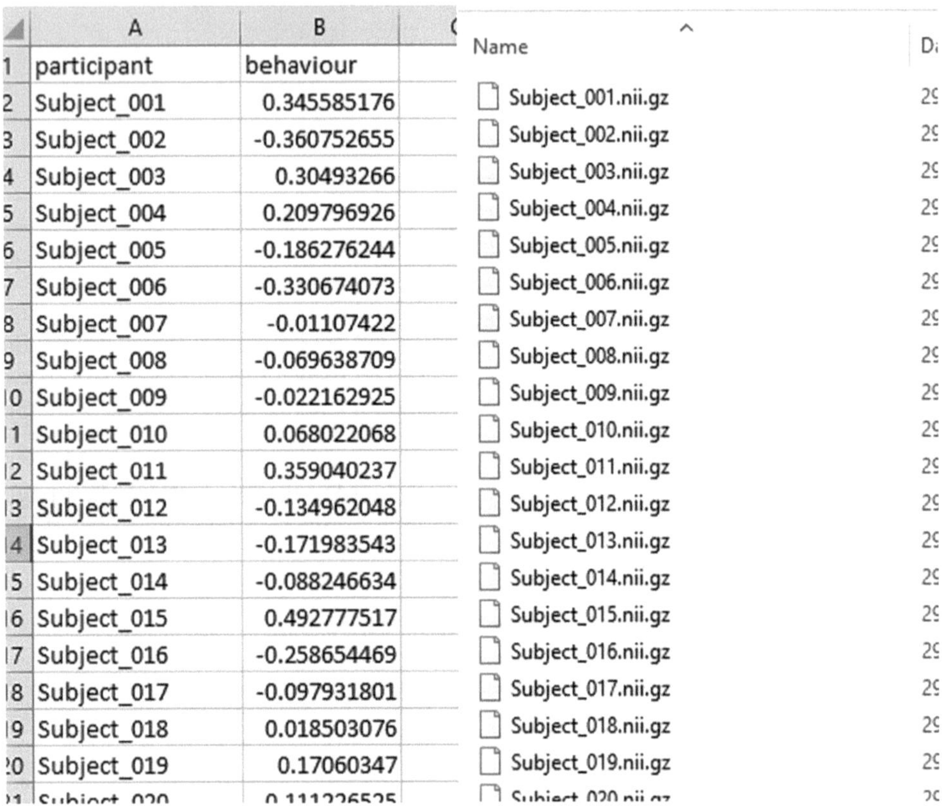

	A	B
1	participant	behaviour
2	Subject_001	0.345585176
3	Subject_002	-0.360752655
4	Subject_003	0.30493266
5	Subject_004	0.209796926
6	Subject_005	-0.186276244
7	Subject_006	-0.330674073
8	Subject_007	-0.01107422
9	Subject_008	-0.069638709
10	Subject_009	-0.022162925
11	Subject_010	0.068022068
12	Subject_011	0.359040237
13	Subject_012	-0.134962048
14	Subject_013	-0.171983543
15	Subject_014	-0.088246634
16	Subject_015	0.492777517
17	Subject_016	-0.258654469
18	Subject_017	-0.097931801
19	Subject_018	0.018503076
20	Subject_019	0.17060347
21	Subject_020	[illegible]

Name
Subject_001.nii.gz
Subject_002.nii.gz
Subject_003.nii.gz
Subject_004.nii.gz
Subject_005.nii.gz
Subject_006.nii.gz
Subject_007.nii.gz
Subject_008.nii.gz
Subject_009.nii.gz
Subject_010.nii.gz
Subject_011.nii.gz
Subject_012.nii.gz
Subject_013.nii.gz
Subject_014.nii.gz
Subject_015.nii.gz
Subject_016.nii.gz
Subject_017.nii.gz
Subject_018.nii.gz
Subject_019.nii.gz
Subject_020.nii.gz

Then you must rename the data tab sheet to "NiiStat" (mind spelling and capitalization). Save this as an .xlsx file. This file must be in the same folder as your lesion .mat files generated by nii_nii2mat!

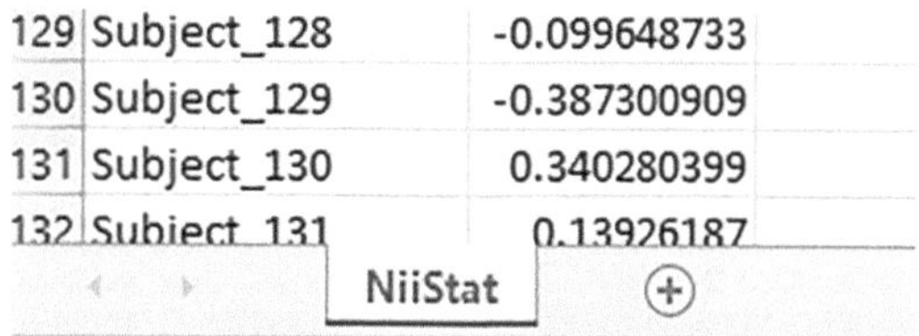

129	Subject_128	-0.099648733
130	Subject_129	-0.387300909
131	Subject_130	0.340280399
132	Subject_131	0.13926187

NiiStat

The Analysis Itself

If you want to use the GUI, you can start by typing **NiiStat-GUI** in the command line. Then you should get the window presented below. (Note that some may experience issues when using NiiStatGUI, if you do, there are other possibilities further below.)

In **Data Dir**, you indicate the directory where you have your lesion data as well as your behavioral .xlsx file.

In **Output Dir**, you provide the directory where NiiStat should create the results folder.

In **Excel File**, you are prompted to select your behavioral data file.

Minimum overlap provides the minimum amount of patients that need to have damage to a certain voxel for it to be included in the analysis. In this example, we picked 5.

Then you need to choose your method of multiple comparison correction. The suggested "gold standard" on the NITRC website is the **Normal Permutation** approach with at least 5000 permutations. In this example, we chose 1000 in order to save time, as the analysis can take a while.

In **Corrected p Threshold,** you can provide the threshold against which you would like to test.

You also need to specify the modality of your images. Go to the tab "MODALITY" to see available choices. In our case, we are using lesion data, so we select just that.

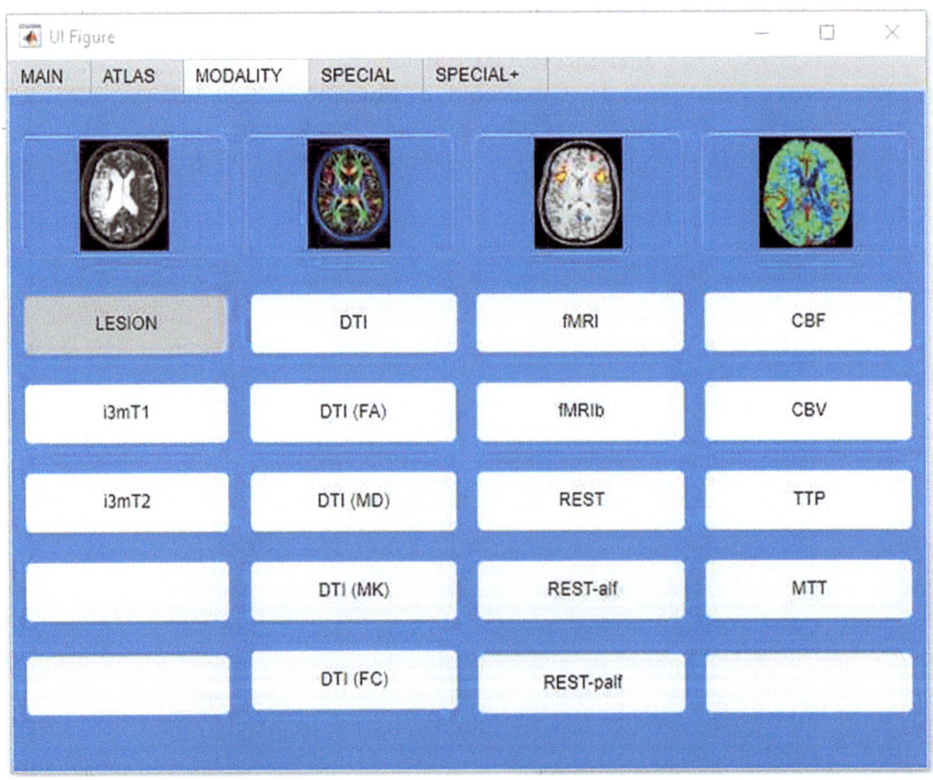

If you are interested in more specialized analyses, such as ROI analyses, regressing over lesion sizes, etc.—you can check the other options and select whatever is relevant to you. When you are ready, you can move back to the “MAIN” tab and click “GO.”

Some people experience issues with the standard GUI; therefore you can also tick the “Use NiiStat Classic” box in the MAIN tab and then click “GO”. Alternatively, you can just type **NiiStat** into the MATLAB command line.

In that case you are prompted to provide the behavioral excel file:

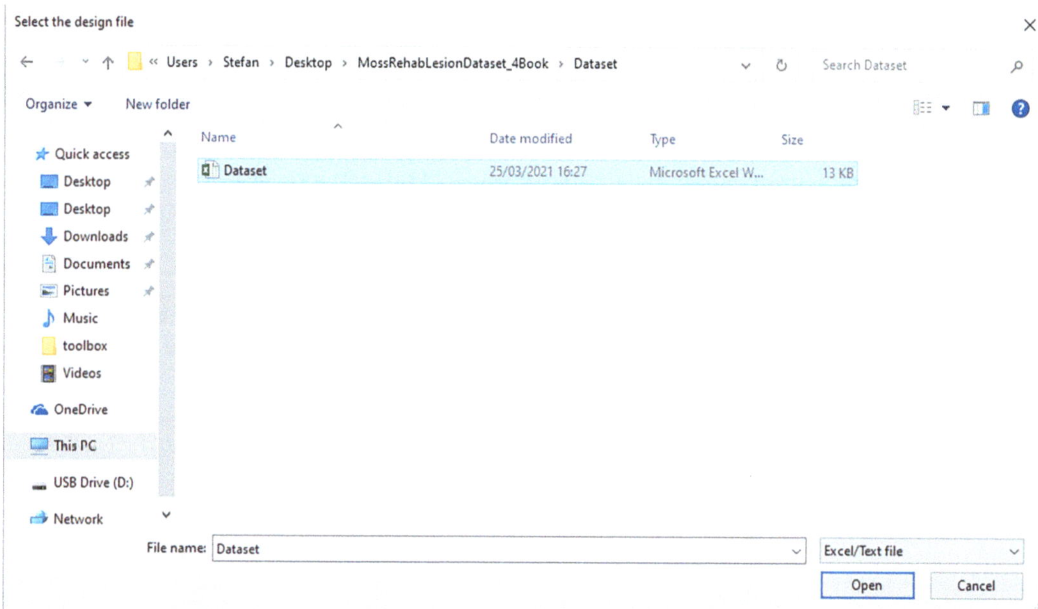

Afterwards you get the following menu, whereby you can enter all the information analogously as above. If you are ready, you can click “OK” to get an output.

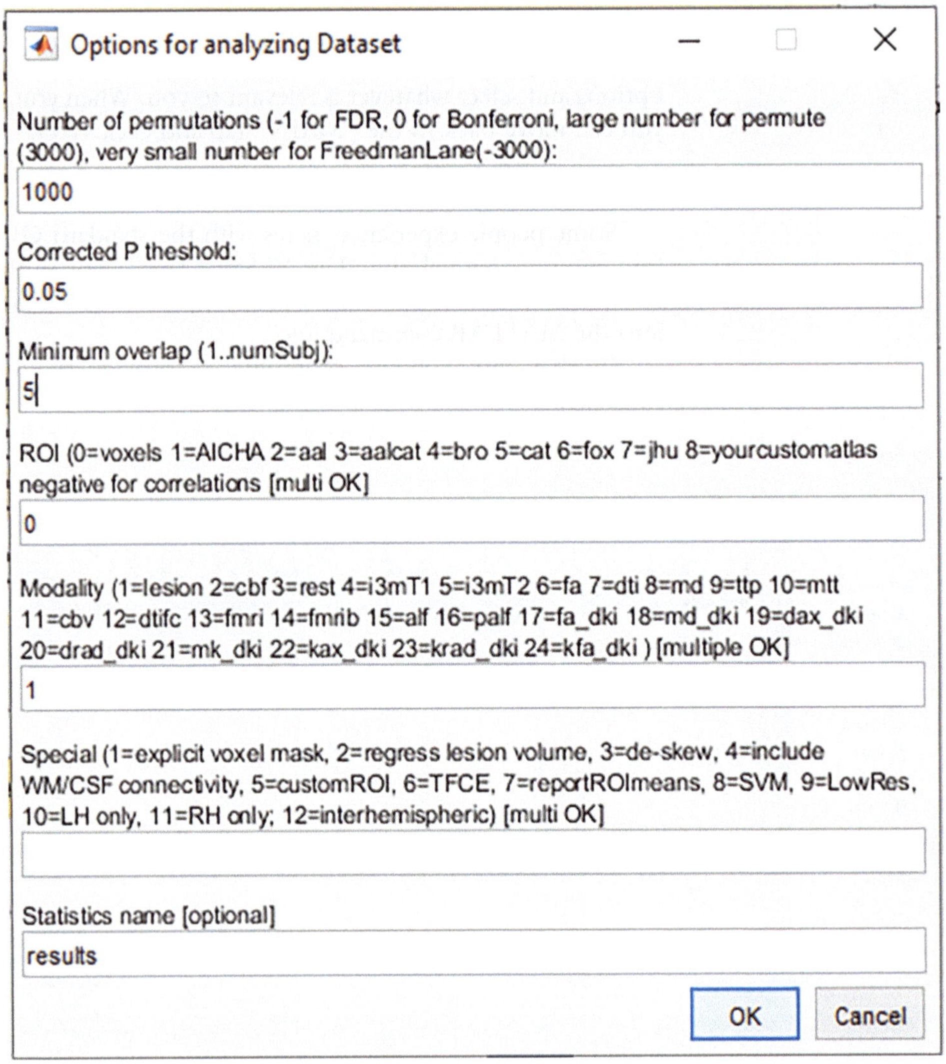

Output

Whichever method you chose, if everything worked out, your MATLAB command line should look kind of like this:

```
Version 3 March 2017 of NiiStat PCWIN64 9.9.0.1570001 (R2020b) Update 4
---------------------------------------------------------------------------------------------
MATLAB Version: 9.9.0.1570001 (R2020b) Update 4
MATLAB License Number:
Operating System: Microsoft Windows 10 Pro Version 10.0 (Build 19041)
Java Version: Java 1.8.0_202-b08 with Oracle Corporation Java HotSpot(TM) 64-Bit Server VM mixed mode
---------------------------------------------------------------------------------------------
MATLAB                                                Version 9.9          (R2020b)
Simulink                                              Version 10.2         (R2020b)
Statistical Parametric Mapping                        Version 7771         (SPM12)
Statistics and Machine Learning Toolbox               Version 12.0         (R2020b)
Symbolic Math Toolbox                                 Version 8.6          (R2020b)
Warning: To enable updates run "!git clone git@github.com:neurolabusc/NiiStat.git"
 or "!git clone git clone https://github.com/neurolabusc/NiiStat.git"
 You will get an error if your Excel file does not have a worksheet named "NiiStat" (case sensitive)
Using regions of interest from folder C:\Users\Stefan\Desktop\NiiStat-master\roi\
Analyzing roi=0, modality=1, permute=1000, design=Dataset
Generating voxel mask for large voxelwise statistics
545385 of 7109137 voxels (7.67161%) show signal in at least 5 participants
The imaging data will require 0.532 gb of memory
Only analyzing voxels non-zero in at least 5 individuals.
**** Analyzing results with 131 participants for behavioral variable behaviour across 7109137 (of 545385) regions/voxels.
Computing glm (pooled-variance t-test, linear regression) for 545385 regions/voxels with analyzing 1 behavioral variables (positive
Expected permutation time is 506.349244 seconds
>>
```

When the analysis is done, you will get this output:

```
1000 permutations required 530.1195 seconds
Thresholds are one tailed (we predict injured tissue will only cause poorer performance not better performance
p<0.050 permutation correction for behaviour is z<-4.54067 z>4.20348
behaviour z=-8.598163..3.211713, 83436 voxels survive threshold
```

If you navigate back to your original folder, you should now have a "niistat_cfg.mat" file as well as a results folder with the exact date and time it finished (e.g., "results_26Mar2021_120254"). Inside your results folder, there should be five files:

results.txt is a text file with the output that MATLAB generated in the command line (see above).

power.mat is a MATLAB file that contains some summary information.

resultssum.nii is an overlap plot of all lesions that were included.

Zresultsbehaviour.nii is a topography plot of the Z-transformed voxel-based results for all voxels that were included in the statistical analysis.

threshZresultsbehaviour.nii is a topography plot of all voxels that were significant.

This is the result that you will end up reporting.

Here you can see a comparison of these three plots produced by NiiStat, each displayed on the ch2bet-template using MRIcron:

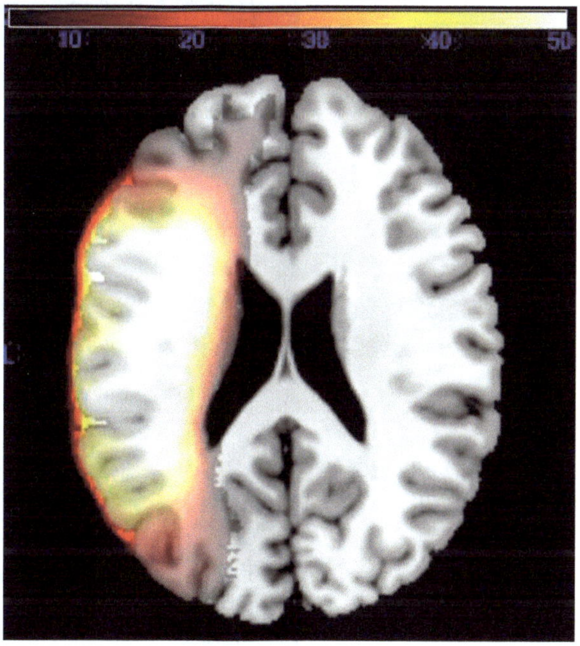

resultssum.nii: brighter colors indicate a higher number of patients having a lesion in a particular voxel

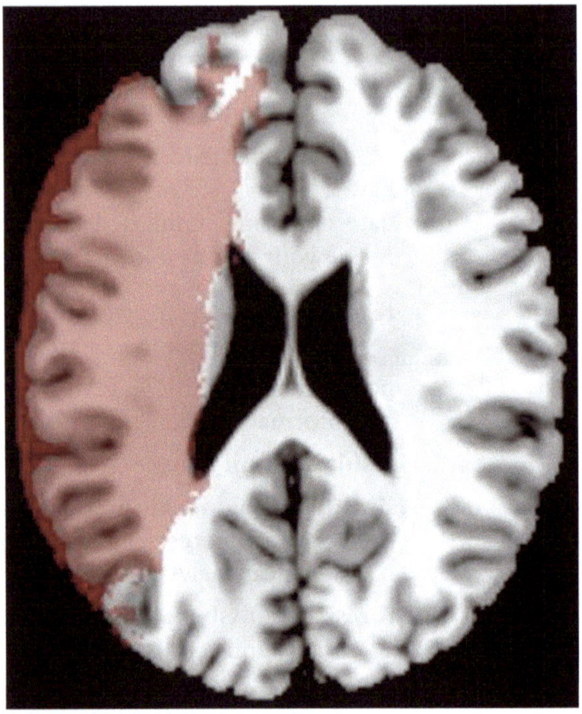

Zresultsbehaviour.nii: All voxels that were included in the analysis are colored red

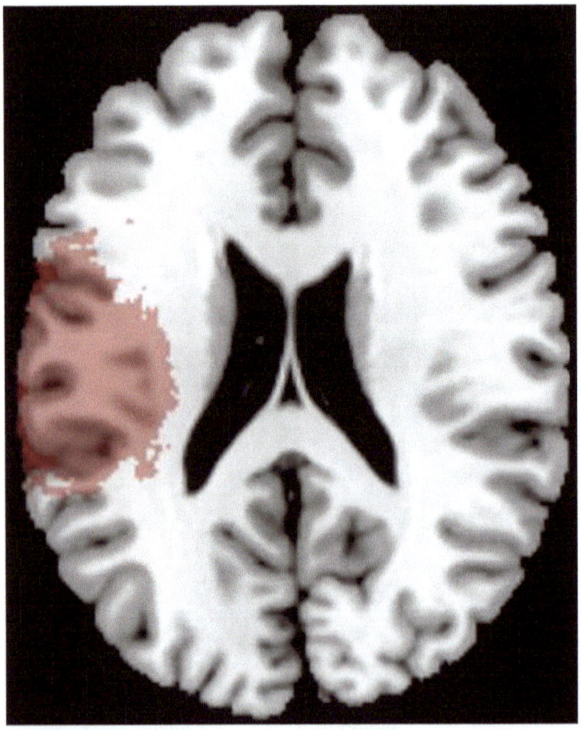

threshZresultsbehaviour.nii: All voxels that were included in the analysis and were significant are colored red

Extra Notes

- When both the behavioral data and the brain data are binomial (either zero or one), NiiStat automatically computes the Liebermeister test statistic instead of t-tests.
- Extra information can be found under the NiiStat Documentation page on NITRC: https://www.nitrc.org/plugins/mwiki/index.php/niistat:MainPage.

Lesion-Symptom Mapping Analyses Using LESYMAP

Getting Started

The analyses described here require the following packages:

- ANTsR (https://github.com/ANTsX/ANTsR): A general neuroimaging data analysis packagc for loading, saving, and manipulating NIfTI files.
- LESYMAP (https://github.com/dorianps/LESYMAP): Lesion-symptom mapping analysis package with both univariate and multivariate analysis options.

The linked GitHub pages provide guidance on how to install these packages across operating systems. Once installed, open RStudio and load the packages using the code provided below.

```
# Load required packages

library(LESYMAP)
```

Read in Data

The binarized, template space lesion masks and the corresponding behavioral scores for each participant are used as inputs for LSM analyses. Example code for reading in the Moss dataset is provided below. This code assumes that your current working directory is one level above the following folders: *lesions*, *template*, and *behavior*. Lesion size is also calculated and saved alongside the behavioral data as this is often useful information for characterizing participants.

```
# define subject files

filename <- list.files(path="~/lesions", full.names=TRUE,

pattern="Subject")


# read in lesion data

lesions <- list() # list of lesions

for (i in 1:length(filename)){

  lesions[[i]] <- antsImageRead(filename[i])

}


# read in behavioral data

behav <- read.delim("~/behavior/behavior.txt", header=FALSE,

col.names="score")


# calculate lesion sizes

behav$lesion_size <- getLesionSize(lesions)
```

Lesion Overlap Map

A lesion overlap map shows the lesion masks of all participants overlaid on a template and is a useful way to visualize the lesion coverage across the sample. The overlap map for the Moss data is shown in Fig. A.1, and the code used to generate this figure is provided below.

```
# create average lesion mask

lesion_avg <- antsAverageImages(lesions) # average all lesions


# plot overlap map

plot(ch2, lesion_avg,

     axis=3,

     alpha=0.7,

     slices=seq(70, 100, by=10),

     colorbar=TRUE,

     window.overlay=range(c(0, 1)))
```

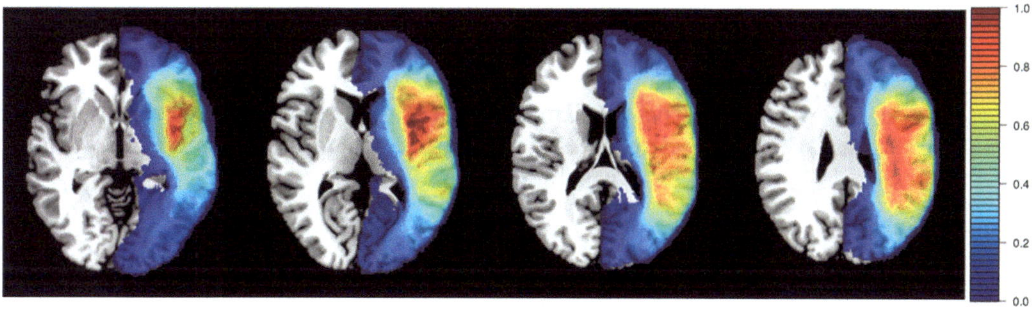

Fig. A.1 Lesion overlap map. Note that the left hemisphere is shown in the right side of the template brain

Mass-Univariate Analysis

The mass-univariate, or voxel-based lesion-symptom mapping (VLSM), approach tests the behavior-lesion association at each voxel within the lesion territory. For all analyses using LESYMAP, the lesion territory is internally defined as voxels where at least 10% of the sample has damage, but a manual value can be specified using the *minSubjectPerVoxel* argument.

The code for running a mass-univariate LSM analysis controlling for lesion volume and correcting for multiple comparisons is provided below. In this example, the following arguments are passed to LESYMAP:

- **lesions**: the list of participant lesion files.
- **behav$score**: the vector of participant behavioral scores.
- **method**: the method type is specified as "*regresfast*" which is a fast linear regression approach that allows for permutation based multiple comparison correction. There are several other mass-univariate and multivariate methods that can be specified.
- **correctByLesSize**: a lesion volume correction method which weights lesioned voxels from smaller lesions more than voxels from larger lesions. For more information, *see* [1]. Lesion size can also be controlled by using residualized behavior scores that have had the effect of lesion size removed ("*behavior*" option) or by applying both methods of correction ("*both*" option).
- **multipleComparison, v, pThreshold**: the default multiple comparison correction is "*fdr*", but permutation-based continuous FWER is selected in the example. This correction controls for the number of false-positive results specified by the *v* value. For more information, *see* [2].

After running the analysis, all output files can be saved using the **save.lesymap()** function demonstrated below. This will save each of the objects in the output list as a separate file in the specified directory. This method of saving the results is useful because, in addition to saving the statistical map containing the voxels which surpassed the FWER threshold and other image files, it also saves text files with the console output and information about how the analysis was run. A single image can be saved using the **antsImageWrite()** function. This can be a useful function to save an image file that has been modified in some way (i.e., it was binarized).

```
# run mass-univariate analysis

vlsm_output <- lesymap(lesions, behav$score, method="regresfast",

correctByLesSize="voxel", multipleComparison="FWERperm", v=100,

pThreshold=0.05)

# save output files in folder called vlsm_results

save.lesymap(vlsm_output, saveDir="~/vlsm_results")

# could alternatively only save the stats map

antsImageWrite(vlsm_output$stat.img, "vlsm_output.nii.gz")
```

SCCAN Analysis

In addition to the traditional mass-univariate approach, LESYMAP provides options for running multivariate LSM analyses. The sparse canonical correlation analysis for neuroimaging (SCCAN) method is demonstrated below, but SVR-LSM can also be run by indicating "*svr*" as the method in the call to LESYMAP. Note, however, that the arguments passed to LESYMAP will necessarily be different for this method and can be further investigated by reading the LESYMAP documentation.

The code for running a standard SCCAN analysis is provided below (*see* [3] for a comprehensive description of SCCAN). The arguments in the provided example that are specific to SCCAN are as follows:

- **method**: the method type is specified as "*sccan*".
- **optimizeSparseness**: SCCAN relies on a sparseness parameter which determines the extent of voxels retained in the output. This sparseness value is optimized using fourfold cross-validation by default, but a sparseness value can also be directly specified using the *sparseness* argument if the optimization algorithm has been previously run. There was no a priori

sparseness value calculated for this sample, so sparseness was optimized.

- **directionalSCCAN**: SCCAN can produce both negative and positive associations which may be of interest depending on the research question. In this example, we only expect negative associations (increased lesion damage associated with decreased behavioral scores), so this parameter is set to *FALSE*.

SCCAN can take several hours to run if the sparseness value is optimized. Similar to the mass-univariate analysis, SCCAN will output a statistical map of voxels where lesion damage is associated with the behavioral score. In addition, SCCAN will output other method specific summary statistics such as the optimized sparseness value and the CV correlation (the correlation between the trained and tested data from the cross-validation procedure). By default, this CV correlation must have a *p*-value less than 0.05 or a null map is returned (although this p-threshold can be manually adjusted with the *pThreshold* argument).

```
# run sccan analysis

sccan_output <- lesymap(lesions, behav$score, method="sccan",

correctByLesSize="voxel", optimizeSparseness=TRUE,

directionalSCCAN=FALSE)


# examine optimized sparseness value

sccan_output$optimalSparseness # 0.2049748


# examine CV correlation

sccan_output$CVcorrelation.stat # 0.7832094


# save output files in folder called sccan_results

save.lesymap(sccan_output, saveDir="~/sccan_results")
```

Visualizing Results

The results of both analyses are shown in Fig. A.2, and the code used to generate these figures is provided below. This ANTsR plot function has been demonstrated previously but not described, so the functionality is described briefly below.

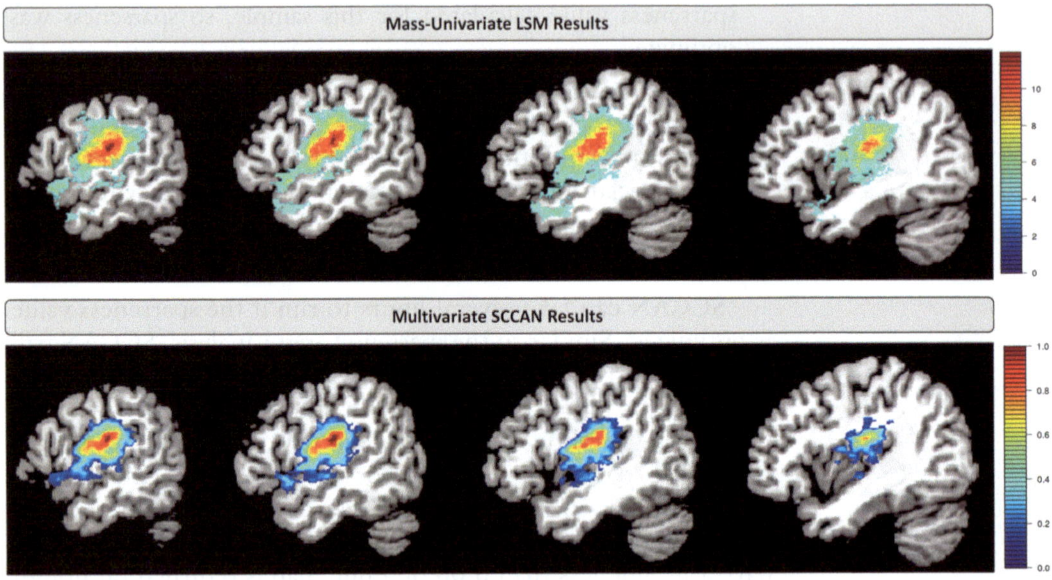

Fig. A.2 Results for the mass-univariate analysis (top panel) and multivariate SCCAN analysis (bottom panel)

A single image or two images can be passed to the plot function. If a second image is provided, it will be plotted on top of the first image.

- **axis**: determines which anatomical plane will be shown in the figure (sagittal slice—1; coronal slice—2; axial slice—3).
- **colorbar**: whether a colorbar will be included next to the plot.
- **slices**: the slices to display in the figure. Specifying slices is optional, but is often necessary if plotting a specific result or slice of the brain is desired. A single slice or sequential slices with a pre-specified gap can be plotted. The *by* argument determines the number of slices between the next displayed slice.
- **window.overlay**: the minimum and maximum values to display for an overlay image.

It is important to note that the **plot()** function demonstrated below assumes that the data have been processed with ANTsR functions and therefore may not work correctly (i.e., may plot a result in the wrong orientation or direction) for NIfTI image files that have been processed with a different application. This error occurs because the orientation information in the header is ignored and the function assumes that the image being plotted has the default antsImage orientation. In these cases, using a general visualization tool such as ITK-SNAP is more appropriate.

```
# VLSM results

plot(ch2, abs(vlsm_output$stat.img),

     axis=1,

     colorbar=TRUE,

     slices=seq(35, 50, by=5),

     window.overlay=range(abs(vlsm_output$stat.img)))


# SCCAN results

plot(ch2, sccan_output$stat.img,

     axis=1,

     colorbar=TRUE,

     slices=seq(35, 50, by=5),

     window.overlay=c(0, 1))
```

References

1. Zhang, Y., Kimberg, D. Y., Coslett, H. B., Schwartz, M. F., & Wang, Z. (2014). Multivariate lesion-symptom mapping using support vector regression. *HumanBrain Mapping, 35*(12), 5861–5876. https://doi.org/10.1002/hbm.22590

2. Mirman, D., Landrigan, J.-F., Kokolis, S., Verillo, S., Ferrara, C., & Pustina, D. (2018). Corrections for multiple comparisons in voxel-based lesion-symptom mapping. *Neuropsychologia, 115*, 112–123. https://doi.org/10.1016/j.neuropsychologia.2017.08.025

3. Pustina, D., Avants, B., Faseyitan, O. K., Medaglia, J. D., & Coslett, H. B. (2018). Improved accuracy of lesion to symptom mapping with multivariate sparse canonical correlations. *Neuropsychologia, 115*, 154–166. https://doi.org/10.1016/J.NEUROPSYCHOLOGIA.2017.08.027

Network Modification Tool 2.0

Overview

The Network Modification (NeMo) Tool takes a white-matter lesion mask and estimates how that lesion will disrupt structural connectivity. We compute the intersection of the lesion mask with a reference database of streamlines estimated using whole-brain tractograms of 420 unrelated healthy young adults from the Human Connectome Project. Brain regions or gray-matter voxels containing the endpoints of these streamlines are assigned a change-in-connectivity (ChaCo) score according to the fraction of streamlines disrupted, assigning a value from 0 (no disruption) to 1 (complete disconnection). ChaCo scores can be calculated at a voxelwise level in the entire gray matter or for regions in a user-selected gray matter atlas (called "chacovol"), where a single value for each voxel/region conveys the total disruption to that voxel/region. Pairwise outputs (called "chacoconn") are square matrices with a value describing the disconnectivity between each pair of voxels/regions.

We used MRtrix3 to compute 5 million streamlines for each HCP subject (2.1 billion streamlines total) using both probabilistic (iFOD2+ACT) and deterministic (SD_STREAM) tractography approaches. More information on the data and tractography options can be found in the GitHub documentation.

The NeMo Tool 2.0 runs in the cloud via a user-friendly web interface. Users upload one or more lesion masks as volumes in MNI152 space, select from a few configuration options, and receive an email with a link to download results after processing is complete.

- The web interface can be found here: https://kuceyeski-wcm-web.s3.us-east-1.amazonaws.com/upload.html.

- Source code and additional documentation can be found here: https://github.com/kjamison/nemo.

Uploading a Lesion Mask

Begin by pointing your web browser to the web interface: https://kuceyeski-wcm-web.s3.us-east-1.amazonaws.com/upload.html.

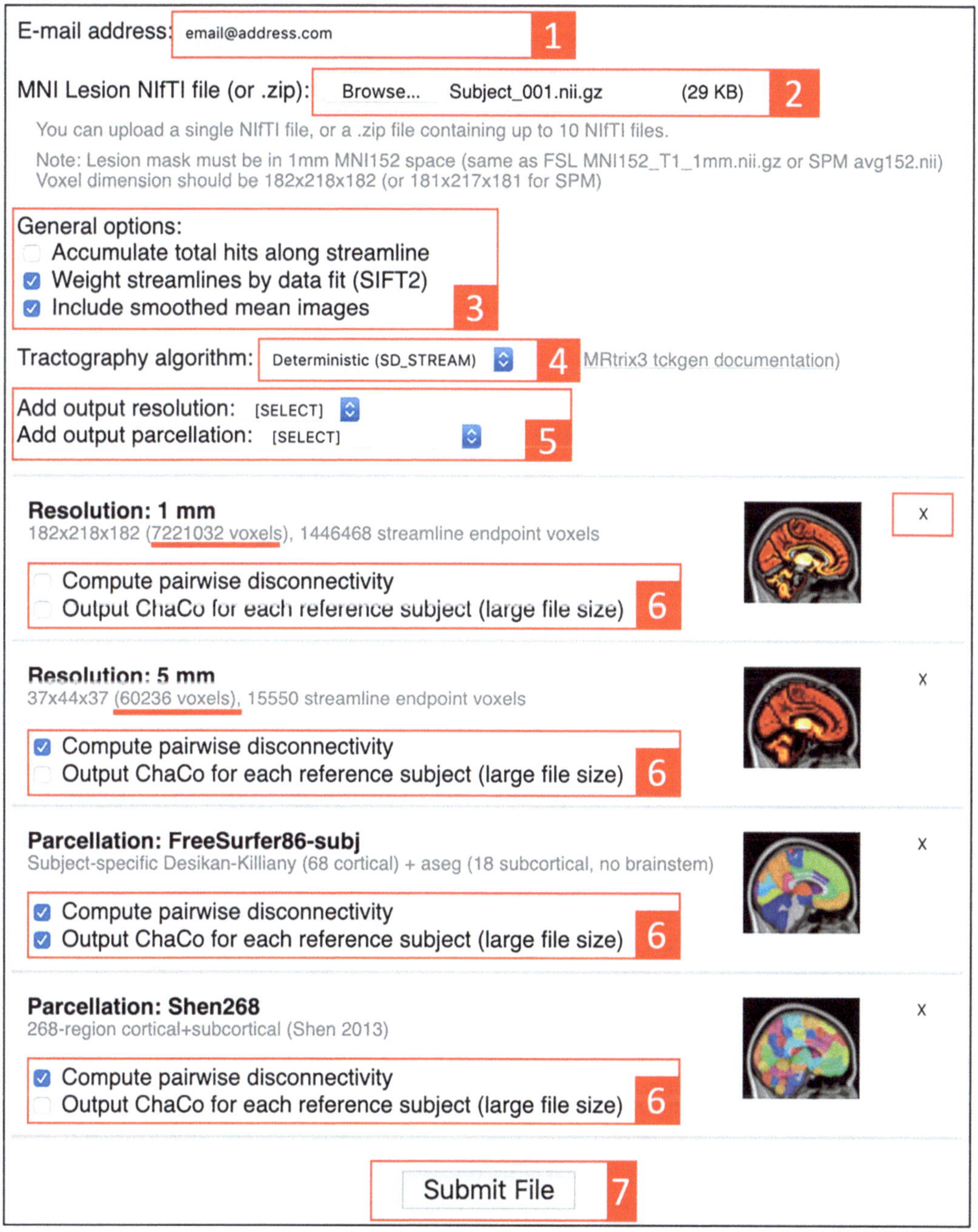

1. Enter a valid email address where results can be sent.
2. Click the "***Browse...***" button and select your lesion masks
   a. Input can be a single NIfTI volume (.nii, .nii.gz) or a .zip archive containing up to 10 separate NIfTI files to be run sequentially as a batch.
   b. Input volumes must already be transformed to 1mm MNI152 v6 space. This is the same as the MNI152_T1_1mm.nii.gz volume distributed with FSL (182 × 218 × 182 voxels) or the avg152.nii volume with SPM (sometimes 181 × 217 × 181 voxels).
   c. To confirm your volume is in the appropriate space, you can download the MNI152_T1_1mm_brain.nii.gz volume here and overlay it in your favorite visualization software (e.g., fsleyes, MRIcron): https://github.com/kjamison/nemo/blob/master/website/atlases/MNI152_T1_1mm_brain.nii.gz.
   d. OR you can download and run nemo_save_average_glassbrain.py, which will save an image with which to visually confirm lesion placement and print out the voxel dimensions:

```
> python nemo_save_average_glassbrain.py --out Subject_001_-
glass.png Subject_001.nii.gz

181x217x181
```

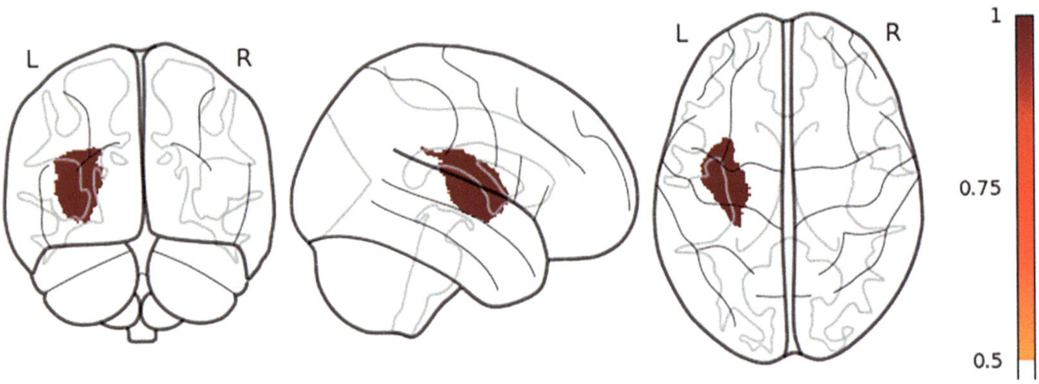

3. Select checkboxes from the general configuration options:
   a. ***Accumulate total hits along streamline***: If selected, a streamline will be "more disconnected" if it intersects multiple voxels in the lesion mask. Otherwise, streamlines are considered completely disconnected if they intersect at least one lesion voxel (default).

   b. ***Weight streamlines by data fit (SIFT2)***: If selected, each streamline is assigned a weight to better match the global tractography estimate for each subject to the observed diffusion signal.

   c. ***Include smoothed mean images***: Voxelwise outputs will additionally include volumes that have been smoothed by 6mm before averaging across the 420 HCP reference subjects, to increase intersubject overlap.

4. Select a tractography algorithm: Select either "***Probabilistic (iFOD2+ACT)***" or "***Deterministic (SD_STREAM)***".

   a. Probabilistic tractography is generally more liberal and more prone to false-positive connections.

   b. Deterministic tractography is more conservative and sparse and may miss certain true connections.

5. Select one or more output resolutions or parcellations.

   a. You can add different output resolutions using the "***Add output resolution***" dropdown. The default output will include the 1mm voxelwise (non-pairwise) output.

   b. You can add any of the provided atlases for parcellated results by selecting options from the "***Add output parcellation***" dropdown.

   c. To upload your own custom parcellation, select "[Upload atlas]" from the parcellation dropdown. A new entry will be added below where you can assign your parcellation a name and select the appropriate NIfTI volume to upload (must be 1mm MNI space).

   d. Resolutions and parcellations can be removed from your request by selecting the "x" for the appropriate entry.

   e. Details about the provided atlases can be found in the GitHub documentation: https://github.com/kjamison/nemo#parcellations.

6. For each resolution or parcellation, you can select a set of output options:

   a. ***Compute pairwise disconnectivity***: If selected, output matrix of pairwise ChaCo scores for each pair of regions or voxels. Output will **always** include the regionwise or voxelwise "chacovol" output, whether the "pairwise" option is selected or not.

      For example, for the "FreeSurfer86-subj" atlas, outputs will include an 86 × 86 disconnectivity "chacoconn" matrix.

      For 1mm resolution, this would create a 7 million × 7 million voxel sparse matrix, which is not particularly useful and is disabled by default.

For 5mm resolution, this output would include a more manageable 60,000 × 60,000 voxel sparse matrix.

b. ***Output ChaCo for each reference subject***: If selected, output will include a vector and/or matrix for each of the 420 reference subjects. If not selected, output only includes the mean and standard deviations across the reference subjects.

These subject-wise outputs can be **much** larger and are not generally recommended, but could be useful if you want to accurately re-parcellate voxelwise outputs at some later time or for non-parametric analyses of ChaCo variability across the population.

7. Click "***Submit File***" at the bottom to submit your request.

a. After uploading, inputs are validated to ensure proper format and dimensions. After a few minutes, you should see an image appear on the web interface showing your lesion mask (or the average of several lesion masks if you submitted a .zip file) in a glass brain presentation in MNI template space. ***If this output does not appear to have the correct orientation or brain location, you will need to correct your MNI registration and resubmit.***

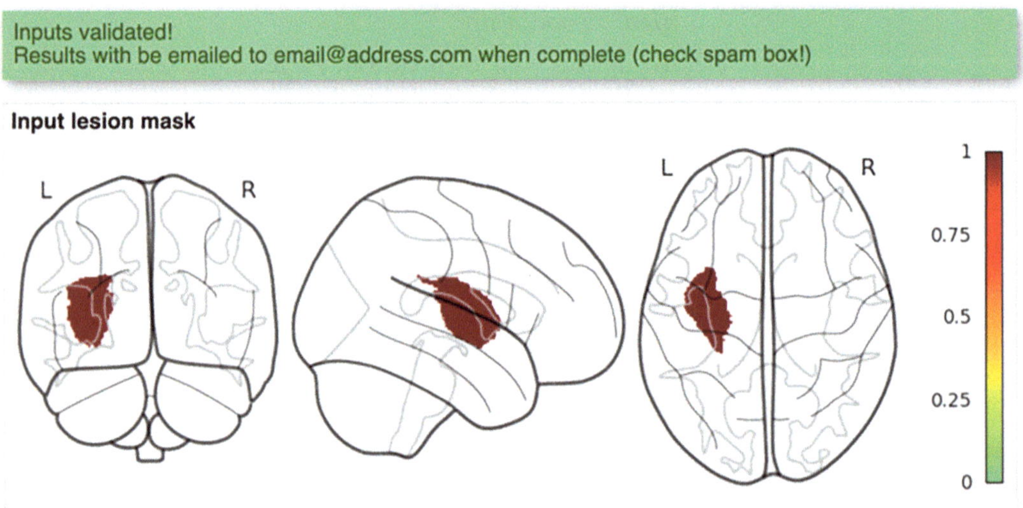

b. You are now free to submit additional inputs.

c. You will receive an email with a download link when the results for each submission are ready. This can take anywhere from 5 min to several hours, depending on the size of your lesion masks, how many masks were uploaded, how many output spaces were selected, etc.

d. Results can be downloaded for 7 days, after which the outputs will be deleted from our storage system.

Viewing Outputs

Depending on your input selections, your outputs may contain a number of different files. For the request described above, the output is a 15MB .zip archive including:

Outputs for 1 mm resolution (No pairwise outputs requested)

```
1.1M Subject_001_nemo_output_sdstream_chacovol_res1mm_mean.nii.gz
1.1M Subject_001_nemo_output_sdstream_chacovol_res1mm_stdev.nii.gz
2.0M Subject_001_nemo_output_sdstream_chacovol_res1mm_smooth6mm_mean.nii.gz
3.2M Subject_001_nemo_output_sdstream_chacovol_res1mm_smooth6mm_stdev.nii.gz
```

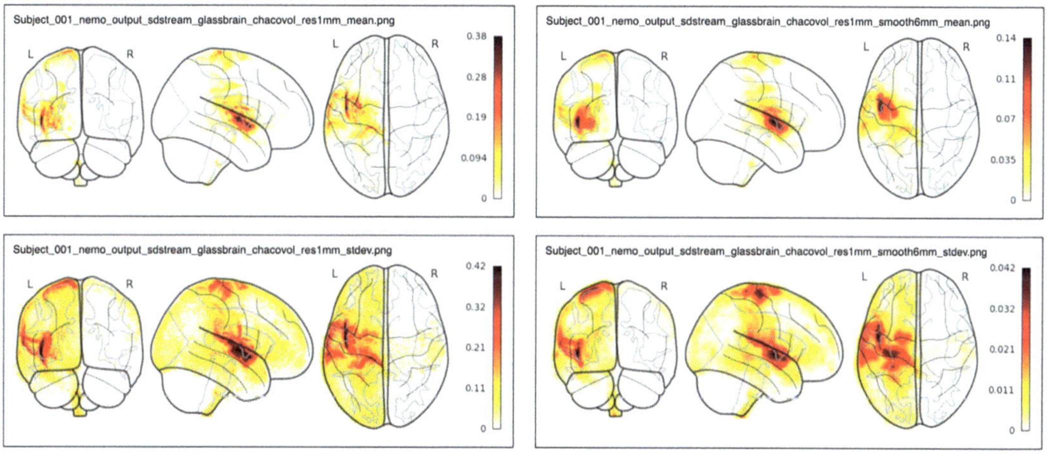

Outputs for 5 mm resolution (Includes pairwise outputs)

```
19K Subject_001_nemo_output_sdstream_chacovol_res5mm_mean.nii.gz
23K Subject_001_nemo_output_sdstream_chacovol_res5mm_stdev.nii.gz
24K Subject_001_nemo_output_sdstream_chacovol_res5mm_smooth6mm_mean.nii.gz
33K Subject_001_nemo_output_sdstream_chacovol_res5mm_smooth6mm_stdev.nii.gz
5.0M Subject_001_nemo_output_sdstream_chacoconn_res5mm_mean.pkl
5.0M Subject_001_nemo_output_sdstream_chacoconn_res5mm_stdev.pkl
```

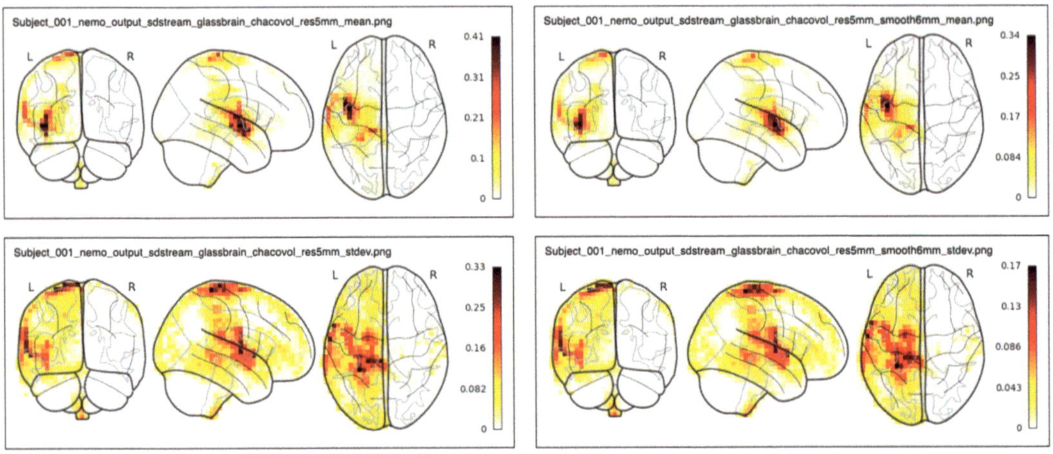

Outputs for Shen268 parcellation (Includes pairwise outputs)

```
1.3K Subject_001_nemo_output_sdstream_chacovol_shen268_mean.pkl
1.3K Subject_001_nemo_output_sdstream_chacovol_shen268_stdev.pkl
45K Subject_001_nemo_output_sdstream_chacoconn_shen268_mean.pkl
45K Subject_001_nemo_output_sdstream_chacoconn_shen268_stdev.pkl
1.5K shen_268_parcellation_networklabels.csv
```

Outputs for FreeSurfer86 parcellation (Includes pairwise outputs and "all ref")

```
504 Subject_001_nemo_output_sdstream_chacovol_fs86subj_mean.pkl
504 Subject_001_nemo_output_sdstream_chacovol_fs86subj_stdev.pkl
131K Subject_001_nemo_output_sdstream_chacovol_fs86subj_all-ref.pkl
131K Subject_001_nemo_output_sdstream_chacovol_fs86subj_all-ref_denom.pkl
9.8K Subject_001_nemo_output_sdstream_chacoconn_fs86subj_mean.pkl
9.8K Subject_001_nemo_output_sdstream_chacoconn_fs86subj_stdev.pkl
621K Subject_001_nemo_output_sdstream_chacoconn_fs86subj_all-ref.pkl
6.1M Subject_001_nemo_output_sdstream_chacoconn_fs86subj_all-ref_denom.pkl
5.4K fs86_FreeSurferLUT.txt
```

Outputs are either NIfTI volumes (*.nii.gz) or python "pickle" format (*.pkl). To read pkl files, you will need Python 3.6+ as well as NumPy and SciPy. NiBabel and Nilearn are also recommended. More details on the Python requirements, as well as example scripts for converting outputs into plaintext or Matlab .mat format, can be found in the GitHub documentation: https://github.com/kjamison/nemo.

Loading and Viewing Voxelwise Outputs

For voxelwise, non-pairwise "chacovol" outputs, the results are saved as NIfTI volumes (and glassbrain figures) for the mean and standard deviation across the 420 HCP reference subjects. Examples above include ***_chacovol_res1mm_mean.nii.gz**. These volumes can be viewed in any NIfTI viewer or read into python using NiBabel:

```
import nibabel as nib
import numpy as np
Vnii=nib.load('Subject_001_nemo_output_sdstream_chacovol_-
res1mm_mean.nii.gz')
voldata=Vnii.get_fdata() #this is a 182x218x182 3D array

#Use Nilearn to display volumetric data:
from nilearn import plotting
plotting.plot_glass_brain(Vnii,output_file='Subject_001_exam-
ple_chacovol.png',cmap='hot_r',colorbar=True)
```

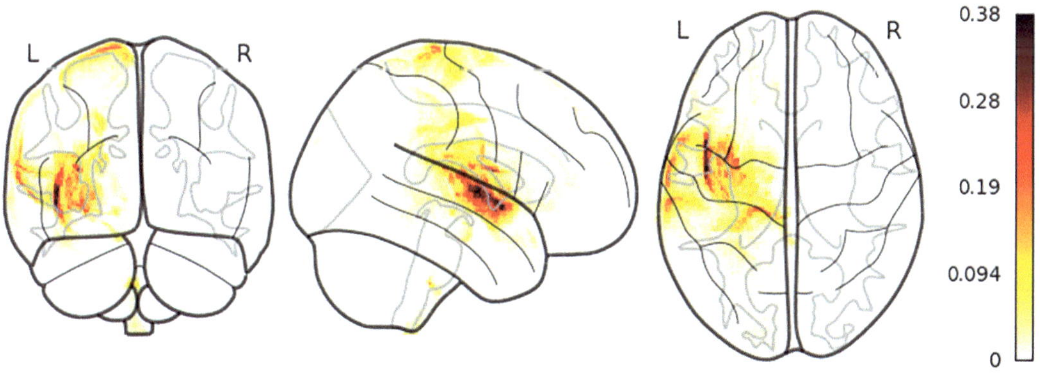

To load voxelwise chacovol results from multiple lesion masks, you might do:

```
import nibabel as nib
import numpy as np
import glob
meanfiles=sorted(glob.glob('Subject_*_nemo_output_sdstream_-
chacovol_res1mm_mean.nii.gz'))
```

```
#create a 182x218x182x(lesionfiles) 4D array
voldata=np.stack([nib.load(f).get_fdata() for f in meanfiles],
axis=3)
volmean=np.mean(V,axis=3) #182x218x182 mean across all lesion
masks

#create a new NIfTI structure to hold the new 3D matrix for
plotting
Vnii=nib.load(meanfiles[0])
Vnii_mean=nib.Nifti1Image(volmean,affine=Vnii.affine, head-
er=Vnii.header)
plotting.plot_glass_brain(Vnii,output_file='all_lesions_exam-
ple_chacovol.png',cmap='hot_r',colorbar=True)
```

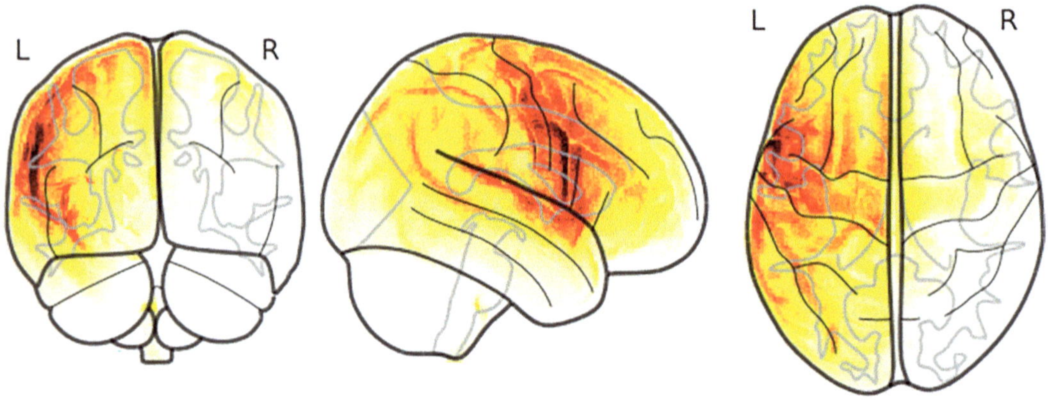

Loading and Viewing Regionwise Outputs

To load the regionwise, non-pairwise "chacovol" outputs for multiple lesion masks:

```
import numpy as np
import glob
import pickle
meanfiles=glob.glob('Subject_*_nemo_output_sdstream_chacov-
ol_fs86subj_mean.pkl')
meanfiles=sorted(meanfiles) #sort alphanumerically
roidata = [pickle.load(open(f,'rb')) for f in meanfiles]
roidata = np.concatenate(roidata,axis=0) #create a
(lesionfiles)x86 matrix

#Now create a heatmap showing regionwise connectivity for each
lesion mask
from matplotlib import pyplot as plt
fig=plt.figure()
ax=plt.imshow(roidata,cmap='hot')
plt.xlabel('ROI')
plt.ylabel('lesion mask')
```

```
fig.colorbar(ax)
fig.savefig('all_lesions_chacovol_fs86subj.png')
```

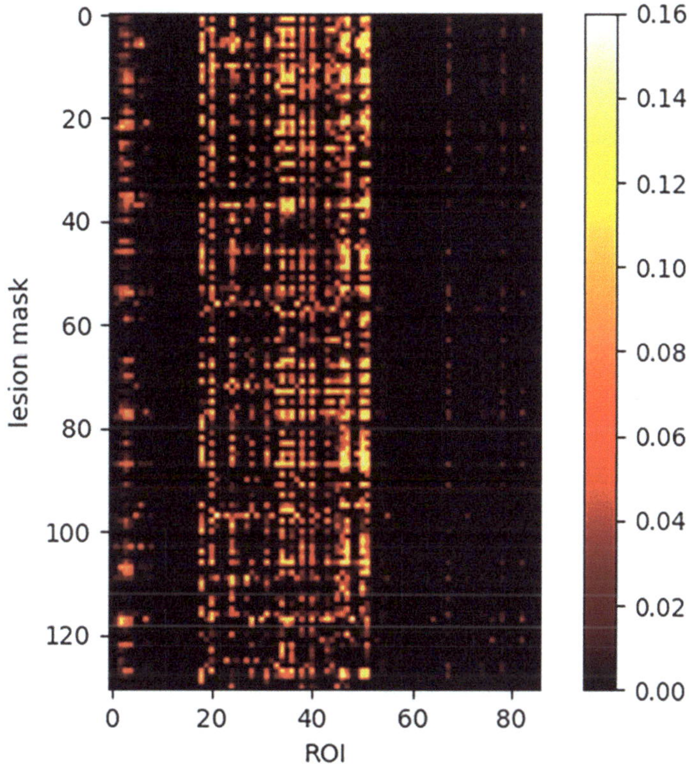

To load the pairwise "chacoconn" outputs for a single lesion mask:

```
import numpy as np
import pickle
connfile='Subject_001_nemo_output_sdstream_chacoconn_fs86-
subj_mean.pkl'
conndata = pickle.load(open(connfile,'rb'))
#chacoconn files are SciPy sparse matrix format, and must be
converted to dense arrays for most uses
conndata = conndata.toarray() #this is now an 86x86 matrix
#Now create a heatmap showing pairwise disconnectivity for
this lesion mask:
from matplotlib import pyplot as plt
fig=plt.figure()
ax=plt.imshow(conndata,cmap='hot',clim=[0,1])
plt.xlabel('ROI')
plt.ylabel('ROI')
fig.colorbar(ax)
fig.savefig('Subject_001_chacoconn_fs86subj.png')
```

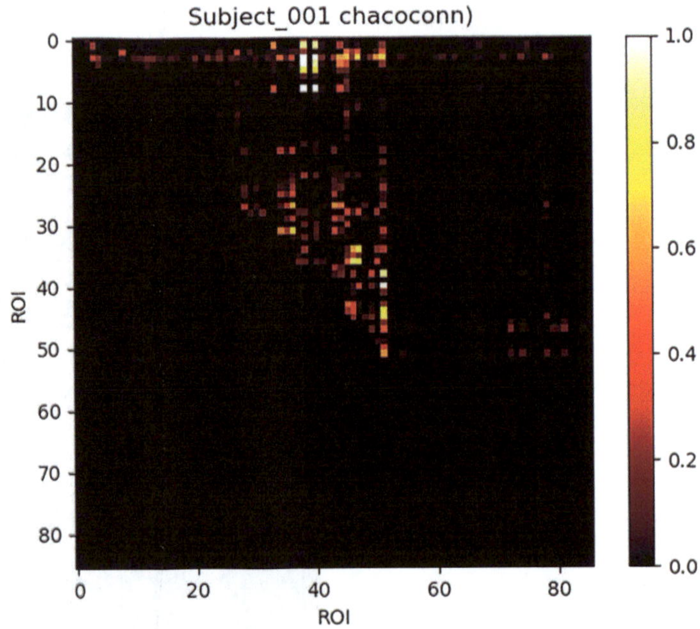


```
#Or load the chacoconn files for each lesion mask and display
the mean:
connfiles=sorted(glob.glob('Subject_*_nemo_output_sdstream_-
chacoconn_fs86subj_mean.pkl')
conndata = [pickle.load(open(f,'rb')).toarray() for f in con-
nfiles]
conndata = np.stack(conndata, axis=2) #create an 86x86x(lesion
files) array
connmean = np.mean(conndata,axis=2)
fig=plt.figure()
ax=plt.imshow(connmean,cmap='hot')
plt.xlabel('ROI')
plt.ylabel('ROI')
fig.colorbar(ax)
fig.savefig('all_lesions_chacoconn_fs86subj.png')
```

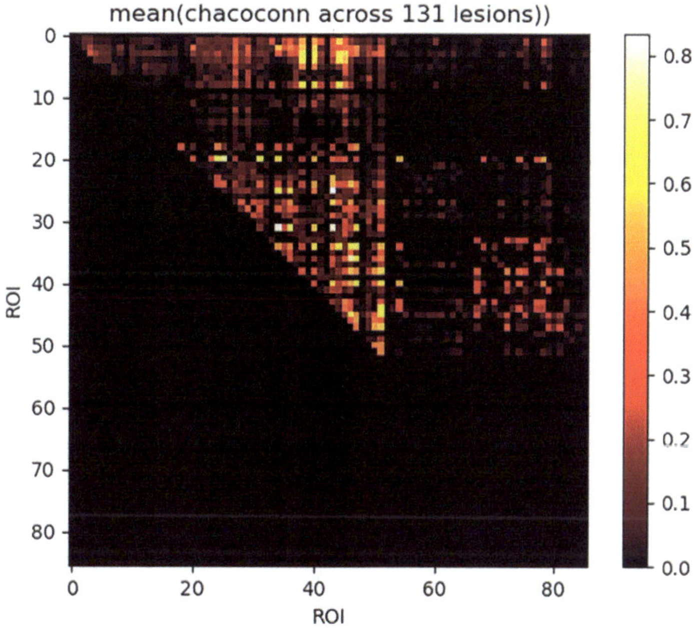

Loading and Viewing Voxelwise, Pairwise Outputs

The output from a voxelwise, pairwise "chacoconn" analysis is a VOXELSxVOXELS pairwise matrix, where chacoconn_mean[i,:] represents a VOLUME of disconnectivity to from voxel i to all other voxels. Thus, we can view the "disconnectivity" volume for a specific gray matter "seed":

```
import pickle
import numpy as np
import nibabel as nib
from nilearn import plotting

#This file contains a 60236x60236 sparse matrix
volconnfile='Subject_001_nemo_output_sdstream_chacoconn_-
res5mm_mean.pkl'
volconndata = pickle.load(open(volconnfile,"rb"))

#We need the corresponding chacovol_resXmm_mean.nii.gz output
which defines the output dimensions
volreffile='Subject_001_nemo_output_sdstream_chacovol_-
res5mm_mean.nii.gz'
Vref = nib.load(volreffile)

#for this demonstration, find the voxel with the largest total
disconnectivity, convert that row
#of the chacoconn output to a volume, and visualize that volume
volconndata_sum=np.sum(volconndata,axis=0)+np.sum(volconnda-
ta,axis=1).T #data is upper triangular so we need to sum across
```

```
both dimensions
voxel_index=np.argmax(data_sum)
newdata=volconndata[voxel_index,:]+volconndata[:,voxel_in-
dex].T
Vnew=nib.Nifti1Image(np.reshape(newdata.toarray(),Vref.
shape),affine=Vref.affine, header=Vref.header)

#convert voxel index to coordinates we can include the position
in the output filename
voxel_ijk=np.unravel_index(voxel_index,Vref.shape) #convert
to (i,j,k) coords
voxel_xyz=(Vref.affine @ np.array(voxel_ijk+(1,)).T)[:3] #can
also convert to x,y,z mm using Vref affine
voxel_ijk_string="%d_%d_%d" % (voxel_ijk)

#save glassbrain image of this volume using nilearn plotting
tools
plotting.plot_glass_brain(Vnew,output_file="mylesion_chaco-
conn_res5mm_voxel_%s.png" % (voxel_ijk_string),colorbar=True)
```

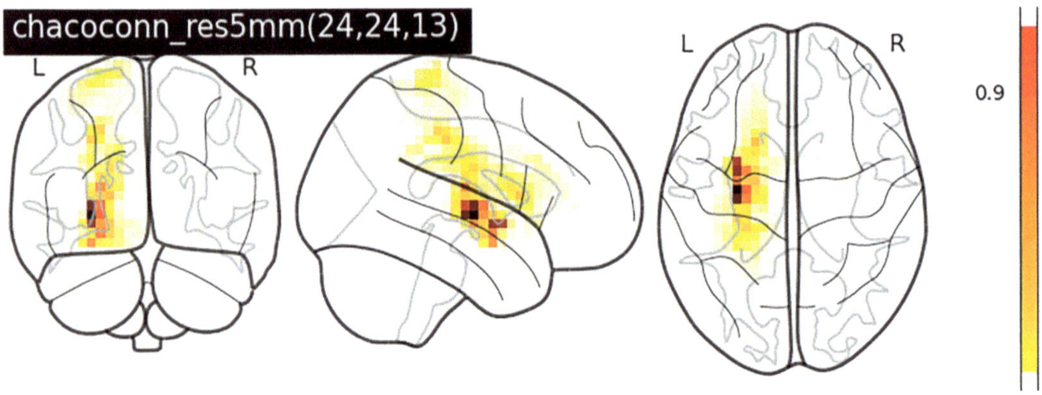

Appendix 2: Automated Segmentation—Tools and Examples

Introduction

This appendix will provide two software usage examples for automated lesion segmentation. The single case being segmented is part of the ATLAS dataset (Native space ID: c0005s0007t01, Left MCA embolism, lesion size 56 ml). The ATLAS dataset can be downloaded at https://fcon_1000.projects.nitrc.org/indi/retro/atlas.html. The order of the examples is the following (authors listed under each example):

1. LINDA software
   Melissa Thye, University of Edinburgh, Edinburgh, United Kingdom
2. Lesion_gnb software
   Joseph C. Griffis, Washington University, St. Louis, MO, USA

Automated Lesion Segmentation Using LINDA

Getting Started

The analyses described here require the following packages:

- ANTsR (https://github.com/ANTsX/ANTsR): A general neuroimaging data analysis package for loading, saving, and manipulating NIfTI files
- LINDA (https://github.com/dorianps/LINDA): Lesion segmentation package which implements a supervised segmentation algorithm called lesion identification with neighborhood data analysis (LINDA)

Dorian Pustina and Daniel Mirman (eds.), *Lesion-to-Symptom Mapping: Principles and Tools*, Neuromethods, vol. 180,
https://doi.org/10.1007/978-1-0716-2225-4, © Springer Science+Business Media, LLC, part of Springer Nature 2022

The linked GitHub pages provide guidance on how to install these packages across operating systems. Once installed, open RStudio and load the packages using the code provided below.

```
# Load required packages
library(LINDA)
```

Automated Lesion Segmentation with LINDA

Automated lesion segmentation algorithms (1) make lesion-based research more accessible for researchers without direct access to a professional trained to identify lesions via manual tracing, (2) reduce the time and effort needed to process participant data which is a significant limitation for large studies, and (3) provide a reproducible method for identifying lesion territory. The lesion identification with neighborhood data analysis (LINDA) approach implemented in R is demonstrated here. A comprehensive description of LINDA has been provided previously (*see* [1]).

The code below shows how LINDA can be applied to the T1 data used to demonstrate automated lesion segmentation in Chapter 3. A hypothetical example for processing several participants is also provided below. In this example, it is assumed that there is a directory called *MRI* in which each participant (i.e., sub-###) has a folder containing their chronic stage T1 MRI data. As a note, if your data include right hemisphere lesions, you will need to flip the orientation of the lesion prior to segmentation with LINDA.

```
# Lesion segmentation for one participant
linda_output <- linda_predict("c0005s0007t01.nii.gz")

# List MRI data filenames
MRI <- list.files(path="~/MRI", pattern="sub", full.names=TRUE, recursive=TRU
E)

# Lesion segmentation for several participants
for (i in 1:length(MRI)){
  linda_output <- linda_predict(MRI[i])
}
```

Lesion segmentation can take up to several hours depending on your computer. A *linda* folder containing the output files will be generated within each participant's folder. The table below provides an overview of the output files. Several of these files are visualized in Fig. A.3. For those files, the figure panel letter is listed next to the file name.

Dorian Pustina and Daniel Mirman (eds.), *Lesion-to-Symptom Mapping: Principles and Tools*, Neuromethods, vol. 180, https://doi.org/10.1007/978-1-0716-2225-4, © Springer Science+Business Media, LLC, part of Springer Nature 2022

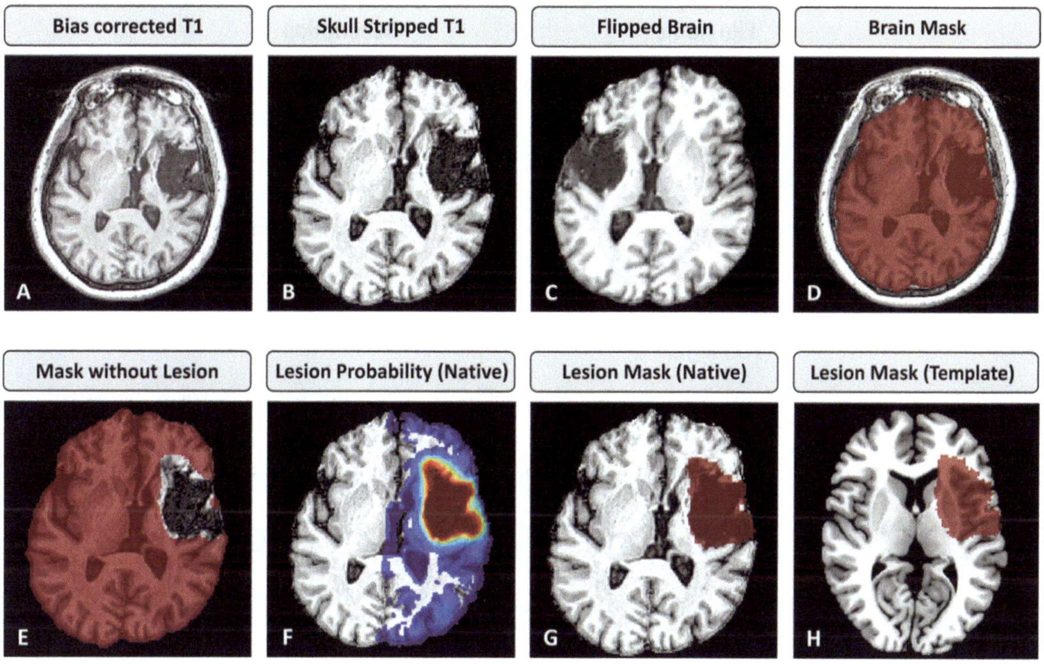

Fig. A.3 Files produced after lesion segmentation with LINDA for the ATLAS dataset case. The ANTsR plotting function uses radiological convention (as opposed to neurological convention), so the left hemisphere is shown on the right side of the brain slice, and the right hemisphere is shown on the left

File name	Description
Native space	
N4corrected (**A**)	Bias corrected T1 image
N4corrected_Brain (**B**)	Bias corrected, skull-stripped T1 image
N4corrected_Brain_LRflipped (**C**)	Flipped brain (used to compute asymmetry mask)
BrainMask (**D**)	Native space brain mask
Mask.lesion(1-4) (**E**)	Brain mask without lesion for registration
Prediction3_probability_native (**F**)	Final lesion probability in native space
Prediction3_native (**G**)	Final binarized lesion mask in native space
Template space	
Reg3_registered_to_template[a]	Bias corrected and skull-stripped T1 in template space
Prediction(1-2)[a]	Lesion prediction after first and second round

(continued)

File name	Description
Prediction3_probability_template[a]	Final lesion probability in template space
Prediction3_template[a]	Final binarized lesion mask in template space
Subject_in_MNI	T1 normalized to template space
Lesion_in_MNI (**H**)	lesion mask normalized to template space
Other files	
Reg3_sub_to_template_(affine/warp)	Transformation matrices (native to template)
Reg3_template_to_sub_(affine/warp)	Transformation matrices (template to native)
Console_Output	Log of console display
Session_Info	Information on R session and package versions

Note.[a]These files are in template space, but have different voxel dimensions than the default ch2 template

Quality Assurance

Automated lesion segmentation is not perfect especially for small lesions, and it can be challenging to detect potential errors in segmentation and fix them when they occur. Further, the value of using a reproducible lesion segmentation method is partially negated if manual, individualized edits are required for each participant. The best way to assess the quality of the segmentation is to visually inspect the output to ensure that the lesion territory was accurately captured. This can be easily done by plotting the data within R. The most useful files to visually examine are (1) the native space lesion prediction overlayed on the native space T1 image and (2) the template space lesion prediction overlayed on the template brain (example code for plotting this is provided below and produces panel **H** from Fig. A.3).

```
# read in files for plotting
ch2 <- antsImageRead(getANTsRData("ch2")) # Colin/ch2 template
lesion_mni <- antsImageRead("Lesion_in_MNI.nii.gz")

# plot
plot(ch2, lesion_mni,
     axis=3,
     alpha=.7,
     slices=80)
```

Consistent errors in segmentation are most notable when several participants have a lesion pattern that is either not likely or not

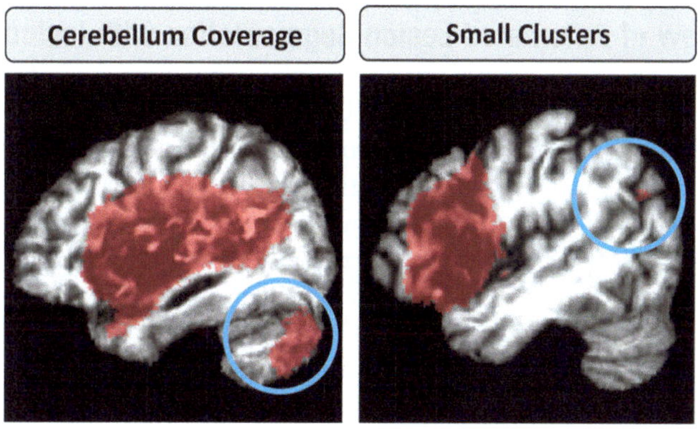

Fig. A.4 Example of errors in lesion segmentation

probable. As an example, in a study of semantic and letter fluency in participants with post-stroke aphasia [2], LINDA produced lesion masks with cerebellum coverage or with small non-contiguous clusters for several participants (Fig. A.4).

None of the participants in the study had cerebellar lesions, and the underlying vasculature made the small, distal clusters unlikely to be lesioned. To address these consistent errors, a dilated mask of the cerebellum was first subtracted from the lesion masks. This had the effect of removing any cerebellum coverage. Second, small clusters (<2000 contiguous voxels) were removed from the lesion masks which effectively removed any distal, artifactual clusters. This approach to fixing consistent errors in segmentation was reproducible and less time-consuming than manual edits, but it worked because the participants in the study had left hemisphere MCA territory lesions that were larger than 2000 voxels. These errors in segmentation and the solution used to address them may be useful for other researchers, but it is critical to consider the properties of the data and the participants when conducting any quality assurance protocol.

References

1. Pustina, D., Coslett, H. B., Turkeltaub, P. E., Tustison, N., Schwartz, M. F., & Avants, B. (2016). Automated segmentation of chronic stroke lesions using LINDA: Lesion identification with neighborhood data analysis. *HumanBrain Mapping, 37*(4), 1405–1421. https://doi.org/10.1002/hbm.23110
2. Thye, M., Szaflarski, J. P., & Mirman, D. (2021). Shared lesion correlates of semantic and letter fluency in post-strokeaphasia. *Journal of Neuropsychology,* 15(1), 143–150.

Overview of Automated Lesion Segmentation with lesion_gnb

This is a brief overview of how to perform automated lesion segmentation using the lesion_gnb toolbox for SPM12. Here, I will focus on the procedure for using MATLAB batch scripts, which can be employed to process data from multiple patients, but this overview will use just the single example lesion shown in the main text of Chapter 3. Please note that *lesion_gnb* also does feature a basic graphical user interface (GUI), and the procedure for automated lesion segmentation through the GUI is explained in detail in the user guide included with the *lesion_gnb* software package.

Within the *lesion_gnb* folder, which should be placed within SPM12's toolbox folder, you will find the following set of files:

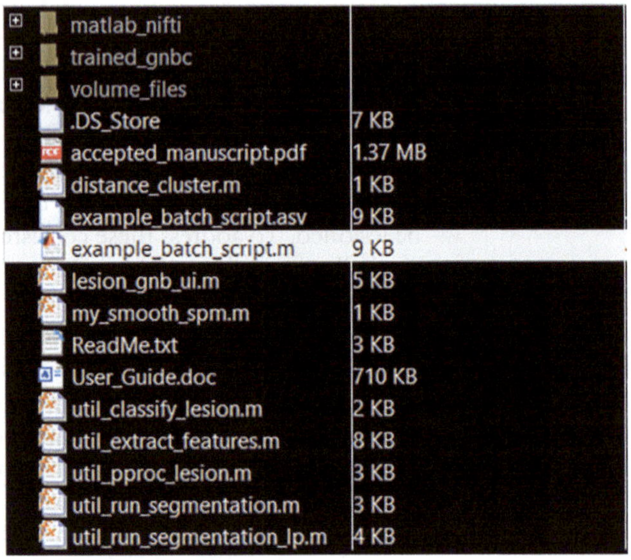

The highlighted file (example_batch_script.m) contains a pre-written template batch script that, with a few minor modifications, can be applied to your data. As noted above, this overview will focus on applying *lesion_gnb* via batch scripts, but the User_-Guide.doc file contains detailed instructions for how to apply the toolkit via the GUI. The ReadMe.txt file also contains other relevant details for use and should be read prior to running the toolkit.

Once the example_batch_script.m file is opened in the MATLAB editor, you will see that it is organized into several blocks of code, each of which corresponds to the implementation of a different processing step or set of steps.

The first step is to apply the SPM12 New Segment routine to the T1-weighted scans in order to obtain the initial tissue probability estimates:

```
%% SPM12 segmentation of T1w scans
clc
clear all

% set appropriate paths
my_spm_path = ('/home/usr/griffisj/Scripts_and_Tools/spm12'); % path to spm
addpath(my_spm_path); % add spm path
addpath(fullfile(my_spm_path, 'toolbox', 'lesion_gnb')); % add toolbox path

% define directory containing patient folders with t1w scans
data_dir = '/data/LesionSeg/ATLAS_R12/ATLAS R 1.2/native_1/c0005'; % directory containing T1w scans in folders
cd(data_dir); % navigate to directory

my_patients = dir('c0005s*0007t01'); % use wildcard to identify patient folders

for i = 1:length(my_patients) % loop through patient folders
    cfg.data = fullfile(data_dir, my_patients(i).name); % cfg field for directory containing the patient's T1w scan
    cd(cfg.data); % navigate to directory
    my_t1 = dir('*.nii'); % use wildcard to identify scan volume (assumes that T1w scan is only scan in directory)
    cfg.t1 = my_t1(1).name; % cfg field for t1w scan file
    util_run_segmentation(cfg); % run new segmentation with default parameters.
end
```

Here, three variables need to be modified for your analysis. The first is the variable my_spm_path, which should contain the path to the SPM12 software package on your machine. The second is the variable data_dir, which should contain the path to the directory the T1-weighted scan data. Here, it is assumed that within data_dir, there exist patient folders that each contain the T1-weighted scan data (in nifti format) for each patient (i.e., see how the variables cfg.data and my_t1 are defined within the for loop). If your data are organized differently, then you may need to also modify these pieces to match your data organization scheme. The third variable that you will need to modify (if working with data other than the example shown here) is the variable my_patients, which should be modified based on the naming scheme of folders within data_dir. These changes will need to be made in every subsequent code block, but will only be explicitly discussed here.

Information about the location of the data is stored in the variable cfg, which is a MATLAB struct, and this variable is then input to util_run_segmentation() (i.e., at the end of the for loop), which runs the SPM12 New Segment routine.

The second step is really several steps all packed into one. Here, we use the tissue probability maps that were created in the first step to define feature maps that encode "missing" and "abnormal" tissue.

```
%% Create feature maps, assign voxel class labels, and apply post-processing
clc
clear all
% define directory containing patient folders with t1w scans
data_dir = '/data/LesionSeg/ATLAS_R12/ATLAS R 1.2/native_1/c0005'; % directory containing T1w scans in folders
cd(data_dir); % navigate to directory

my_patients = dir('c0005s*0007t01'); % use wildcard to identify patient folders

for i = 1:length(my_patients) % loop through patient folders
    cd(fullfile(data_dir, my_patients(i).name)); % navigate to patient directory
    cfg.data = fullfile(data_dir, my_patients(i).name); % cfg field for directory containing the patient's segmented T1w scan
    cfg.out = 'Lesion_Mask_TPM_L'; % cfg field specifying output directory (will be created in each patient folder)
    cfg.fwhm = 8; % cfg field specifying smoothing kernel width (FWHM) to apply to TPMs/PPMs during feature map creation
    cfg.im_mask = 0; % use implicit mask when smoothing -- default is off for feature map creation
    cfg.priors = [0.89 0.11]; % [0.5 0.5]assumes equal prior probability for each tissue class. Original priors are [0.8936, 0.1064] for non-lesion, lesion.
    cfg.affected = 'L'; % cfg field specifying affected hemisphere (L or R)
    cfg.ppm_only = 'N'; % cfg field specifying whether to construct feature maps using both unaffected hemisphere TPM and PPM volumes (N) or using only PPM volumes (Y)
    util_extract_features(cfg); % run feature map creation
    util_classify_lesion(cfg); % use trained/cross-validated GNB classifier to predict class labels
    cfg.fwhm = 8; % cfg field specifying smoothing kernel width (FWHM) to apply to binary lesion class labels
    cfg.clust = 100; % cfg field specifying minimum cluster size to retain in final lesion mask
    cfg.im_mask = 0; % use implicit masking when smoothing -- default is off for post-processing.
    util_pproc_lesion(cfg); % run post-processing (outputs post-processed lesion mask volumes.
end
```

Here, there are multiple parameters that may be changed within the for loop, but in practice only three of these are likely to be relevant for most scenarios. The first is the variable cfg.out, which is simply the name of the folder (which will be created within the patient folder) in which the outputs will be saved. Here, it is defined as "Lesion_Mask_TPM_L," which is intended to indicate that the algorithm is using patient tissue probability maps (TPMs) to compute the features as described below (i.e., indicated by the suffix "_TPM"), with the lesion in the left hemisphere (i.e., indicated by the suffix "_L"). The second is the variable cfg.affected, which specifies the hemisphere that is affected by the lesion. For this example patient, the lesion is in the left hemisphere, and so this variable is set to "L." If it were in the right hemisphere, it would be set to "R" (and you would want to change the suffix of cfg.out to "_R" instead of "_L"). The third is the variable cfg.ppm_only, which is set to "N," meaning that we want to use the patient TPMs in addition to SPM12's prior probability maps (PPMs) to compute the features. This is typically recommended unless the patient has significant bilateral lesions, since such a scenario will render the assumption that there exists an "unlesioned" hemisphere invalid. In such a scenario, we might want to instead use only the PPMs to compute the features by setting cfg.ppm_only to "Y," and we would also want to change the name of the output folder set by cfg.out to "Lesion_Mask_PPM_L" to reflect this choice. It is typically not recommended that the other parameters set by the other fields of the cfg variable be changed, although this is left as optional since in some cases it may be reasonable. Please refer to the user guide, the *lesion_gnb* source code, and SPM12's documentation for more details on these parameters. You can also send me an e-mail (jcgriffis@wustl.edu) or message on a social media platform such as ResearchGate with more detailed questions.

Three functions are implemented in this step. The first is util_extract_features(), which produces the "missing" and "abnormal" tissue feature maps. The second is util_classify_lesion(), which applies the trained GNB classifier to the voxel-level feature maps. The third is util_pproc_lesion(), which applies post-processing to the feature maps. At the end of this step, you will have the feature maps (i.e., f1.nii, f2.nii) as well as the classifier outputs (i.e., lesion_posterior.nii, lesion_labels.nii) and the post-processed outputs (e.g., s8lesion_labels_clustered_100.nii), where the "s8" prefix indicates the full-width half maximum of the Gaussian smoothing kernel applied to the lesion labels (i.e., 8mm in this case) and the "clustered_100" suffix indicates the cluster threshold applied during post-processing (i.e., 100 voxels in this case). The post-processed lesion images obtained from this step are the outputs originally described in the initial paper.

```
%% Re-run segmentation using smoothed lesion probability map as additional prior
clc
clear all
% define directory containing patient folders with t1w scans
data_dir = '/data/LesionSeg/ATLAS_R12/ATLAS R 1.2/native_1/c0005'; % directory containing T1w scans in folders
cd(data_dir); % navigate to directory

my_patients = dir('c0005s*0007t01'); % use wildcard to identify patient folders

for i = 1:length(my_patients) % loop through patient folders
    cd(fullfile(data_dir, my_patients(i).name)); %  navigate to directory
    t1_c1 = dir('c1*.nii'); %  get c1 segmentation of grey matter (it has a consistent naming scheme, so it's easy to wildcard)
    my_t1 = t1_c1(1).name(3:length(t1_c1(1).name)); % get T1w scan name by removing prefix
    cfg.data = fullfile(pwd); % cfg field for directory containing the patient's segmented T1w scan
    cfg.t1path = fullfile(pwd, my_t1); % cfg field with path to t1w scan file
    cfg.t1name = my_t1; % name of T1w scan  file
    cd(fullfile(pwd,'Lesion_Mask_TPM_L')); % directory containing extra prior
    my_smooth_spm('s8lesion_labels_clustered_100.nii', [8 8 8]); % smooth the final post-processed lesion mask to use as a tissue prior
    cfg.lpdir = pwd; % cfg field specifying output directory from lesion mask creation
    cfg.lp = 's8s8lesion_labels_clustered_100.nii'; % additional tissue prior (i.e. smoothed final mask or lesion probability map (final mask reccomended))
    cfg.out = 'LP_Seg'; % output directory
    util_run_segmentation_lp(cfg); % run second segmentation using cfg.lp as tissue prior
end
```

The third (optional) step is to re-run the segmentation routine while using the post-processed lesion segmentation as a prior for an extra tissue class, analogous to the procedure used by Seghier et al. [1] and Sanjuan et al. [2] for the ALI method. Prior to assigning the post-processed lesion segmentation as an additional tissue prior, it may be smoothed and rendered continuous (i.e., by setting the [8 8 8] to the desired smoothing kernel FWHM—in this case, it is 8mm). In addition, you will want to make sure that the file assigned to the variable cfg.lp corresponds to the file that you want to use as the extra tissue class prior. Here, it is the output of applying the function my_smooth_spm() to the post-processed lesion segmentation (in practice, you could alternately use the raw lesion labels, the lesion posteriors, or smoothed versions of these if desirable, but my experience has been that using a smoothed version of the post-processed lesion tends to work best), which is the same file with another "s8" suffix appended. The output of the segmentation will be in the folder defined by cfg.out, which will be inside of the original output folder (i.e., in this case, Lesion_Mask_TPM_L).

This step will output a new set of tissue segmentations that include files "c7*.nii" (i.e., the native-space lesion probability map) and "wc7*.nii" (i.e., the template-normalized lesion probability map). The lesion probability maps output by the second segmentation tend to be more anatomically detailed than the original *lesion_gnb* outputs, and they can be manually thresholded and further processed as desired for your intended application. It is also worth noting that this extra segmentation step can help to improve registration performance and also reduces errors in the segmentation of other tissue classes (as shown in the main text). This is the final step, although additional code blocks are provided to facilitate the application of the atlas-space transformations to other files (e.g., other modalities aligned to the T1w image) and/or to allow for the reverse normalization of other outputs back into the patient's native space (e.g., if you wanted to visualize the feature maps on the original scan). However, these blocks simply employ standard SPM12 atlas normalization procedures, and so they will not be discussed here (the interested reader is directed to the SPM12 manual).

References

1. Seghier, M. L., Ramlackhansingh, A., Crinion, J., Leff, A. P., & Price, C. J. (2008). Lesion identification using unified segmentation-normalisation models and fuzzy clustering. *Neuroimage*, 41(4), 1253-1266.

2. Sanjuán, A., Price, C. J., Mancini, L., Josse, G., Grogan, A., Yamamoto, A. K. & Seghier, M. L. (2013). Automated identification of brain tumors from single MR images based on segmentation with refined patient-specific priors. *Frontiers in Neuroscience*, 7, 241.

Index

A

Advanced Normalization Tools (ANTs) ... 80, 81, 83, 91, 92
Advanced Normalization Tools for R (ANTsR) ... 62, 73, 80, 83, 84, 317, 321, 322, 337, 339
ANTsPY ... 80
Aphasias ... 96, 100, 110, 111, 120, 127–129, 136, 143, 144, 150, 159, 168, 169, 174, 177, 190, 191, 193, 220, 226, 228, 229, 231, 239, 242, 244, 249, 252, 284, 302, 341
Automated lesion segmentation ... 53–55, 58, 60–62, 74, 337–346

B

Behavioral neurology ... 142
Brain-behavior mapping ... 20–21, 95, 108–112, 135–145, 150, 160
Brain lesions ... 1, 8, 20, 21, 53, 54, 66, 155, 156, 171, 181, 182, 184, 188, 190, 191, 193–195, 225, 239, 240, 246
Brain mapping ... 1, 7, 8, 14, 16–18, 20, 84, 91, 142, 257, 260, 264, 266–268, 322, 341
Brain networks ... 14, 131, 151, 161, 174, 177, 182, 191, 193, 247, 266, 268

C

Causal mechanism ... 200
Cognition ... 136, 200, 263
Computational neuroimaging ... 188
Connectome ... 150–161, 169–175, 182–185, 187–189, 241, 324
Cortical mapping ... 261
Cost masking ... 85–87, 89
Cross-validation ... 70, 72, 109, 221–224, 228, 231, 306, 307

D

Diaschisis ... 8, 155, 169, 181, 225, 252
Direct electrical stimulation (DES) ... 257, 258, 262, 264–268

E

Enantiomorphic ... 87–89, 92

F

Focal brain injury ... 155
Functional reorganization ... 17, 20, 129, 226, 227, 233, 285

I

Inferences ... 27, 36, 49, 90, 91, 119, 195, 200, 203, 209–216, 219–233, 239, 279
Inter-individual variability ... 262–264

L

Language ... 2, 7, 8, 12, 14–17, 19, 21, 100, 106, 108–111, 113, 127, 136, 143, 144, 150, 159, 160, 168, 169, 174, 216, 228–231, 243, 244, 257–266, 282–285
Lesion analyses ... 54, 96, 101, 112, 119, 131, 168–169, 181, 184, 188, 189, 308–317
Lesion-behavior mapping ... 95–96, 119–132, 168, 169, 177, 199–217, 290–295
Lesion-deficit mapping ... 7, 8, 201, 202, 212
Lesion filling ... 3, 9, 87–89, 92
Lesion-network mapping ... 156–158, 161, 181–195
Lesion studies ... 60, 96, 111, 112, 168, 244, 248
Lesion symptom mapping (LSM) ... 1–21, 27, 62, 66, 85, 95–113, 136, 167–178, 219, 237–253, 289, 295, 302, 317–322
Localizations ... 150, 184, 190, 191, 193, 200, 209, 211, 212, 214, 216, 244, 248, 249, 259, 261, 276, 278, 281

M

Magnetic resonance imaging (MRI) ... 3–8, 10, 13–15, 18, 32–34, 36, 38, 45, 46, 53, 54, 63–71, 80, 84, 95, 97, 100, 136–139, 144, 150, 152–156, 158, 170, 174, 181–195, 229, 230, 233, 258, 259, 261, 264, 266, 284, 337
Manual segmentations ... 28–32, 36, 39, 47, 50, 56, 58, 60, 65, 72
Mass-univariate ... 119, 200, 202, 205, 207–209, 213, 248, 250, 290, 304, 319–322

Dorian Pustina, and Daniel Mirman (eds.), *Lesion-to-Symptom Mapping: Principles and Tools*, Neuromethods, vol. 180, https://doi.org/10.1007/978-1-0716-2225-4,

N

Neurodegenerative diseases 12–14, 20, 21, 62, 135–145, 225
Neuroimaging 15, 21, 32, 53, 55, 58, 62, 64, 80, 83, 95, 96, 101, 111, 136, 140, 142, 144, 171, 188, 225, 226, 229, 230, 232, 264, 296, 317, 320, 337
Neuroimaging methods...................................... 101, 176
Neuroplasticity 227, 264, 266–268
Neuropsychology 244, 341

P

Perception................ 127–129, 259, 266, 271, 275, 282
Predictions......................59, 70–72, 131, 150, 159, 160, 203, 210–212, 216, 219–233, 239, 306, 339, 340
Primary progressive aphasia (PPA).................................. 12, 13, 136, 142–144
Psychometrics............................210, 238, 242, 243, 252

R

Reproducibility................................ 49, 54, 62, 140, 306
Resting state functional connectivity MRI (rs-fcMRI)...............................182, 184, 187, 189

S

Spatial inference 100, 200, 201, 206, 208, 211
Statistical inference...122
Statistical parametric mapping (SPM) 62, 65, 66, 68, 69, 119–123, 125, 129–131, 137, 138, 172, 200, 202, 304, 326
Strokes 2–9, 12, 14, 20, 21, 28, 29, 32, 38, 40, 54, 55, 62–66, 68, 71, 72, 79–92, 95, 100–102, 104–107, 109–111, 125, 127–129, 131, 135, 136, 138, 144, 145, 155, 156, 158–161, 168, 169, 171, 172, 175–177, 182, 184–186, 190, 193, 204, 208, 225–228, 230, 231, 233, 239, 252, 341
Structural connectivity.......................112, 149, 212, 324
Structural MRI 12, 15, 40, 56, 136, 144
Subcortical mapping 263
Surface-based morphometry (SBM) 137, 138, 140, 142–144

T

Transcranial magnetic stimulation (TMS) 18–21, 144, 159, 193, 194, 271–286
Traumatic brain injuries....................................... 2, 8–10, 18, 38, 100, 160, 186
Tumors 14–16, 18, 20, 54, 101, 112, 155, 156, 172, 184, 186, 188, 257–259, 261, 262, 264–267, 284, 285

V

Voxel-based lesion symptom mapping (VLSM) 27, 96–112, 121, 168, 169, 172–174, 178, 188, 190, 209, 239, 245, 248–250, 290, 291, 294–299, 319
Voxel-based morphometry (VBM)................. 12, 66, 96, 136–140, 142–144

MIX
Papier aus verantwortungsvollen Quellen
Paper from responsible sources
FSC® C105338
www.fsc.org

If you have any concerns about our products,
you can contact us on
ProductSafety@springernature.com

In case Publisher is established outside the EU,
the EU authorized representative is:
Springer Nature Customer Service Center GmbH
Europaplatz 3, 69115 Heidelberg, Germany

Printed by Libri Plureos GmbH
in Hamburg, Germany